The Way It Works

MAN AND HIS MACHINES

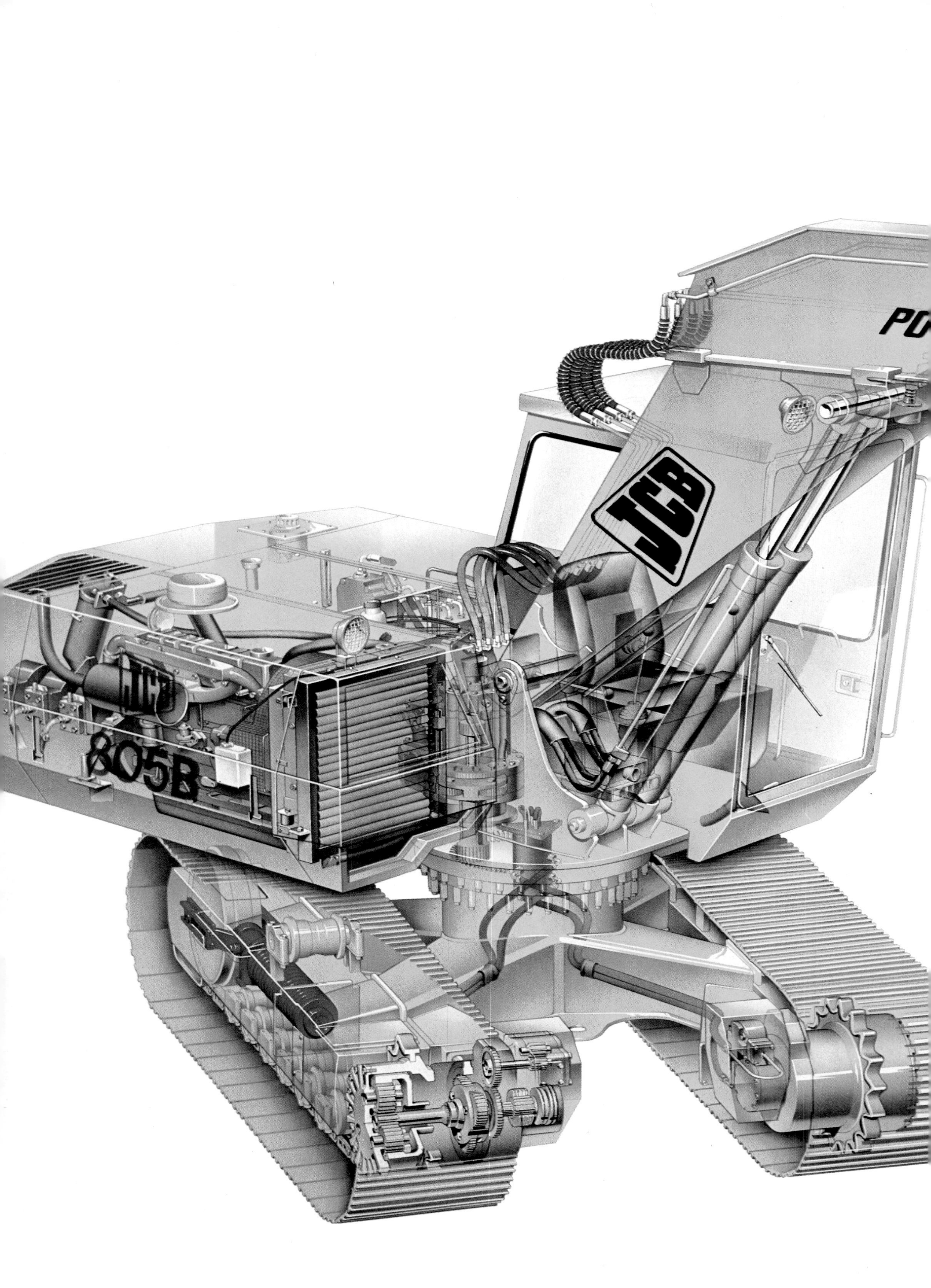
JCB
805B

The Way It Works

MAN AND HIS MACHINES

Robin Kerrod

FOREWORD BY
Professor Meredith Thring
Queen Mary College, London

Written and designed by
Robin Kerrod

Illustrated by
Bryan Foster
Ian McIntosh **Mike Atkinson**

First published in 1980 by
Octopus Books Limited,
59 Grosvenor Street, London W1

ISBN 0 7064 0986 8 (*hardback*)
ISBN 0 7064 0985 X (*paperback*)

Produced by Mandarin Publishers Limited,
22a Westland Road, Quarry Bay, Hong Kong
Printed in Hong Kong

Contents

Foreword

When Aladdin rubbed his magic lamp the Genie appeared and offered to do his bidding. As this book shows all the requests that he made have been given to the people of the developed countries by technology based on the heavy consumption of fossil fuels (oil and coal but particularly oil) to the extent that each person in these countries consumes the equivalent of some four tons of oil every year (or six tons of coal).

Our equivalent of the magic carpet of eastern legends is the aeroplane, or for shorter distances the car and train, and for sea transport the ship. We can fly halfway round the world in a subsonic aeroplane in 24 hours or in Concorde which flies twice as fast.

We can talk to people, and see what is happening on the other side of the world in a fraction of a second. One man can talk to several million or write a book which goes into a million homes. We can erect buildings tens of storeys high. We can live comfortably in the Arctic, we can move mountains (small ones) and dam the largest rivers. Every worker can have ten horsepower at his elbow to supplement his own physical capacity of less than one-quarter horsepower and some make use of several hundred horsepower to move earth or lift machinery. Finally, we have even achieved the greatest flight of earlier literary imagination; we have landed a man on the moon and brought him back safely.

After a very interesting discussion of the very beginnings of technology the second chapter explains clearly and in simple terms the various ways in which man can obtain the energy on which all his technology achievements are based. This covers the fossil fuels as well as nuclear energy and one renewable form of energy, namely hydro-electricity. The next two chapters deal with one of the factors causing the biggest change in human life, the possibility of travel, local and world wide. Many people believe that as the oil runs out we shall return to sailing ships, coal fired trains and airships, all of which are clearly described in these chapters. The next two chapters dealing with communications and people at work are very valuable in explaining the reader's understanding of the way in which technology has affected humans' lives, both by extending our possibilities for receiving information and by changing the kind of work we do. The last chapter deals with the very glamorous subject of space exploration and gives us a very interesting account of some of the results of this work.

Coming, as it does, at a time when the importance of mechanical engineering and technology is being stressed in schools and colleges, *The Way It Works* makes a fine contribution to our library of books on the subject.

M. W. Thring
Professor of Mechanical Engineering
Queen Mary College
London

Foundation Stones

Man, now the most dominant, most adaptable and most wide-ranging creature on the face of the Earth is a comparative newcomer on the evolutionary scene. A mere 5-million year old upstart, he co-exists with species, like the coelocanth and dragonfly, that have remained virtually unchanged for hundreds of millions of years.

The species of modern Man, *Homo sapiens* ('wise man'), was firmly established 35,000 years ago. He became a skilful tool user and competent artist. By using fire and making clothing he was able to survive the chilly climates of the glacial periods.

By 5000 BC in several regions Man had begun to forsake the nomadic existence of a hunter and gatherer for the more settled life of a farmer. This led soon to the development of the first complex societies, or civilizations, in which people had time to devote to things other than the struggle for existence.

The first civilization grew up in Mesopotamia, the then very fertile region between the rivers Euphrates and Tigris. Not 1500 km (1000 miles) away arose the great Egyptian civilization, whose colossal edifices and tomb relics still amaze us today. By 2000 BC these civilizations were adept at smelting metals. Sailing ships were plying the seas. Writing in the form of pictographs and hieroglyphics was developing rapidly.

The period between 2000 and 500 BC saw the gradual decline of the civilizations in Mesopotamia and Egypt and the rise of the Greeks, who scaled new heights in science, art and literature. The millennium between 500 BC and AD 500 saw the rise, decline and fall of the mighty Roman Empire. With the fall of this Empire, the West entered a period of decline which historians usually term the Dark Ages. But light could be seen in the East, where the Chinese and Arabs made spectacular advances.

By the 1400s a re-awakening of interest was occurring in the West, giving rise to the period we call the Renaissance ('rebirth'). By the end of this period (about 1600) the Western world was poised to make the great leap forward into a Machine Age.

Above: Metal-working in the Middle Ages, using an undershot waterwheel to drive a tilt-hammer and bellows.

Opposite: A wooden model of a plough, one of Man's half-dozen key inventions, left as a funerary offering in an early Egyptian tomb.

Below: Two of the three main pyramids at Giza, dating from about 2500 BC, which stand as a vivid reminder of the architectural achievements of the ancient world.

Early Days

Tracing our earliest ancestors is no easy task, and discoveries in recent years have tended to confuse rather than clarify the issue. Early finds in the 1920s and 1930s of fossils of the man-like ape *Australopithecus* in Southern Africa suggested that it could be the 'missing link' between ape and man. The Australopithecine remains were judged to be about 1 million years old.

But between 1972 and 1974 anthropologists working at Lake Rudolf in Kenya, Laetolil in Tanzania, and Hadar in Ethiopia discovered the remains of much more advanced creatures than *Australopithecus*, yet up to 3¾ million years old. They appear to have walked erect, and their remains are so much more distinctly human than ape-like that they have been assigned to the genus *Homo*. They are classified as early forms of *Homo erectus*. From the sites in which the fossils were found, it is clear that men-like apes and ape-like men co-existed together for millions of years.

The sites also offer evidence that both *Australopithecus* and the early *Homo erectus* used simple tools. At first they simply picked up handy sized rocks to use as tools and weapons. Later they deliberately struck pebbles to break off flakes and leave a sharp cutting edge. The tool-user became a tool-maker. Man had taken a giant leap, beginning to use his brains to extend his physical capabilities and shape the environment for his own ends, something that other species had singularly failed to do.

Contemporary remains of early tool-users outside Africa are lacking, and the reason is plain. About 2 million years ago the world climate began cooling as the Pleistocene Ice Age began. Primitive man, naked to the elements, could survive only in tropical regions.

There were, between the bitterly cold glacial periods, warmer times when Man would have been tempted out of the tropics in search of game, only to beat a retreat at the next glacial advance. By about half a million years ago, however, the taming of fire put a different complexion on things. With a means of keeping warm, Man could live outside the tropics, and so he did. Remains of advanced forms of *Homo erectus* have been found in Java (Java Man), China (Peking Man) and Germany (Heidelberg Man), all dating to about 400,000–500,000 years ago.

Gradually early men began making better stone tools, particularly hand axes. They were pear-shaped tools obtained by chipping flakes off both sides of a pebble with a hammer stone, stick or bone. Apparently as good as a modern knife for skinning an animal, the hand axe could also be used for digging and hammering.

By about 70,000 BC Man had made further progress in overcoming his environment. He lived in caves and knew how to *make* fire. He kept himself warm by wearing skin clothing. The dominant species at the time was Neanderthal Man, of whom many fossils survive. His stone tools were made from thick

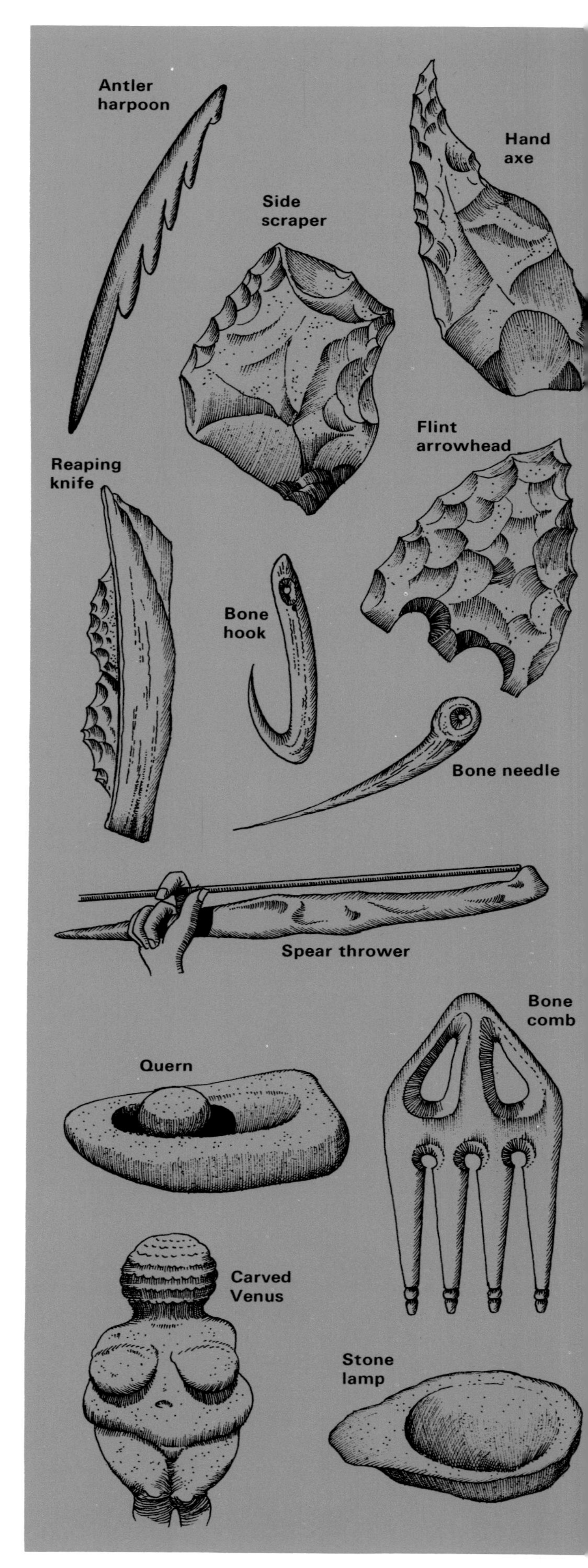

flakes of stone, finished by fine chipping. In their burial of the dead, the Neanderthals showed a great cultural advance.

Wise Man Cometh

The short, thickset and beetle-browed Neanderthals were ousted from the evolutionary scene by the direct ancestors of modern man, a breed we call the Cromagnons. Tall and well built, the Cromagnons had a long skull, narrow nose and prominent chin. Their brain capacity was, if anything, greater than our own. Their stone tools were the most sophisticated yet. They struck thin, narrow blades from a prepared core and fashioned them into numerous tools, such as knives, scrapers and awls. More important was their manufacture of specialist tools, or burins – sharp-edged chisels and gravers – for working antler, bone and ivory. It has been said that these burins were the forerunners of machine tools – tools made to shape other implements.

With greater guile than their forebears, the Cromagnons found hunting easy on the steppe and tundra, which teemed with bison, reindeer and mammoth. This allowed them more leisure, which they soon put to good use. They radically improved their weaponry, adding flint tips to their spears and making spear throwers to launch them with greater force. They invented the bow and arrow, which revolutionized hunting, for it enabled the hunter to kill his prey from a safer distance.

These early men extended their mastery over their environment when they learned how to make lamps. These lamps were hollowed-out stones or shells filled with animal oil or fat, into which dipped a crude form of wick. When you consider how important artificial light is to us today, the invention of the lamp can be seen in true perspective.

Increased leisure also allowed the advanced hunters ample time for artistic pursuits. They became skilful carvers, competent sculptors and consummate painters. The breathtakingly beautiful cave paintings of Lascaux in France and Altamira in Spain represent the summit of their achievement. Although they are difficult to date, these cave paintings are now believed to be the work of the Magdalenians, a race who flourished from about 15,000–10,000 BC.

By 10,000 BC the last Ice Age was drawing to a close. So was the Paleolithic, or Old Stone Age

Above: Part of the magnificent painting adorning the 'Room of the Bulls' in the Lascaux caves, near Montignac, in the Dordogne Valley of southern France. The astonishing colour and vigour of the Lascaux paintings belie their age, which is up to 15,000 years. Discovered in 1940, the Lascaux caves were open to the public for several years. But they began to deteriorate owing to attack by airborne moulds and bacteria, and they are now closed except for academic research.

Opposite: The tools that prehistoric man developed to improve his lot, using the materials nature provided. But very early on his artistic bent began to show itself, in painting and carving. The carved 'Venus' depicted here, found at Willendorf, Austria, is one of the earliest representations of the human form, dating from perhaps 20,000 BC.

Overleaf: The march of civilization, *ca* 8000–2000 BC.

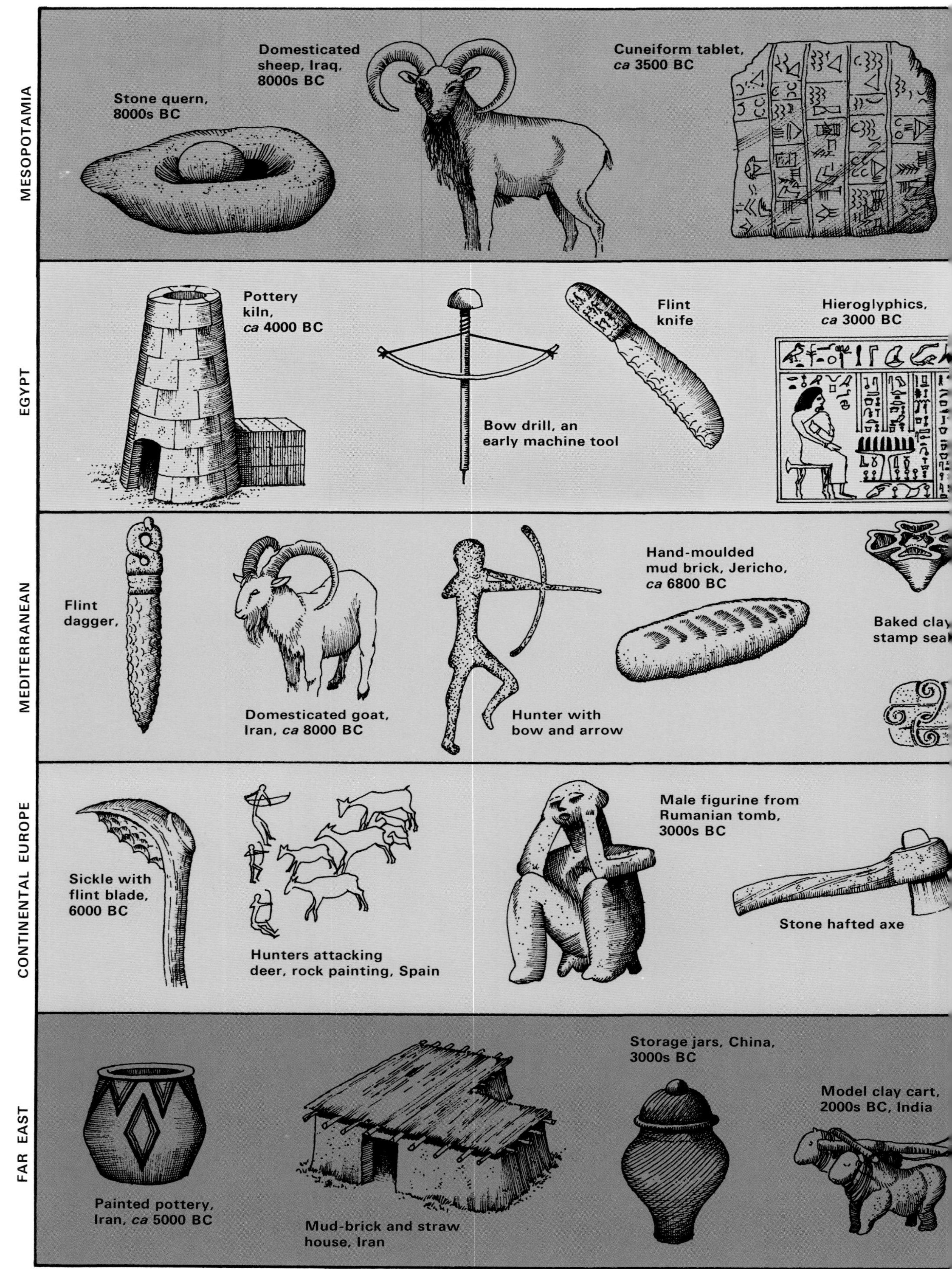
8000–2000 BC
MESOPOTAMIA
Stone quern, 8000s BC
Domesticated sheep, Iraq, 8000s BC
Cuneiform tablet, ca 3500 BC
EGYPT
Pottery kiln, ca 4000 BC
Bow drill, an early machine tool
Flint knife
Hieroglyphics, ca 3000 BC
MEDITERRANEAN
Flint dagger,
Domesticated goat, Iran, ca 8000 BC
Hunter with bow and arrow
Hand-moulded mud brick, Jericho, ca 6800 BC
Baked clay stamp seal
CONTINENTAL EUROPE
Sickle with flint blade, 6000 BC
Hunters attacking deer, rock painting, Spain
Male figurine from Rumanian tomb, 3000s BC
Stone hafted axe
FAR EAST
Painted pottery, Iran, ca 5000 BC
Mud-brick and straw house, Iran
Storage jars, China, 3000s BC
Model clay cart, 2000s BC, India

8000–2000 BC

lid-wood
heel, ca 3000 BC
Ziggarut of Ur,
ca 2200 BC
Brick-arch
vault, Ur
Cast copper
goddess, Ur,
2000s BC
Shaduf
Step pyramid,
Sakkara,
ca 2700 BC
Pyramids of Giza
ca 2600–2300 BC
'Beehive' house,
Cyprus, ca 5000 BC
Copper casting,
2000s BC
Ox-drawn plough,
Crete, 3000 BC
Cretan ship,
3000s BC
Rectangular
long-house,
Germany, 3000s BC
Copper axeheads,
2000s BC
Grimes 'graves', Britain,
flint mine, from ca 2300 BC
Seal stones,
India
Timber and thatched
houses, Japan, 2000s BC
Bronze statuette
of dancing girl,
Mohenjo-Daro, India
2000s BC

period of prehistory, to be followed by the Mesolithic, or Middle Stone Age period. This was characterized by the use of tiny flints called microliths, which were set in bone and wood to make tools such as reaping knives and sickles. On Mesolithic sites such tools are often found together with querns and pestles and mortars, which were used to grind grain. Man the hunter was well on the way to becoming a farmer, though as yet he was only harvesting cereals that grew wild.

The next step was for man to start growing his own crops from gathered seed, which first happened in about 8000 BC. To prevent him having to leave his growing crops to hunt, he began herding wild sheep and goats around his settlement. Then, by becoming selective in his choice of animals, he began true domestication.

The more settled life of the farmer led to more permanent dwellings of stone or mud brick. Craft industries grew up. The art of firing pottery became known, as did spinning and weaving. Mankind had entered into the last Stone-Age period, the Neolithic, or New Stone Age.

Until about 8000 BC the ways of life of men everywhere had been more or less the same. But after that date they began to diverge, region by region. The Near and Middle East began taking great strides forward, while in the Far East, Continental Europe and the Americas the old ways of life lingered on.

For this reason we have adopted in the illustrations appearing on pages 12–13, 16–17, 20–21, a regional viewpoint of Man's progressive mastery over materials and developing technology that laid the foundations for our Machine Age.

The 'War' side of the famous 'Standard of Ur', a decorated wooden box, thought to be the sounding box of a lyre. The figures are in shell and limestone, set on a background of lapis lazuri. Found in a tomb of the Royal Cemetery of Ur, it dates from about 2500 BC. Horse-drawn vehicles were already in widespread use by then.

Foundling Civilizations

The two great centres for the Neolithic Revolution as it can be called were Mesopotamia and Egypt. They had in common a warm climate that favoured agriculture and annual river flooding that deposited fertile silt over the land and revitalized the soil. Mesopotamia was situated between the Rivers Euphrates and Tigris, while Egypt lay on the River Nile.

By 5000 BC Mesopotamia and Egypt had many farming settlements. Over the next 2000 years these settlements blossomed into townships and then extensive cities, with the complex societies that went with them. In Mesopotamia the first great civilization arose – the Sumerian. The cities were dominated by massive mudbrick temples built to appease the gods, which towered above the simple rectangular houses of the ordinary folk. In Egypt, however, monumental architecture was confined to funeral tombs.

Among the growing band of artisans generated by city life were the coppersmiths, who first worked with native specimens of this attractive metal. Soon, whether by design or accident, they discovered how to smelt ores into copper and cast the molten metal. The furnaces they used for smelting were undoubted-

An almost life-size cast bronze head depicting the first great king in history – Sargon of Akkad, who ruled over the Sumerians in *ca* 2350 BC. It was found at Nineveh.

ly devised from the potter's kiln. The science of metallurgy began. But the budding metallurgists found that, like the other native metals – gold and silver – copper was really too soft to be made into really effective tools and weapons. So they began smelting other mineral deposits to try to produce a harder metal.

By 3000 BC the great breakthrough had occurred, not only in Mesopotamia but also farther west. The art of smelting bronze from copper and tin ores was discovered. The world was poised for another gigantic leap forward from a Stone into a Bronze Age.

The invention without which machines could not function – the wheel – had been made a little earlier. And the wheeled cart came into its own as a major form of transport, again first in Mesopotamia. Earlier, loads had been carried on sledges or moved with the aid of rolling logs or stones.

Transport speeded up on water too as men discovered how to harness the wind. Simple dugouts and reed boats propelled by paddles gave way to larger ocean-going vessels. They had a square sail to run with the wind, and oars to use when the wind was unfavourable. The Cretans were probably the first to go down to the sea in ships.

The wheel found another useful outlet – in pottery making. By shaping clay on a spinning wheel pots could be made with greater ease. The usefulness of rotary action was further realized in the bow drill, the first power tool.

The increasingly complex society of the Sumerians in Mesopotamia made it necessary for people in authority to keep records of such things as crop yields and numbers of livestock on estates. This prompted the greatest invention of all – that of writing. At first it was in the form of pictograms – pictures representing the objects being recorded. The next step was to use signs to stand for ideas (ideograms) or sounds (phonograms). Eventually the pictures became stylized and evolved into the wedge-shaped writing we call cuneiform, executed on clay tablets.

The Egyptians were a little later (3300 BC) in developing writing in the form of attractive hieroglyphics. But they were more adept at the art than the Sumerians, and soon they had invented a better material to write on – paper made from papyrus. This was a great advance over the bulky clay tablets previously used.

The Great Pyramids

The period 3000–2000 BC saw the full flowering of the Sumerian and Egyptian civilizations. Bronze became the dominant metal for tools and weapons. The standard bronze contained about one part tin to nine parts copper. With better tools craftsmen were able to produce better artefacts.

The Sumerian architects built the distinctive

Overleaf: The march of civilization, *ca* 2000–500 BC.

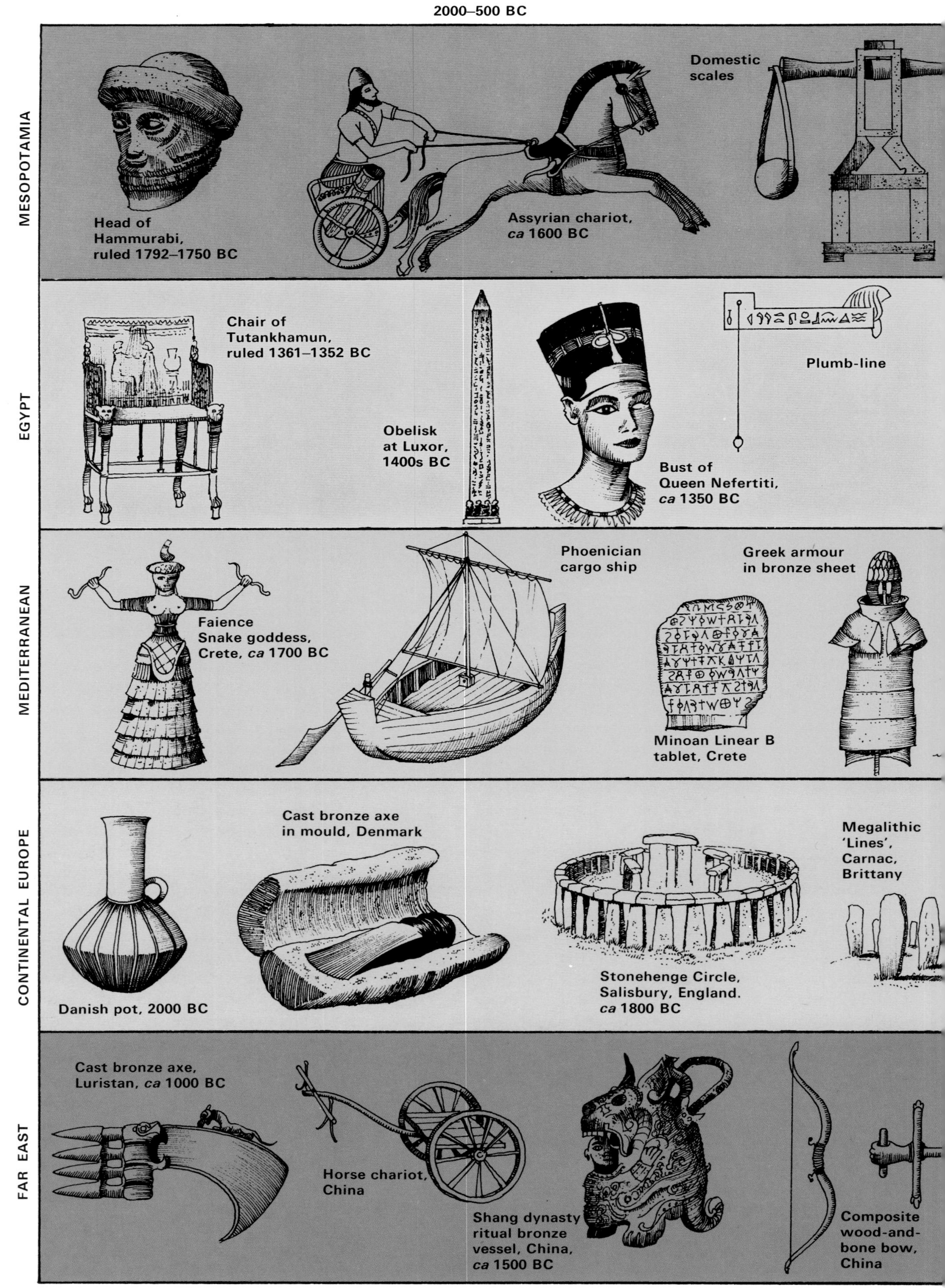
2000–500 BC
MESOPOTAMIA
Head of Hammurabi, ruled 1792–1750 BC
Assyrian chariot, ca 1600 BC
Domestic scales
EGYPT
Chair of Tutankhamun, ruled 1361–1352 BC
Obelisk at Luxor, 1400s BC
Bust of Queen Nefertiti, ca 1350 BC
Plumb-line
MEDITERRANEAN
Faience Snake goddess, Crete, ca 1700 BC
Phoenician cargo ship
Minoan Linear B tablet, Crete
Greek armour in bronze sheet
CONTINENTAL EUROPE
Danish pot, 2000 BC
Cast bronze axe in mould, Denmark
Stonehenge Circle, Salisbury, England. ca 1800 BC
Megalithic 'Lines', Carnac, Brittany
FAR EAST
Cast bronze axe, Luristan, ca 1000 BC
Horse chariot, China
Shang dynasty ritual bronze vessel, China, ca 1500 BC
Composite wood-and-bone bow, China

2000–500 BC

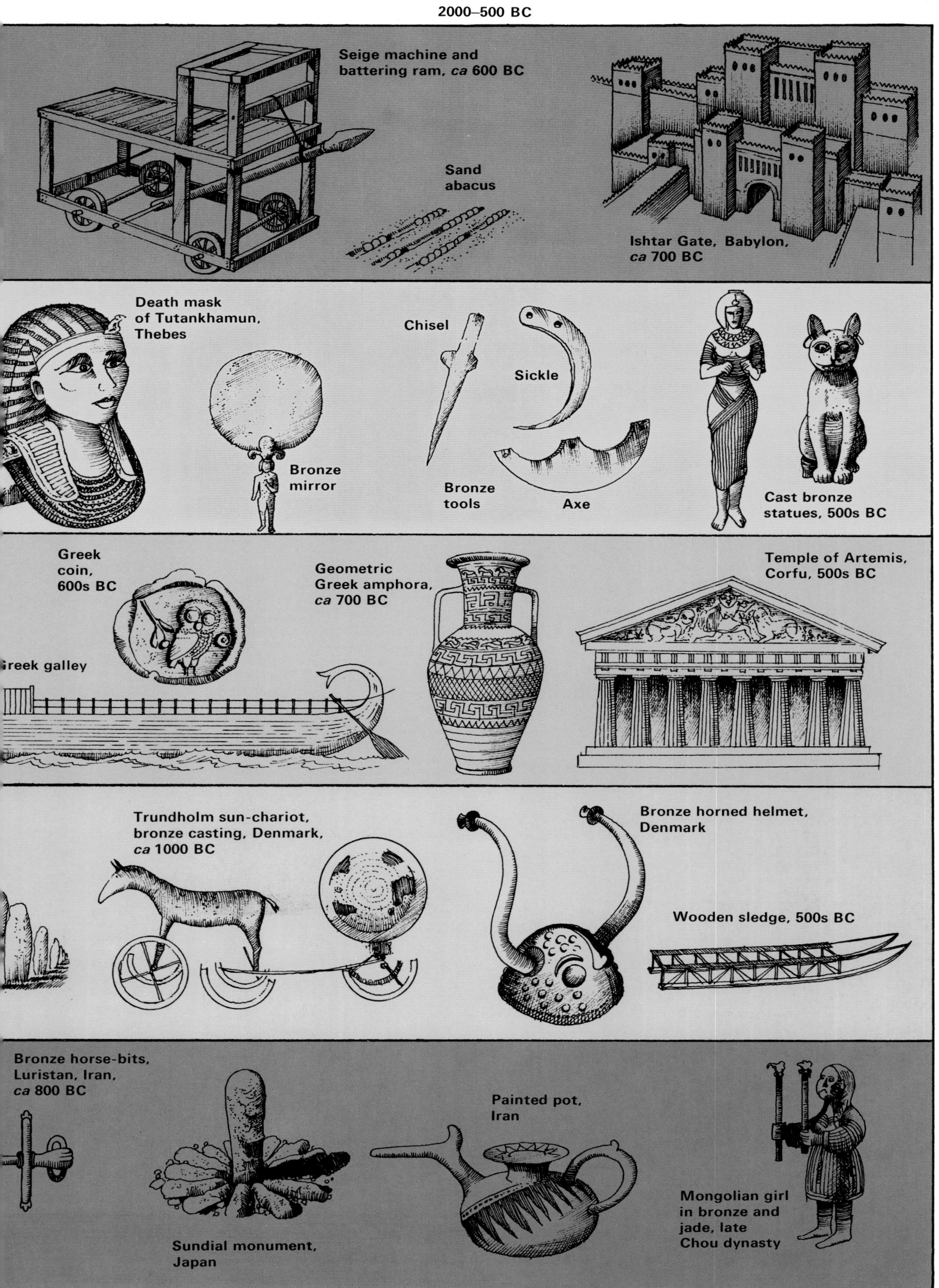
Seige machine and battering ram, *ca* 600 BC
Sand abacus
Ishtar Gate, Babylon, *ca* 700 BC
Death mask of Tutankhamun, Thebes
Chisel
Sickle
Bronze mirror
Bronze tools
Axe
Cast bronze statues, 500s BC
Greek coin, 600s BC
Geometric Greek amphora, *ca* 700 BC
Temple of Artemis, Corfu, 500s BC
ireek galley
Trundholm sun-chariot, bronze casting, Denmark, *ca* 1000 BC
Bronze horned helmet, Denmark
Wooden sledge, 500s BC
Bronze horse-bits, Luristan, Iran, *ca* 800 BC
Painted pot, Iran
Mongolian girl in bronze and jade, late Chou dynasty
Sundial monument, Japan

stepped ziggurats, but their greatest achievement was the invention of the brick arch. In the Nile Valley the Egyptians were building the most imposing tombs of all time for their pharaoh god-kings – the Great Pyramids at Giza, which still stand. The architect-priests who supervised the building of the pyramids were obviously excellent mathematicians, for the geometry of the pyramids is faultless. They must also have been experienced astronomers, for the astronomical alignment of the pyramids was very precise. One of the main passages in the Pyramid of Cheops, for example, pointed at the star Thuban, which at the time was the pole star.

But the thing that most impresses about the pyramids is their sheer size. The Great Pyramid of Cheops stands 137 m (450 ft) high, and its base covers more than 5·2 hectares (13 acres). A permanent work force of 4000 would have been required for nearly 30 years to manoeuvre its 2,300,000 limestone blocks, weighing 2 tonnes each, into position.

Rollers and sledges would have been used to transport the blocks, for strangely enough the wheel had not yet spread to Egypt. To raise the blocks up into position the workers would have used a ramp and balanced beams. These are forms of the simplest machines, the inclined plane (or wedge) and the lever. The balanced beam was already in use in Egypt to raise water, in the form of the shaduf. So was the weighing balance, which used the same principle.

The shaduf consisted of a pole with a bucket at one end and a counterweight at the other, pivoting about a pillar or stake. Because of the counterweight, the water-filled bucket could be raised with little effort. The stone blocks for the pyramids would have

Above: A scene from the colourfully illustrated papyrus 'Book of the Dead', many copies of which have been found in Egyptian tombs. Some texts date from 2400 BC. This scene shows the jackal-headed god Anubis weighing the soul of Anhai, the girl whose hand is held by the falcon-headed Horus, while the ibis-headed Thoth writes down the result. Practically identical scales are still in use in the Middle and Far East today.

Below: The side of a painted and stuccoed casket found in the famed tomb of Tutankhamun. It shows the boy king riding into battle in the newly invented war chariot.

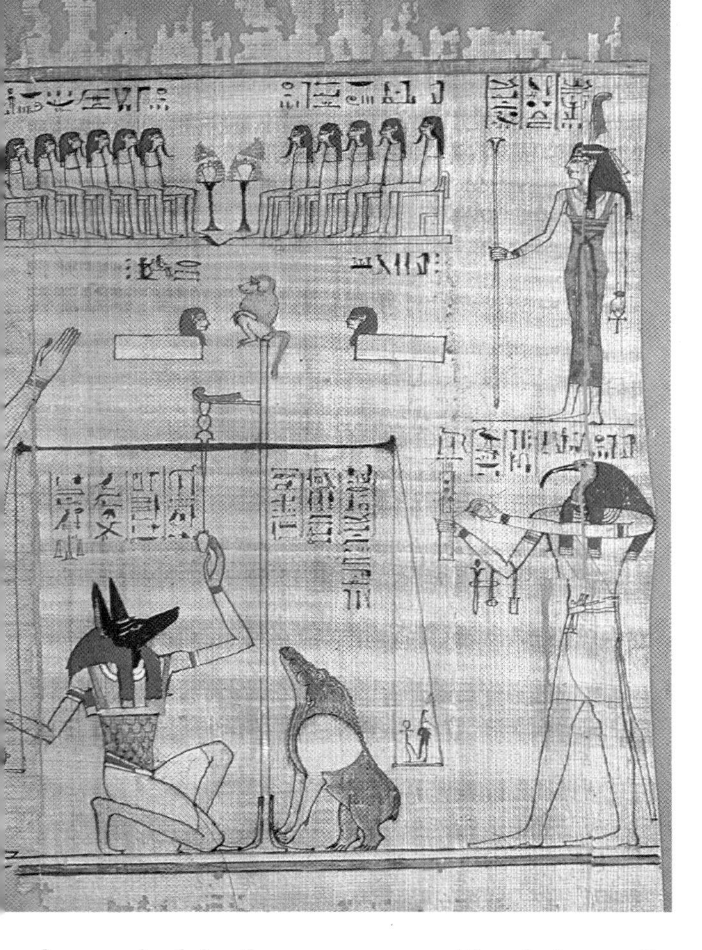

Below: The Rillaton gold cup, a precious relic of the so-called Wessex culture of Bronze Age England. Found in a grave beneath a round barrow at Rillaton, Cornwall, it is beaten from a single sheet of gold. The handle is riveted on. It dates from about 1600–2100 BC.

been raised in the same way, with what was the forerunner of the crane.

The shaduf was used widely for irrigation, which was essential for agriculture in both Egypt and Mesopotamia to distribute and conserve the flood waters. Building dams, dykes and canals was already an accomplished art. Some time before 2000 BC the agricultural scene was revolutionized, first in Mesopotamia, by the invention of the ox-drawn plough.

Model of a sailing boat, a funerary offering found in an Egyptian tomb. No fundamental changes in ship design occurred for 3000 years.

The plough came into use at much the same time in Crete, where the Minoan civilization grew up.

Sword and Chariot

The thousand years between 2000 and 1000 BC was a period of upheaval in Mesopotamia which saw the destruction of the Sumerian Empire and the rise of Babylonia. Other warring parties, including the Hittites, invaded the territory by and by and established supremacy for a time, only to be overthrown in their turn. There was upheaval in Egypt also, though it was mainly internal conflict.

The art of warfare progressed rapidly during this period and was revolutionized by the introduction of the swift horse-drawn chariot with light body and twin spoked wheels. The sword also made its first appearance and was soon widely adopted.

The Hittites, who ruled briefly in Mesopotamia, discovered the art of making iron in about 1500 BC. Iron ores could not be smelted properly to produce molten iron, because the high temperatures necessary could not be produced in the furnaces then in use. Smelting produced a spongy mass, or bloom, of slag and iron. This had to be beaten to expel the slag, producing what we now call wrought iron.

The Hittites jealously guarded the secret of iron-making for centuries and only after 1000 BC did it become widely known elsewhere. Practically every other community in the world, except in the Americas, had by then mastered the art of bronze working. China, who independently discovered the alloy, soon surpassed everyone in casting techniques.

Overleaf: The march of civilization, *ca* 500 BC–AD 500.

500 BC–AD 500

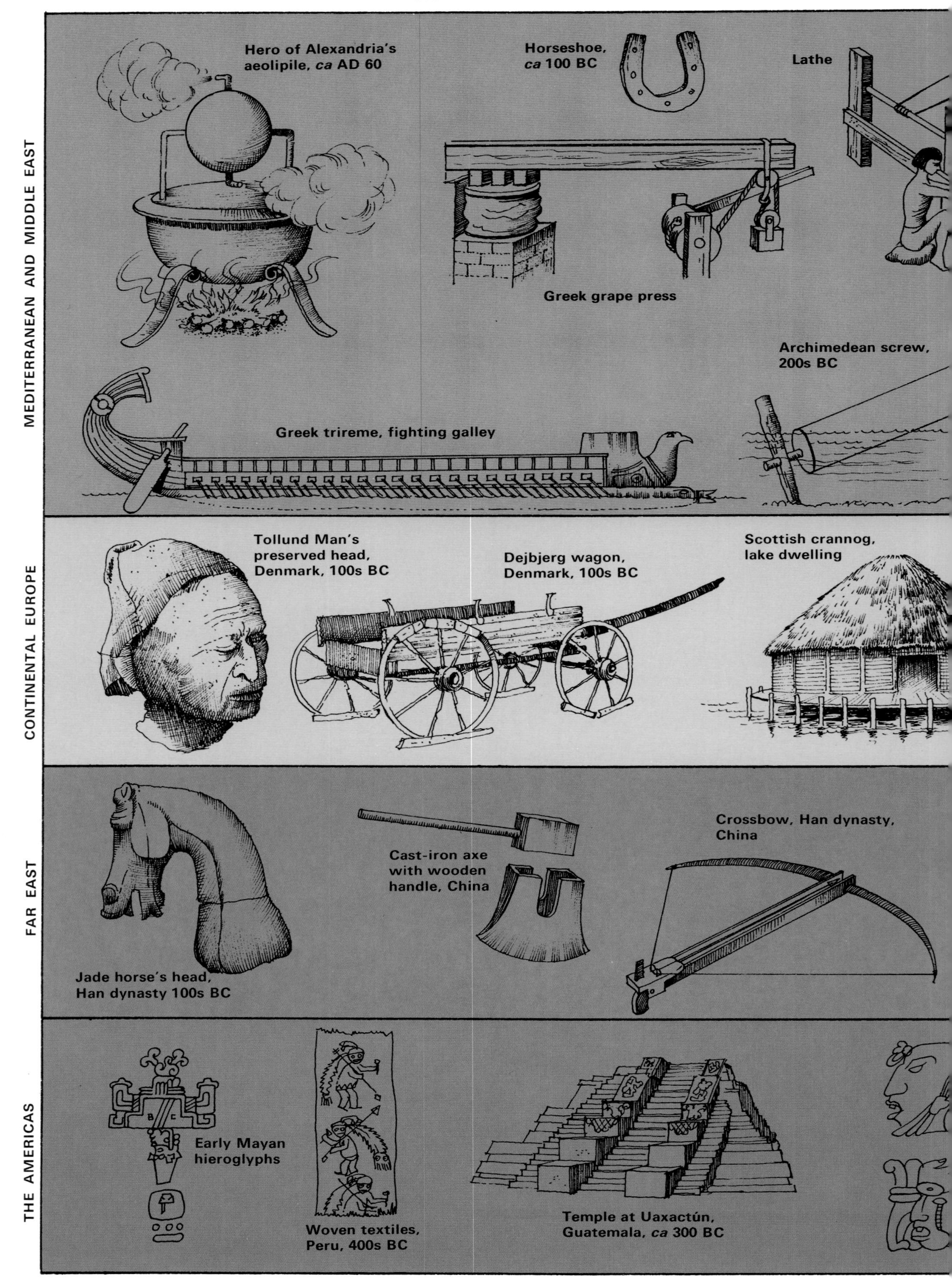
MEDITERRANEAN AND MIDDLE EAST
Hero of Alexandria's aeolipile, *ca* AD 60
Horseshoe, *ca* 100 BC
Lathe
Greek grape press
Archimedean screw, 200s BC
Greek trireme, fighting galley
CONTINENTAL EUROPE
Tollund Man's preserved head, Denmark, 100s BC
Dejbjerg wagon, Denmark, 100s BC
Scottish crannog, lake dwelling
FAR EAST
Jade horse's head, Han dynasty 100s BC
Cast-iron axe with wooden handle, China
Crossbow, Han dynasty, China
THE AMERICAS
Early Mayan hieroglyphs
Woven textiles, Peru, 400s BC
Temple at Uaxactún, Guatemala, *ca* 300 BC

500 BC–AD 500

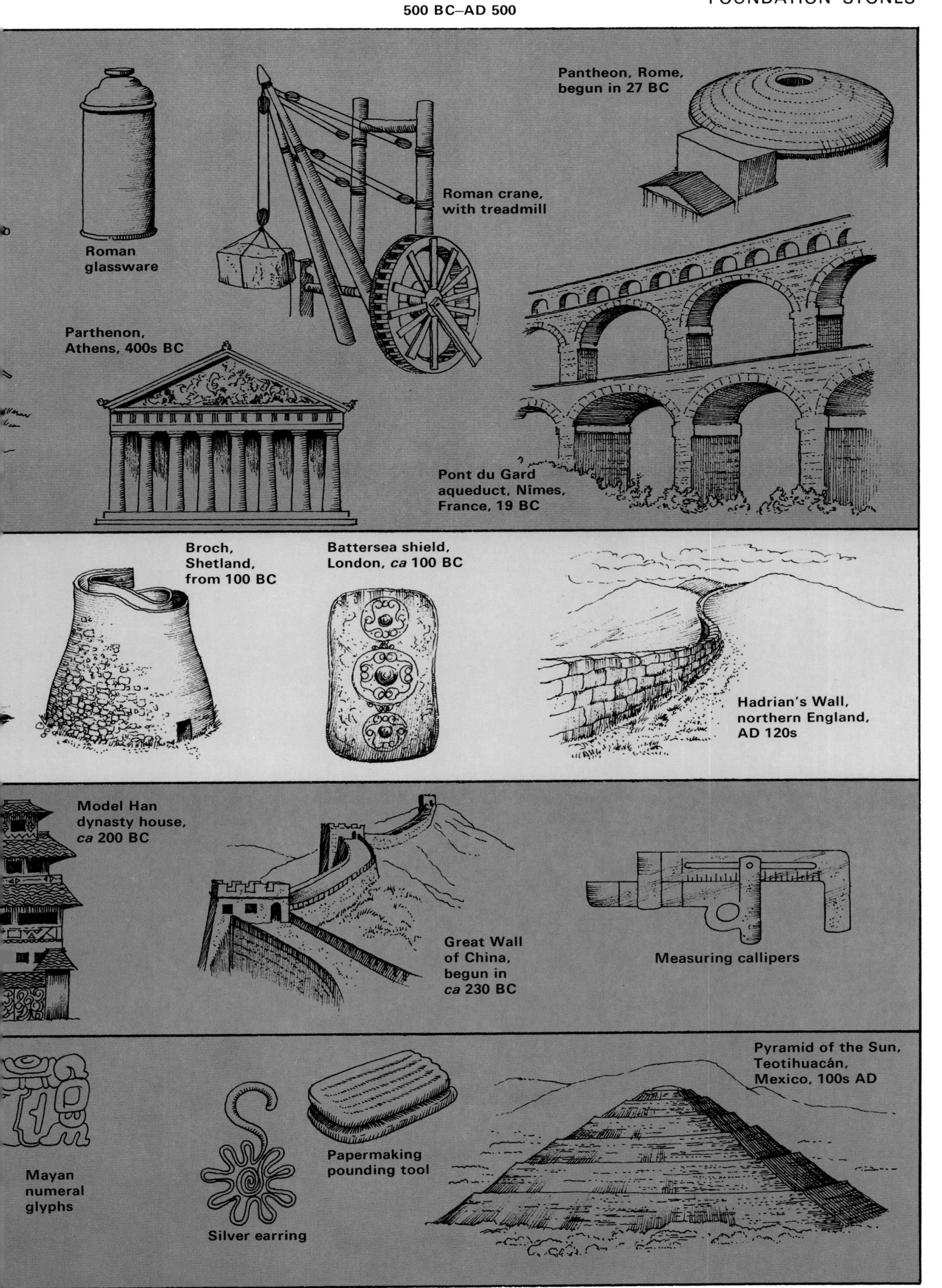
Roman glassware
Roman crane, with treadmill
Pantheon, Rome, begun in 27 BC
Parthenon, Athens, 400s BC
Pont du Gard aqueduct, Nîmes, France, 19 BC
Broch, Shetland, from 100 BC
Battersea shield, London, *ca* 100 BC
Hadrian's Wall, northern England, AD 120s
Model Han dynasty house, *ca* 200 BC
Great Wall of China, begun in *ca* 230 BC
Measuring callipers
Mayan numeral glyphs
Silver earring
Papermaking pounding tool
Pyramid of the Sun, Teotihuacán, Mexico, 100s AD

Megalithic Marvels
The Egyptian civilization achieved its greatest power and glory between about 1500 and 1000 BC, during the period known as the New Kingdom. Thebes was its magnificent capital. Its rulers included the Pharaoh Tutankhamun (1361–1352 BC), whose intact tomb, discovered in 1922, revealed fabulous riches that reflect the unrivalled splendour of the era. In mathematics and astronomy too the Egyptians were no laggards, though they were eclipsed in this respect by the Babylonians. They used a decimal system of counting, had a 365-day calendar, and used shadow clocks to tell the time by day and water clocks by night.

Far away in western Europe the natives boasted no such magnificent culture, and must in general be considered barbarians in comparison with the sophisticated populace of Egypt and Babylon. Yet the strange megalithic monuments these 'barbarians' left behind belie this. The construction of the unique Stonehenge, near Salisbury, for example, must have been the product of an advanced and wealthy society. The erection of the 55-tonne blocks (sarsens) for the main structure must have been a prodigious task; so must the transport of the 4-tonne subsidiary stones (bluestones) from a Welsh quarry 212 km (132 miles) away.

But it is the design of Stonehenge, constructed between about 1800 and 1400 BC, that was so brilliant. It was laid out with incredible precision by architects whose knowledge of astronomy must have equalled if not exceeded that of contemporary Babylonian and Egyptian astronomers. It is thought to have served as a place of worship and as an astronomical calendar which enabled the priest-astronomers of the day to predict eclipses, for example. There is evidence from the architecture that the Bronze Age Britons had contacts with Greeks from Mycenae, who had themselves inherited much from the nearby but now decaying Minoan civilization of Crete, centred on Knossos.

The Rise of the West
The achievements of the Minoans and the Mycenaeans were indicative of the way things were progressing. The centres of cultural influence were beginning to shift westwards. From 1000 BC Egypt went into decline militarily and culturally. The warlike Assyrians dominated Mesopotamia for a while before the Medes and Persians conquered them.

In the Mediterranean the Phoenicians established a great sea-trading empire. Carthage flourished from about 800 BC, and the Greeks became their implacable enemies. This was the time of Homer, of the rise of Greek philosophy at Miletus. In short it was the beginning of classical Greece.

The glory that was Greece, centred on Athens, laid the foundations for the present western civilization. The attainments of the Greeks in the arts, architecture, law, philosophy and sciences were legion, despite the fact that they always seemed to be at war! Mostly it was war between the city-states of Greece, particularly Athens and Sparta.

By 500 BC, to the east, the Persians had accumulated a huge empire which included Babylonia and Egypt, and faced the Greeks across the Aegean Sea. In 490 BC under Darius they invaded Greece in force, only to be decisively defeated at Marathon. Ten years later under Xerxes they crossed into Greece, but had to withdraw when their supporting fleet was routed at Salamis.

In 334 BC Alexander the Great, now ruler of Greece, attacked the Persian Empire and within three years had conquered it. Under him Greek culture spread throughout the Empire even as far as India, bringing in the Hellenistic Age. Among the many cities he founded was Alexandria in Egypt, which rapidly became a centre of learning, science and invention.

Eureka!
It was in Alexandria that the famous Greek mathematician and inventor Archimedes (287–212 BC) studied before settling at his birthplace Syracuse,

Left: The world's most famous prehistoric monument, Stonehenge, near Salisbury, Wiltshire. Though popularly associated with Druid worship and human sacrifice, this unique megalithic structure was almost certainly constructed as an astronomical instrument to serve as a calendar and possibly to predict eclipses.

Right: The Athenians had developed to perfection the art of black-figure vase painting by 500 BC. The vases depict scenes from mythology and also everyday life, such as this one showing blacksmiths at work.

in Sicily. He is most often remembered for his alleged dash naked through the streets after leaping from his bath, having solved a thorny problem in buoyancy. His law of buoyancy, which we know as Archimedes' Principle, is fundamental to hydrostatics and explains, for example, how steel ships can float. But it is in Archimedes' mechanical genius that our interest lies.

Archimedes was the first to study basic mechanics, working out, for example, the 'law of the lever'. 'Give me a place to stand on and I will move the Earth', he said. By use of levers and what appears to have been a primitive winch he enabled his king to move a fully laden ship by hand. For the defence of Syracuse against the Romans, Archimedes built powerful catapults (ballista) and other war machines that kept the Romans at bay for three years. They are reputed to have included a contrivance with convex mirrors that concentrated the heat of the Sun and set fire to the Roman ships. The other invention credited to Archimedes is the Archimedean screw for raising water, which is still used in the Middle East today. But Archimedes was so ashamed of his mechanical gadgets, which he felt were beneath the dignity of a man of science, that he left no record of them.

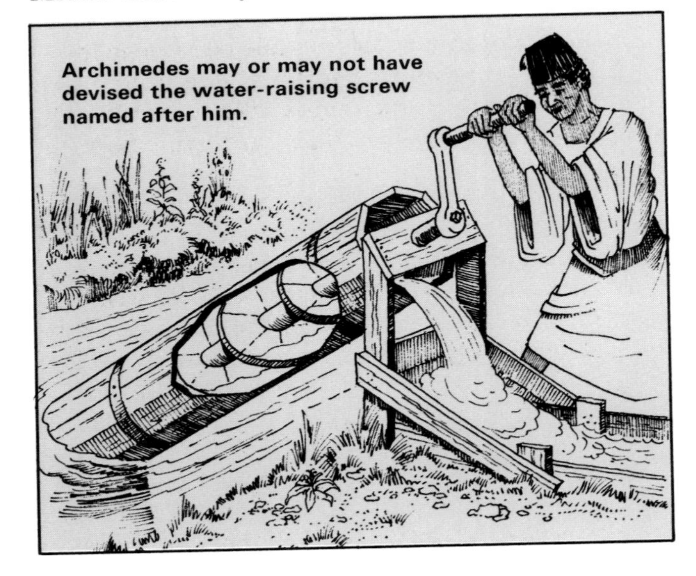

Archimedes may or may not have devised the water-raising screw named after him.

A Greek who must have been living in Alexandria at the same time as Archimedes was Ctesibius. He was a most ingenious fellow who designed many interesting devices. One was a very much improved water clock, or clepsydra, which had a valve to control the water supply into the upper vessel and so keep the water level there constant. With a constant head, the water ran through the exit orifice at a constant rate into a lower vessel, where it made a float slowly rise. By use of spur and bevel gears the motion of the float was transmitted to a pointer which moved over a scale to indicate the time.

Ctesibius's most interesting invention, however, was his hydraulis, which was the forerunner of the music organ. It incorporated several new mechanical devices, including a one-way piston pump. This was the first recorded use of the piston and valve, indispensable components of the modern engine. The piston pump was incorporated in a device that we would now describe as a force pump. It supplied air to a reservoir, which was an open-ended drum upturned in a container of water. Air was admitted to the organ pipes from the top of the reservoir, while it could escape from the bottom. In this way a more or less constant pressure was maintained in the reservoir.

Rome Rules

Syracuse, which, thanks to Archimedes, had withstood Roman attack for years, finally succumbed in 212 BC and Archimedes was killed. Rome was by now one of the two most powerful nations in the western Mediterranean. The other was Carthage in North Africa. Rome eventually in 146 BC sacked Carthage in the last of three lengthy wars (the Punic Wars). In the same year Roman armies destroyed Corinth, and Greece became a Roman province.

Over the next century the Romans, whose fighting

prowess was to become legendary, conquered the whole of the Hellenistic world. They also extended their power in the west. Julius Caesar subdued the Gauls and Franks and invaded Britain (55 BC). In 49 BC he defied the Senate by leading his troops across the Rubicon, and effectively became dictator. Five years later he was murdered as a result of the best-known conspiracy in history, an act intended to preserve the Republic. But it was not to be. In 31 BC Caesar's adopted son Octavian gained supreme power. In 27 BC he took the name Augustus and became Rome's first Emperor.

For 200 years peace reigned throughout the Roman Empire – the Pax Romana – for no country was strong enough to threaten it. At its height the Empire stretched from Britain to the Persian Gulf, from the Danube to North Africa.

The Romans did not sweep away the Greek culture of the regions they dominated. Rather they adopted and extended it. So imperceptible is the transition from Greek to Roman that the thousand years of history from 500 BC is called the Greco-Roman period.

Power to the People

The Romans, considering the wealth, security and longevity of their empire, contributed remarkably little in the way of new inventions. Their great contribution to posterity was in the way they applied and developed the technology they had inherited from their forebears.

They were magnificent engineers whose bridges, buildings and aqueducts were of unrivalled beauty. Their method of building roads, essential for rapid communications throughout the Empire, was masterful. To give permanence to their roads the Romans built them of several layers, incorporating a form of cement that set hard. By adding gravel to it, they discovered that vital building material – concrete.

Above: A reconstruction, in Malibu, California, of a Roman villa, the Villa dei Papiri, once located south of the city of Herculaneum, which was destroyed in the AD 79 eruption of Vesuvius. The recreated villa forms the J. Paul Getty Museum.

Details of the architecture and construction methods of Roman times have come down to us through the book *De Architectura*, written by Vitruvius near the beginning of the Augustan Age. This handbook covers not only every aspect of contemporary architecture and building machines, but such things as water clocks, sundials and, most important, waterwheels.

The upright waterwheel is often called the Vitruvian in his honour, though he almost certainly did not invent it. It was undershot – turned by water flowing beneath it. It was a natural development from the treadmill, used to power cranes or water scoops for irrigation. The importance of the introduction of the waterwheel cannot be over-emphasized for it ushered in the age of power-driven machinery and as a power source was unrivalled until the advent of the steam engine.

Outstanding Hero

If steam was harnessed in ancient times, it was not for a practical purpose. One of the most talented of the Alexandrian Greeks, named Hero (1st century AD), built an interesting device called the aeolipile, which was spun by steam. It can be regarded as the forerunner of the steam turbine and jet engine.

Hero's device (see diagram, page 20) consisted of a boiler on top of which a hollow sphere could rotate on supports. Attached to the sphere were two right-angled tubes. Steam from the boiler was piped via one of the supporting arms into the sphere and escaped from it through the angled tubes. Reaction to the escaping jets caused the sphere to spin.

Hero's mechanical genius must be reckoned on a par with that of Leonardo da Vinci and, like Leonardo, Hero left many books detailing his work. His aeolipile was described in a book called *Pneumatica*, which also included descriptions of siphons, a fire engine, a windmill-operated organ and, believe it or not, the first vending machine – a coin-operated device to dispense holy water! Many of Hero's other books are devoted to theoretical and practical mechanics. And he even wrote one on automata (*Automatopoietica*), describing how such devices could be activated by water, falling weights and steam. He is reputed to have made mechanical birds that chirped, drank and flew.

Below: One of the two most impressive remaining Roman aqueducts, the Pont du Gard, near Nîmes, southern France. The other is in Segovia, Spain. Nîmes is particularly rich in Roman remains, having been one of the wealthiest towns in Roman Gaul.

Decline and Fall

The Pax Romana, the two centuries of Roman peace, drew to a close in AD 180, when the mighty Roman army began to suffer defeats by invading barbarian tribes in the east and west. Over the next 200 years Rome's power dwindled and the Empire shrank. The decline was temporarily halted by the Emperor Diocletian (ruled 284–305), who reorganized the Empire, in particular dividing it into east and west parts. His successor was Constantine (305–337), deservedly known as the Great. He made Christianity legal and established a new capital at Byzantium, renaming it Constantinople (modern Istanbul).

This Pompeiian wall painting shows the typical ship of the period in a mythological scene depicting Ulysses.

The astrolabe had developed into a refined navigational instrument by 1572, when this one was made in the Netherlands.

After Constantine's death the Empire began to fall to pieces again and by 395 had formally split in two. The West Roman Empire, based on Rome, grew weaker and weaker. Rome itself was captured and looted by the Gauls in 410 and sacked and burnt by the Vandals in 455. The Roman Empire in the west was finally overthrown in 476. The East Roman, or Byzantine Empire, however, survived and its Greco-Roman culture flourished for nearly a thousand years longer. Not until 1453 did it succumb to the growing might of the Turks.

Eastern Promise

The thousand years in Europe following the collapse of the West Roman Empire is usually termed by historians the Middle Ages. The first half is gloomily called the Dark Ages, because for many the light of classical civilization had been extinguished. In technology too things appeared to stagnate. But if lull there was in the West, there certainly wasn't in the East, in China in particular.

The Chinese civilization had developed in comparative isolation and had independently evolved technology that was in many respects superior to that anywhere else in the world. For example, it discovered a method of casting iron in 600 BC – 1800 years before Europe did. Yet in other respects China was very backward – for example, in not developing an alphabet. The enigma of China persists to the present day.

China went through a particularly creative period during the Han dynasty (200 BC–AD 200). During this period the Chinese perfected the crossbow, which had a particularly ingenious trigger mechanism.

They introduced the horse collar and stirrup, and the wheelbarrow. The horse collar was particularly important for it allowed horses to be used as draught animals for the first time. With the stirrup, horse soldiers could stay mounted more easily in battle. In AD 105 an official in the Emperor's court, Ts'ai Lun, discovered how to make paper by beating vegetable fibres into a pulp and pressing this into sheets.

But the ingenuity of the Han Chinese is perhaps better illustrated by the invention by the mathematician and astronomer Chang Heng of a seismograph. It consisted of a bronze pot containing a heavy pendulum. Surrounding the pendulum were eight arms, each connected by lever to a decorative dragon with a ball in its mouth. As the pendulum started to swing in an earthquake tremor, it knocked one of the arms, which caused a ball to fall from the dragon's mouth, into the gaping mouth of a bullfrog. By this means the seismograph not only detected earthquakes, but also indicated their direction.

The Chinese jealously guarded their many 'trade secrets' for centuries. The art of papermaking, for example, did not become known outside China until the AD 750s, when Arabs learnt the secret from some Chinese prisoners. It eventually filtered through to the West via the Moorish conquests of Spain.

It is to the Chinese too that credit is usually given for the invention, by AD 1000, of gunpowder, that explosive mixture of sulphur, charcoal and saltpetre. But it is probable that gunpowder was invented independently in Europe as a by-product of the alchemy that many people then practised. The English philosopher Roger Bacon, knowledgeable about things alchemical, published a recipe for gunpowder in 1242. And within a century the cannon had established itself on the battlefield, particularly as a siege weapon.

The Chinese initially used gunpowder in their fireworks, but by 1232 they had almost certainly begun using it in warfare, though not in cannons. In that year a contemporary account speaks of them having used 'arrows of flying fire' in their fight against the Mongols. It seems more than probable that these missiles were rockets with gunpowder as propellant.

Arabian Advance

The Arabs in the Middle Ages were instrumental in transmitting many elements of Eastern technology to the West. But they also had their own contributions to make, particularly in mathematics and science. They were skilled chemists who knew all about such things as distillation and crystallization. They were expert astronomers who built on the observations of the great Greek astronomers Hipparchus and Ptolemy. It is through an Arabian translation published in AD 827 that we know Ptolemy's great encyclopedic mathematical and astronomical work *Almagest* ('The Greatest'). The Arabian influence on astronomy is evident in such star names as Algol, Deneb, and Betelgeuse.

To improve the accuracy of their astronomical observations, the Arabs refined an instrument used by Hipparchus – the astrolabe. The astrolabe started life as a simple device for observing the position and altitude of heavenly bodies. By the late Middle Ages, under Arab influence, it had become a sophisticated navigational instrument, invaluable to the new breed of intrepid mariners.

Three other advances besides the astrolabe paved the way for the great age of seafaring exploration that occurred in the 1400s. They were the lateen sail, the stern rudder and the compass. The triangular lateen sail had been used by the Arabs on their dhows centuries before it was adopted by European sailors in the 1200s in addition to the traditional square sail. By using the triangular sail, it became possible to sail ships close to the wind. By deft use of the newly adopted stern rudder, which had its origins in the Netherlands, ships could now sail into the wind, by tacking – that is zig-zagging at an angle to the wind direction. Though they might constantly change direction, they still knew, more or less, where they were heading, thanks to the magnetic compass. This was in widespread use by the 1200s, some two centuries after knowledge of it filtered through to Europe from – yes, China!

Power from the Elements

While sailors were harnessing the wind more effectively at sea, some landlubbers realized how they could harness it to drive machinery. And the windmill was born. Although primitive windmills had been used in Persia for several centuries, their sails turned on a vertical axis. The European windmill, which came into widespread use in the 1200s, had sails that turned on a horizontal axis. Windmills

A sixteenth-century post mill, which had to be turned round manually to keep the sails facing into the wind.

were to remain a prime power source until the late nineteenth century.

The early windmills were post-mills. They were mounted on a vertical post and had to be turned round bodily to keep the sails facing into the wind. By the late 1400s, however, many were being built with a fixed body. The only part that had to be turned as the wind shifted was a turret that carried the sails.

The windmill now vied with the watermill as a power source for grinding grain. It found most application in lowland regions like the Low Countries in Europe and the Fens in England where the streams did not flow fast enough to spin waterwheels. By the 1400s most of the waterwheels in use were of the vertical, Vitruvian type, though there were still quite a number of horizontal wheels about. Most of the vertical wheels were undershot, that is, driven from below, but where there was sufficient head overshot wheels found favour.

The millwrights became progressively more skilled in construction, increasing the efficiency of power transmission through improved gearing. And soon their skills were in demand in other industries where there was potential for power-driven machinery. It was the waterwheel that became the more versatile prime mover rather than the windmill. Flowing water is a much more dependable power source than the capricious wind.

Initially it was in the iron, textile and paper industries that water power first found application. The machinery used in each one was essentially the same, typified by the tilt-hammer used to forge iron. The tilt-hammer consisted of a beam which could pivot up and down. On one end was a heavy hammer. The other end was knocked down by a cam fixed to the drive shaft from the waterwheel. As the cam moved on the heavy hammer fell on to an anvil, to be lifted again when the cam came round again. In this way a regular hammering action was maintained. In the wool industry this was used in fulling cloth, that is, shrinking and felting woollen material; and in papermaking in beating rag into pulp. The wool industry also benefited about this time (1500) from the introduction of the spinning wheel.

Above: A page from the most beautiful of books – the 42-line Gutenberg bible, printed in 1455 on a crude wine press.

Top right: The mechanism of the oldest working clock, in Salisbury Cathedral, which dates from 1386. Clearly seen in this photograph is the verge-and-foliot regulator above the train of gears.

Opposite: An overshot waterwheel being used to work a rag-and-chain pump for mine drainage. The illustration appears in Agricola's *De Re Metallica* (1556).

Below: A Chinese tomb-rubbing, showing an early spinning wheel.

The paper industry was increasing in importance because of the recent development of printing from movable type. Printing revolutionized communications. For the first time books could be mass produced, instead of being laboriously copied by hand one by one. Learning, once the prerogative of the privileged few, became available to the masses.

Printing from a single carved block dates from the year 770 in Japan, but the idea of movable type came nearly three centuries later. In about 1040 a Chinaman, Pi Sheng, made movable type characters out of clay, but they were too fragile to be used repeatedly. It was a Korean, Yi Kyo-bo, 200 years later who made the system work, carving movable type from wood. In the early 1400s the Koreans started casting type in bronze.

Not until about 1450 did printing get underway in Europe, pioneered by Johannes Gutenberg in Germany. He set up a printing house in Mainz and in 1456 produced the first printed Bible, one of the most beautiful of all books. The machine he used for printing was a modified wine press, capable of producing 300 sheets a day.

Time for Clocks

The increasing interest in things mechanical in the Middle Ages was also applied to timekeeping, resulting in the mechanical clock. The origins of the clock are obscure, but certainly in the 1300s clocks made an appearance in Europe. The first ones were installed in towers and called turret clocks. They had neither hands nor dial, nor did they strike the hour. They simply sounded an alarm to remind the timekeeper to ring a bell. The word 'clock' is derived from a late Latin word meaning 'bell'. (Compare the modern French word 'cloche', meaning 'bell'.)

The oldest surviving striking clock is that in Salisbury Cathedral which dates from 1386. It is powered by a falling weight and has a verge escapement. This consists of an oscillating arm (foliot) on a spindle. Projections (pallets) on the spindle alternatively engage and release one by one the teeth of a gearwheel powered by the weight. The gearwheel drives the clock mechanism. The Salisbury clock, which worked virtually non-stop until 1884, is now fully restored and still strikes the hours.

The early domestic clocks were smaller versions of the turret clocks. Then in about 1500 a German locksmith, Peter Henlein, made the clock a more portable instrument. He introduced the compact coil spring to drive the mechanism instead of falling weights. The clock now had a dial and an hour hand. As a regulator the foliot gave way to the balance wheel, but this was still not very precise. It was to be many years before an accurate regulator was introduced – the pendulum.

The pendulum is an effective regulator because the time it takes to swing back and forth is constant, no matter how large the swing. The time of swing depends only on the pendulum's length. The person who first observed this law was the Italian scientist

and mathematician Galileo, who is reputed to have timed the swing of lamps in the cathedral of his native Pisa. Later he was to build the first telescope (page 192) and train it on the heavens.

The Reawakening

Galileo (1564–1642) was one of the many gifted men who grew up in the period of history we call the Renaissance. This period, which covers approximately the two centuries between 1400 and 1600, saw the revival of interest in classical art, architecture and learning.

People once again began to question their role in society and seek to change that society for the better. They began to reason things out for themselves instead of blindly following the dictates of others – of the Church in particular. This brought about a revolution in religion, the Reformation, which led to the establishment of Protestantism in many European countries.

Part of Michelangelo's Sistine Chapel ceiling fresco.

The Renaissance let loose a veritable flood of creative genius, in Italy in particular. Among the consummate artists of the period were the sculptor Donatello (1386–1466) and the versatile duo Michelangelo (1475–1564) and Leonardo da Vinci (1452–1519), who must surely tie for the title 'greatest artist of all time'. Michelangelo and Leonardo were not only painters, but also architects and sculptors.

In addition Leonardo was a skilled engineer and prolific inventor of mechanical gadgets. If one person had to be singled out as embodying the spirit of the Renaissance, it would be Leonardo. He represented the Renaissance ideal of the universal man.

By 1600 the foundations of our mechanical age had been laid. In the seventeenth century scientific expertise and technology developed at an ever-increasing rate, preparing the ground for the Industrial Revolution that was to come.

THE GENIUS OF LEONARDO

Leonardo da Vinci was born in 1452 in Vinci, a small village in Tuscany. He was the illegitimate son of a Florentine lawyer and property owner. His artistic bent obviously appeared at an early age for when he was 15 he was apprenticed to the painter Verocchio. In 1472 he was accepted in the painters' guild in Florence, where he remained until 1481.

By then Leonardo's expertise with paint brush and palette, pen and pencil was already well advanced. And among his early drawings were many sketches of mechanical apparatus and weapons, evidence of his interest in, and knowledge of things mechanical.

Between 1482 and 1499 he was employed in the service of the Duke of Milan, to whom he was painter, sculptor, musician and technical adviser on military and engineering matters. His artistic achievements in Milan reached their peak with the mural 'The Last Supper' completed in 1497.

But his creative energies now were turning more and more to scientific and literary pursuits. In the 1490s he began monumental treatises on painting, architecture, human anatomy and mechanics. He set down his observations on these themes in voluminous notes and sketches, which he would later assemble in his notebooks. There remain of his notebooks a prodigious 7000 pages, all in characteristic 'mirror-writing'.

Leonardo returned to Florence in 1499, where he painted that most famous painting 'The Mona Lisa' (1503). He then went back to Milan and entered the service of the French King Louis XII. Later he was to work in Rome with Raphael and Michelangelo on designs for the new church of St Peter. In 1516 he settled in France, at Cloux, near Amboise, where he died three years later.

In whatever subject he studied, Leonardo laid absolute faith in the evidence of his eyes. He was no mere theorist advancing fanciful ideas. He was a practical man, who designed things that would work, because he could *see* how they would work.

And it is in his 'things', his machines, that we are interested in this book. There is no evidence that Leonardo actually built the machines and mechanical devices he sketched and described. And in many cases their practical importance remained unrealized and unrealizable for centuries. There was neither the demand for them nor the technology.

It is almost as if Leonardo was born into the wrong period of history. Had he grown up in the eighteenth century, the emergent technology of the Industrial Revolution would have enabled him to give substance to his technical *tours de force*, to advance the Age of Machines. Imagine a collaboration between Leonardo and James Watt!

In the field of weaponry he designed a breech-loading cannon, mammoth crossbows and a multibarrelled machine gun, the like of which was not seen on the battlefield until the 1800s. And his design for an armoured vehicle foreshadowed the tank of World War I.

As a result of his studies of the flow of fluids, he designed the V-shaped double-leafed lock gates, which he realized would be easier to open and better able to withstand the pressure of the water. In his exhaustive studies of air currents and the phenomenon of flight, Leonardo pioneered what we would call today aerodynamics. He designed a 'flying machine' not unlike that of the early glider pioneers, like Otto Lilienthal. His 'aerial screw' is an early ancestor of the helicopter, and his 'tent roof', of the parachute.

Among Leonardo's other ingenious devices were spinning and weaving machines, power loom, pumps, dredgers, treadle lathe, primitive bicycle, hydraulic jack and gas turbine. He incorporated the turbine in his automatic spit, designed for installation in a chimney. Hot air rising from the fire spun a turbine wheel which, through right-angled gears, turned a cooking spit.

But it is perhaps in his appreciation of the form and function of individual parts of the machine – gears, joints, bearings and so on – rather than the machine itself that Leonardo's mechanical genius lay.

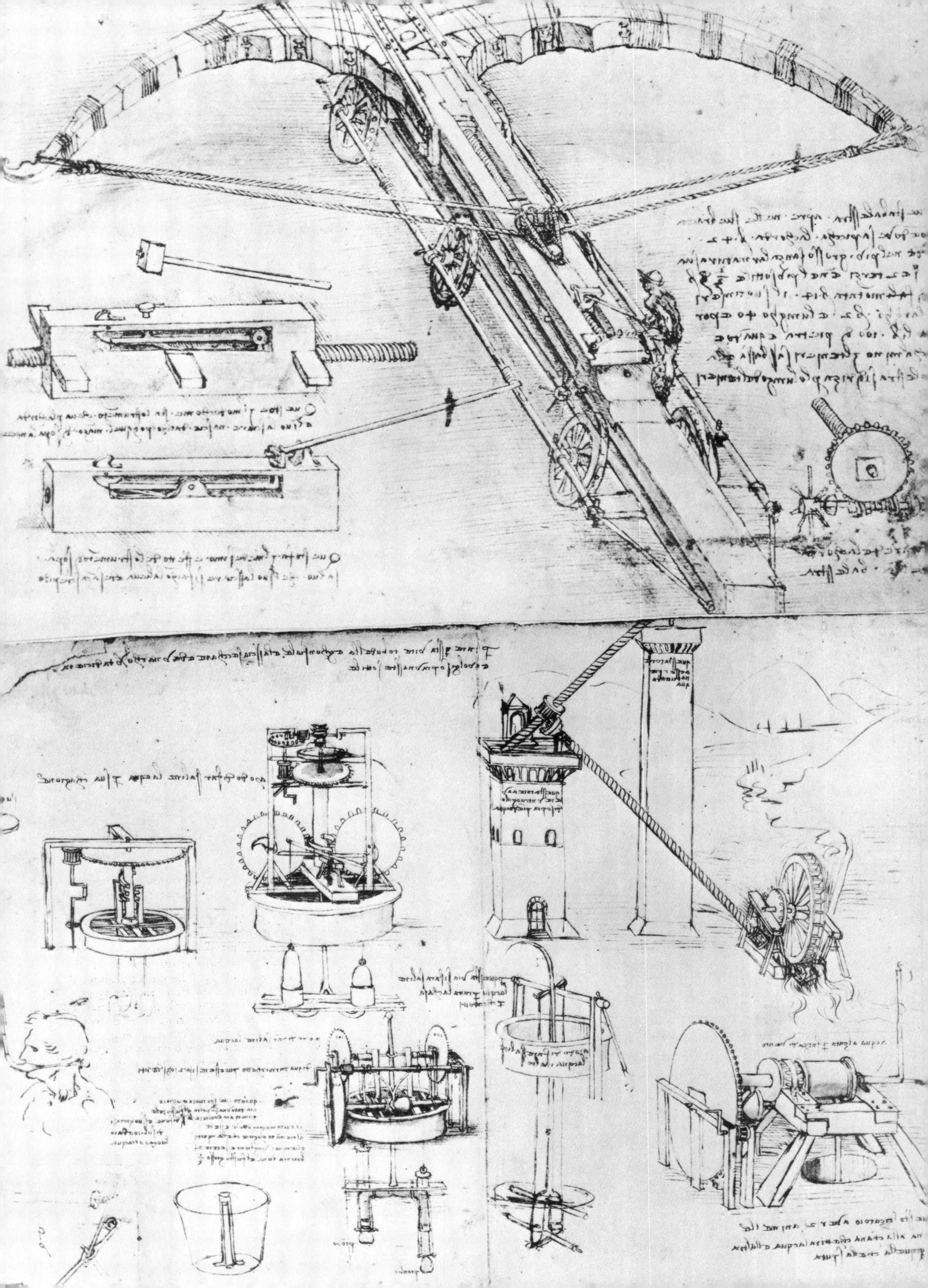

Power to the People

We live in a power-hungry world, for power is the key to industrial progress, better communications, and higher standards of living. Man, being the inventive creature that he is, has over the years created an assortment of power-producing devices, or prime movers, that provide the mechanical muscle to drive the multitude of machines that make the wheels of our industrial society go round.

These prime movers harness the power of flowing water and of expanding steam with windmill-like turbines. Or they exploit the chemical energy locked up in hydrocarbon fuels by exploding them in closed cylinders. Putting prime movers in hulls, on wheels and on wings created means of individual and mass transportation that shaped the geography of the continents and moulded the life-style of the population. Coupling the prime movers to generators produces the most versatile form of energy – electricity, which has the unique advantage of being able to be transmitted long distances.

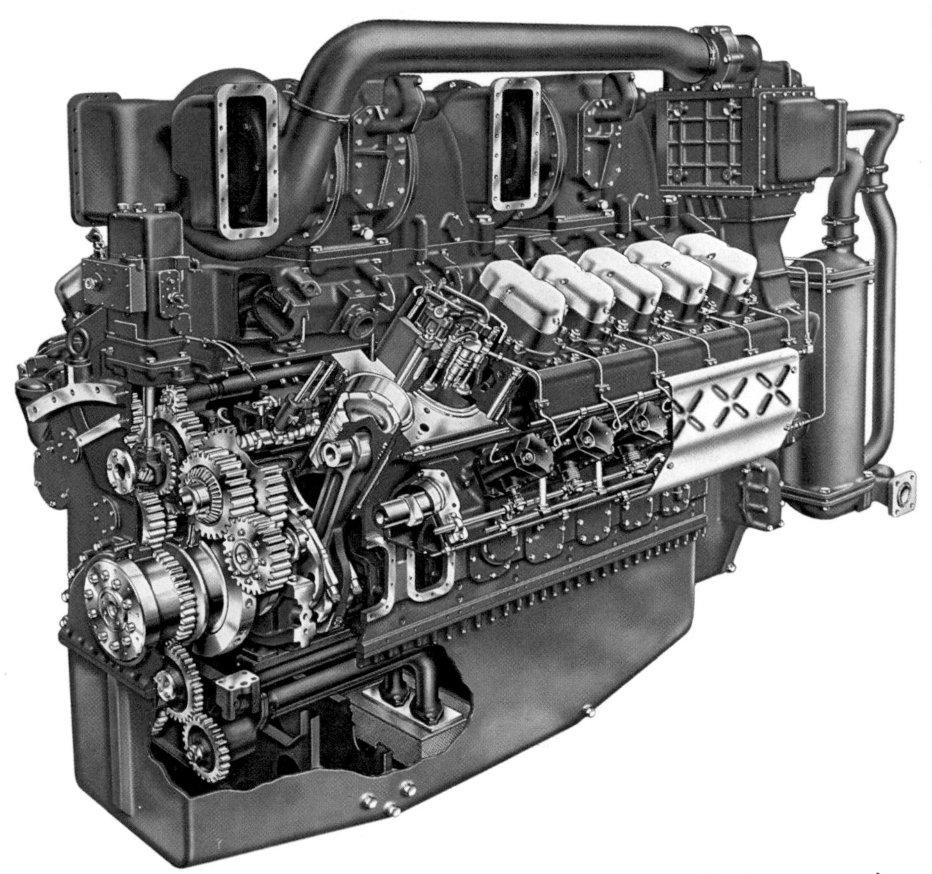

However, we have now entered an era in which power is becoming an increasingly scarce and expensive commodity. The trouble is that the power industry relies overwhelmingly on diminishing assets – deposits of the fossil fuels oil, natural gas and coal. When used up, these deposits cannot be replaced. Despite energy-saving measures, the days of fossil fuels in general, and of oil in particular, are all but numbered.

Fortunately there are alternative energy sources available which will, and must, prevent an energy vacuum next century. Already we are harnessing the energy locked up in the atom, in nuclear reactors. And prospects seem rosy that we shall one day tap, in fusion reactors, the very source of power that drives the universe (see page 218).

Above: Beautiful but deadly is this glow made by radiation in a nuclear 'pool' reactor.

Above left: The all-powerful internal combustion engine, power source par excellence of twentieth-century civilization. Will it survive until the twenty-first?

Top: Refurbished windmills like this could come into their own again as the energy crisis deepens.

Opposite: Pylons at sunset. The picture could become symbolic if we achieve electricity generation from nuclear fusion, for a similar process makes the sun shine.

Steaming Ahead

The key to the rapid industrial expansion we know as the Industrial Revolution was the steam engine. This provided the power that transformed a predominantly rural society into an industrial one.

Thomas Savery in England first put steam to practical use with his steam pump (1698) for draining mines. He called it his 'New Invention for Raiseing of Water and occasioning of Motion to all Sorts of Mill Work by the Impellent Force of Fire'. His engine worked by condensing steam inside a vessel, thus creating a partial vacuum. This sucked in water from the mine through a one-way valve. Steam was again introduced into the vessel to force the water out, and the process was repeated.

Savery's pump had severe limitations, not the least of which was the lack of a safety valve! A great improvement came with fellow-countryman Thomas Newcomen's atmospheric pumping engine of 1712. Steam was condensed by a water spray in a separate cylinder fitted with a piston. Atmospheric pressure forced the piston into the cylinder when the steam beneath it condensed. The motion of the piston rocked a beam which worked the pump.

Scottish engineer James Watt improved the Newcomen engine out of all proportion by condensing the steam in a separate condenser. This brought about a considerable gain in efficiency since the power cylinder remained hot all the time. Watt introduced this important innovation in 1769 but not until Watt met Birmingham businessman Matthew Boulton some years later did it find commercial application. Boulton financed the building of well-engineered machines capable of utilizing the new technology. Between 1775 and 1800 Boulton and Watt engines spurred not only Britain but many other countries to industrial revolution.

Strangely enough Watt baulked at the idea of using high-pressure steam, and it was Richard Trevithick in England and Oliver Evans in the United States who built the first engines to do so. Trevithick introduced his engine onto the railways to create the first locomotive (1804). Eighty years later Sir Charles Parsons used high-pressure steam to power a new type of steam engine, the steam turbine. It was not long before his turbines found their niche on the power scene – powering ships and driving electricity generators.

The prolific American inventor Thomas Edison built the first practical electricity generator and incorporated it in the first power station, which he built to light the Pearl Street district of New York in 1882. Electric lighting had come to stay. But for many years it was rivalled by coal-gas lighting, which had been introduced nearly a century before by William Murdoch, one of Watt's engineers.

Internal Combustion

The potential of coal gas as a fuel for driving engines did not go unnoticed. The first successful gas engine was demonstrated in Paris in 1859 by Étienne Lenoir. Whereas the steam engine was an external combustion engine – it burned fuel outside the power cylinder, the gas engine was an internal combustion engine – it burned fuel inside the cylinder. A mixture of gas and air was introduced to each side of the piston in turn and ignited by an electric spark, driving the piston back and forth. Lenoir showed the way, but it took August Nicolaus Otto to refine the gas engine into a power source that soon began to displace the steam engine in small factories. Otto improved the gas engine by compressing the gas/air mixture before it was burned. He thus introduced the 4-stroke cycle of operations – induction, compression, power, exhaust – now called the Otto cycle.

In the same year that Lenoir showed off his gas engine, Edwin Drake across the other side of the Atlantic struck oil. Soon internal combustion engines burning the vapour of first kerosene (paraffin), then heavy oil and finally petrol were being built. In Germany in 1884 and 1885 Gottlieb Daimler and Karl Benz fitted lightweight petrol engines to two-, three- and four-wheeled vehicles. The motor-cycle and motor-car – bain and boon of the twentieth century – were born.

While everyone else went petrol-engine mad, German engineer Rudolf Diesel stuck with the oil engine. He devised in 1893 a system of injecting oil into hot, highly compressed air, which brought about a great improvement in thermal efficiency. His 'diesel engines' soon became widely used.

The petrol engine meanwhile had been making inroads into another sphere of transportation – aviation. Its high power output, coupled with light weight, tempted Orville and Wilbur Wright in the United States to design the first aeroplane, which made its first successful flight in 1903. Development of the aircraft petrol engine, with its obvious military applications, was rapid. But by 1930 English airman Frank Whittle, almost unnoticed, obtained a patent for a new type of aero engine that was destined to lead the work into a supersonic age. It was the gas turbine, or jet engine. Parallel work in Germany led to the first flight of a jet-powered plane, the Heinkel He-178, in 1939.

Nuclear Promise

In the same year Austrian physicists Lise Meitner and Otto Frisch announced the theory of the nuclear fission of uranium. Three years later, in the unlikely setting of a Chicago squash court, Enrico Fermi and his colleagues operated the first atomic pile, forerunner of the nuclear reactor.

Fermi's demonstration that the nuclear chain reaction could be controlled held promise of power aplenty for future generations. But the other side of the nuclear coin was revealed in 1945 in the mind-numbing holocausts of Hiroshima and Nagasaki in Japan which ended World War 2 in the Pacific. In 1956 the world's first nuclear power station, at Calder Hall in Cumbria, England, began feeding electricity into the national grid.

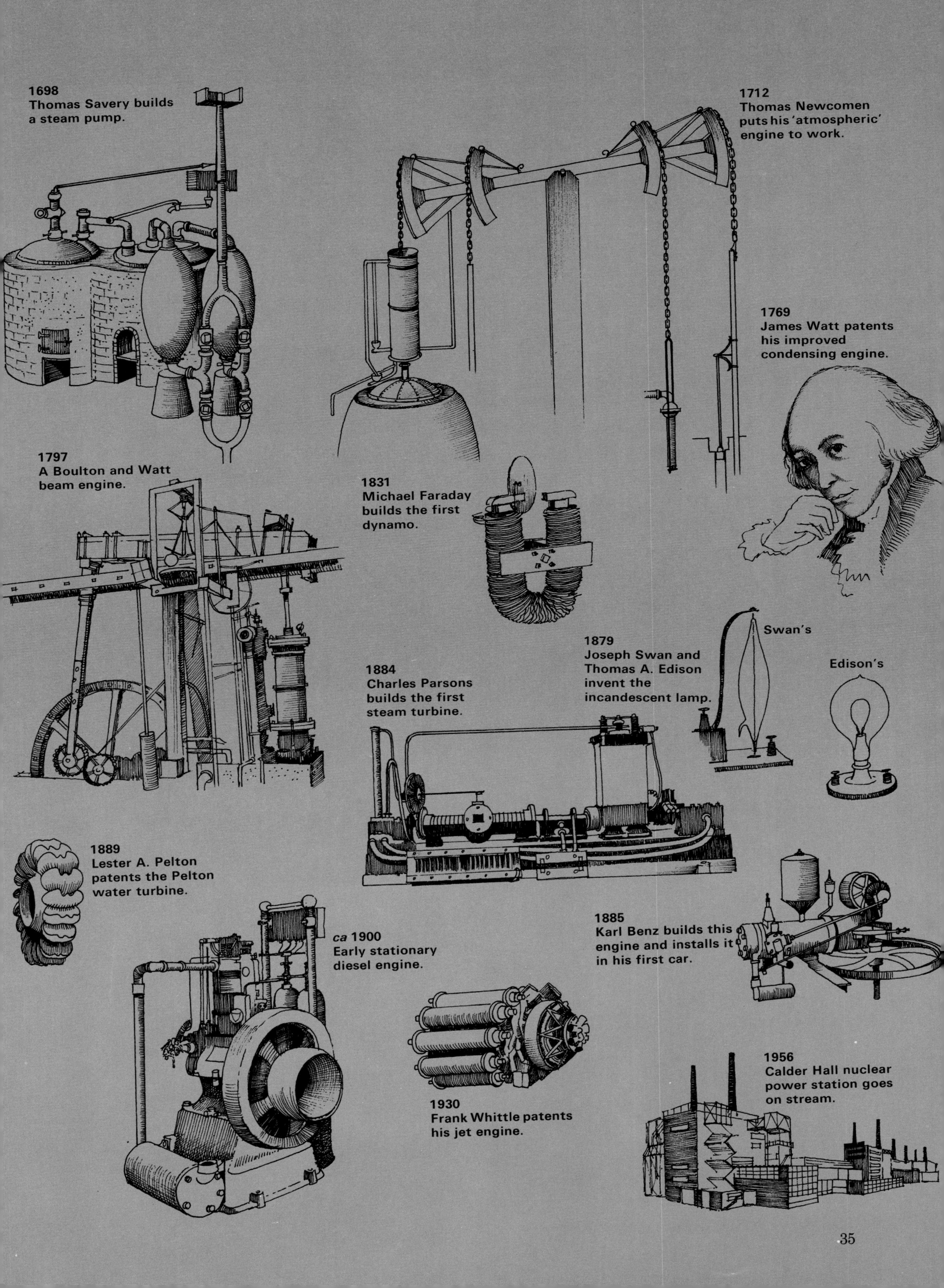
1698
Thomas Savery builds a steam pump.
1712
Thomas Newcomen puts his 'atmospheric' engine to work.
1769
James Watt patents his improved condensing engine.
1797
A Boulton and Watt beam engine.
1831
Michael Faraday builds the first dynamo.
1879
Joseph Swan and Thomas A. Edison invent the incandescent lamp.
Swan's
Edison's
1884
Charles Parsons builds the first steam turbine.
1889
Lester A. Pelton patents the Pelton water turbine.
ca 1900
Early stationary diesel engine.
1885
Karl Benz builds this engine and installs it in his first car.
1930
Frank Whittle patents his jet engine.
1956
Calder Hall nuclear power station goes on stream.

The Petrol Engine

The petrol engine is the most common form of internal combustion engine currently in use. Its greatest application is for powering cars, but it also powers a host of other machines, from chain-saws, lawn-mowers and outboard motors to motorbikes, lorries and piston-engined aircraft. Many lorries and a few cars have a diesel engine (see page 41), while most aircraft have a jet engine (see page 44). The power output from the petrol engine may vary from less than 1 horsepower (HP) to more than 35,000 HP, depending on its size. The engine of an average family saloon car develops around about 100 HP.

All petrol engines derive their power in essentially the same way. They explode with a spark a mixture of petrol and air inside a closed cylinder to produce hot gases. The gases expand and drive a tight-fitting piston down the cylinder. The piston is connected by a connecting rod to a crankshaft in such a way that as it moves down the cylinder, it drives the crankshaft round. What the crankshaft then drives depends on what kind of machine is being powered. In a car engine the rotary motion of the crankshaft is carried to the driving wheels via a transmission system (see page 74). Because of the way it functions, the petrol engine can be described as a reciprocating piston engine which burns its fuel by spark-ignition.

The discussion of the workings of a petrol engine that follows relates primarily to a conventional car petrol engine. But similar considerations apply to most other petrol engines, though they may vary in size, application and detail.

The car petrol engine is a collection of hundreds of separate parts, about 150 of which are in constant motion. They can be grouped together in systems according to their function. There is the main body unit that carries and incorporates all the other parts. Then there is the fuel system, which acts to deliver fuel in the right condition and the right amounts into the engine cylinders. The ignition system provides the spark to burn the fuel. The cooling system removes the excessive heat produced by combustion. And the lubrication system ensures that oil reaches the moving parts so that they are able to move freely.

The Body Unit

The main part of the engine body is the cylinder block, into which holes are bored to form the engine cylinders. It is usually made of cast iron, though aluminium is sometimes used. Fitting on top of the block is the cylinder head. This carries the valves, two to each cylinder, through which fuel is drawn or exhaust gases expelled. The lower part of the cylinder block, the crankcase, carries the bearings that support the crankshaft. Pistons, made close-fitting by piston rings, fit into the bores. The connecting rods linking pistons and crankshaft have their small end linked to the piston by a gudgeon pin and their big end linked to the crankshaft by a bearing.

At one end of the crankshaft is a cog wheel that drives by chain a cogwheel on the end of the camshaft. This is a shaft with raised portions (cams) which work a succession of rods and shafts that operate the valve mechanism. As the cams rotate they alternately raise and lower little cups called tappets, into which fit the push rods. The push rods carry the motion to the cylinder head, where they rock a pivoted rocker arm up and down. The other side of each rocker arm touches the end of a valve stem. When the push rod goes up, the rocker arm pushes the valve open. When it goes down, the valve spring snaps the valve shut again.

To produce power, the engine goes through a repetitive cycle of operations. It is called the four-stroke cycle because the piston in each cylinder has to make four strokes, or up or down movements, to produce power. The cycle is illustrated below.

On the first downward stroke of the piston (induction stroke), the inlet valve opens and lets vaporized fuel be drawn into the cylinder. On the second, upward stroke (compression), the inlet valve closes and the fuel is compressed (typically about 8–9 times). When the piston is near the top of its stroke, a spark from the sparking plug ignites the fuel. The

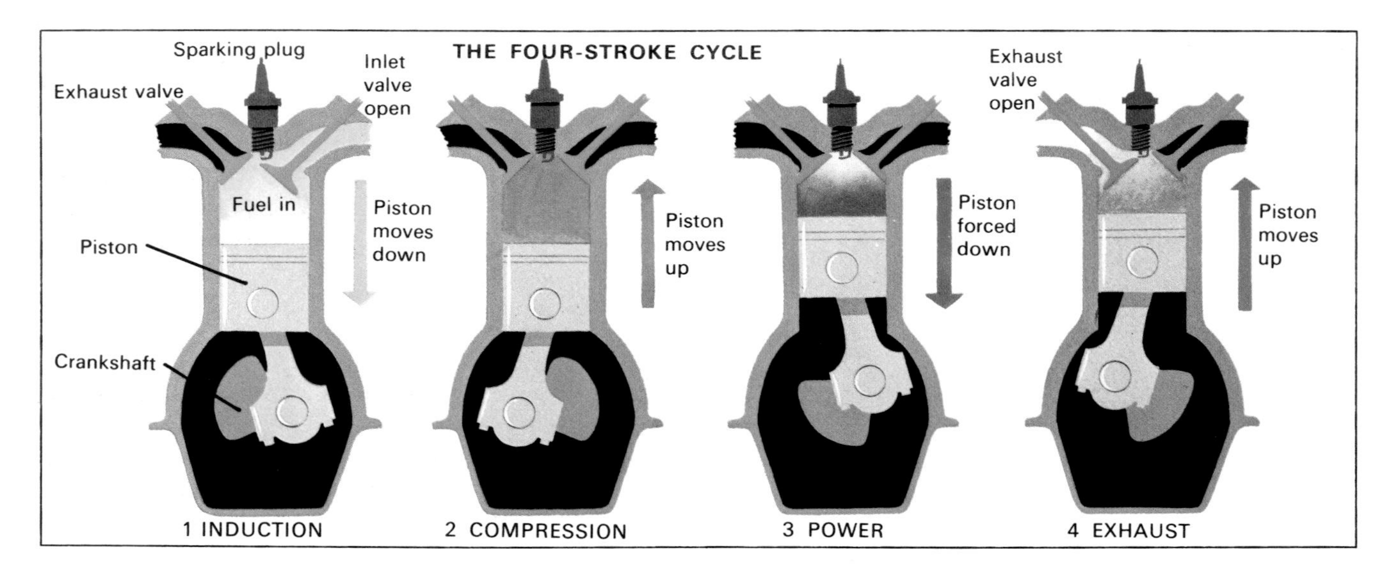

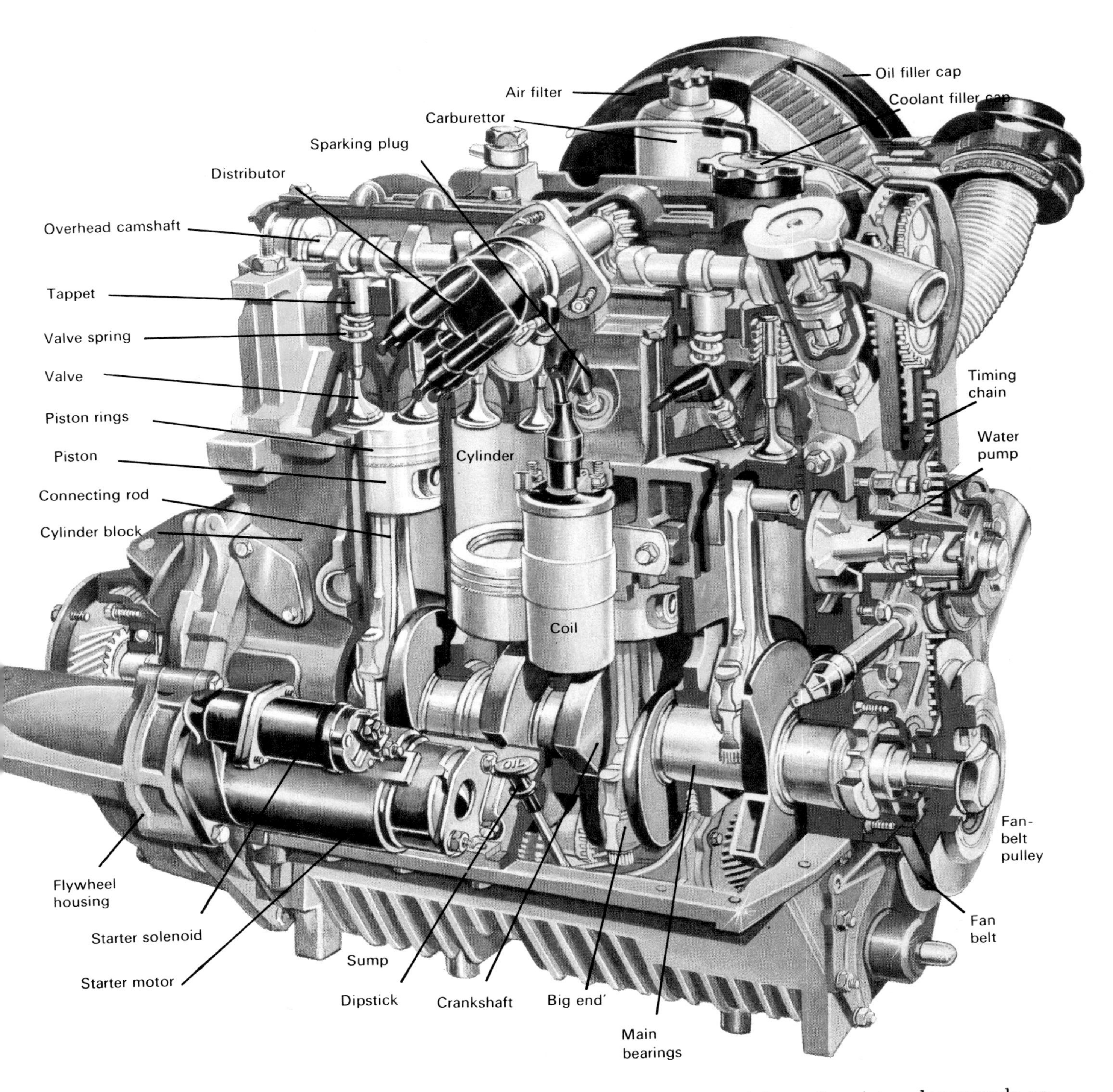

Above: A cutaway drawing of the four-cylinder, 2-litre petrol engine fitted to British Leyland's Princess. It differs in several respects from the standard engine. It has an overhead camshaft, which operates the valves via 'bucket' tappets, and thus lacks the push-rods and rockers of more conventional engines. The pistons are of the hollow-crown type, the hollow forming the combustion chamber. Also, the engine is mounted in a transverse position with the cylinders in-line across the car rather than fore-and-aft as in conventional designs. Integral with the engine is the gearbox, which serves the front-wheel drive transmission. The gearbox is hidden in this view but is visible in the photograph of the same engine shown overleaf.

Left: Representation of the standard 4-stroke cycle of a petrol engine. The mixture of fuel and air from the carburettor is taken in, compressed (typically about nine times), and exploded by an electric spark. The explosion forces down the piston, which drives round the crank-shaft and then returns to expel the exhaust gases from the cylinder.

hot gases produced force the piston downwards on its third (power) stroke. When the piston starts to come up again on its fourth stroke (exhaust), the exhaust valve opens and the spent gases are expelled from the cylinder through it. As the piston reaches the top of its stroke, the exhaust valve snaps shut, the inlet valve opens, and the cycle begins again.

Because the power is delivered intermittently, several cylinders are required to produce a steady power output. A heavy flywheel is attached to the output end of the crankshaft to smooth out power delivery further. It effectively stores the energy imparted to it in 'thumps' and transfers it evenly to the car's transmission. Around the edge of the fly-wheel is a toothed starter ring used for starting the engine. The starter motor spins a toothed wheel against the ring to turn the engine until it fires.

The engine illustrated above is a four-cylinder engine whose cylinders are 'in-line'. This is the most common cylinder arrangement, but others are also used. Some engines have their cylinders arranged in two banks set at an angle and driving a common crankshaft. They have a V-shaped cross-section and are called V-engines. Other engines have the two banks of cylinders opposite one another, which makes for a very flat design. These are called horizontally opposed engines. (A popular aircraft piston-engine design has the cylinders arranged symmetrically in a ring, this being called a radial engine.) The most common car engines have 4 or 6 cylinders in-line, though V-4, V-6, V-8 and even V-12 designs are becoming more widely used. Engine cylinder capacities vary from below 1000 cubic centimetres (cc) to over 5000 cc.

The Fuel System

In order to burn efficiently in the engine cylinder, petrol must be mixed with about 15 times its weight of air, whereupon it forms a highly inflammable mixture. Usually this is done by means of one or more carburettors, but sometimes by direct fuel-injection into the cylinders.

A carburettor works on the same principle as a scent spray. It consists essentially of a tube with a constricted throat (venturi) through which air is sucked into the engine cylinders. A fine jet delivers petrol into the air stream at the throat, where the pressure is low. The petrol is sucked up and immediately vaporizes. The vaporized fuel is then delivered to all the cylinders via the inlet manifold. A steady flow of petrol to the carburettor is maintained by a device called a float chamber.

The carburettor incorporates two flap ('butterfly') valves to control the flow of air and mixture. The choke valve, located upstream of the throat, controls the air flow into the carburettor. The throttle valve, located downstream, controls the mixture flow into the engine.

In practice carburettors must supply richer

(stronger) fuel mixtures at some times than at others. To do this they may use a number of fixed jets which come into use one after the other, or a single variable-flow jet.

The Ignition System

The inflammable fuel mixture from the carburettor enters the cylinder and is compressed. It is then ignited by a spark from a sparking plug. To produce a spark a high voltage must be created, and that is the first main function of the ignition system. The second is to distribute the spark to each cylinder at the right time.

The electricity for the system comes initially from the battery, but this provides low-voltage electricity – only 12 volts. This has to be 'stepped up' to something like 25,000 volts to produce a spark. That is the task of the ignition coil and the contact-breaker in the distributor.

The ignition coil is a simple transformer (see page 62), consisting of two coils of wire wound around a soft iron core. One coil (primary) has only a few turns, while the other (secondary) has many turns. The primary coil is incorporated in a circuit with the battery and the contact-breaker. Inside the distributor a shaft with raised cams on it alternately opens and closes the contacts. This makes the current in the primary coil circuit constantly rise and fall. Every time it does so it sets up, or induces, a high-

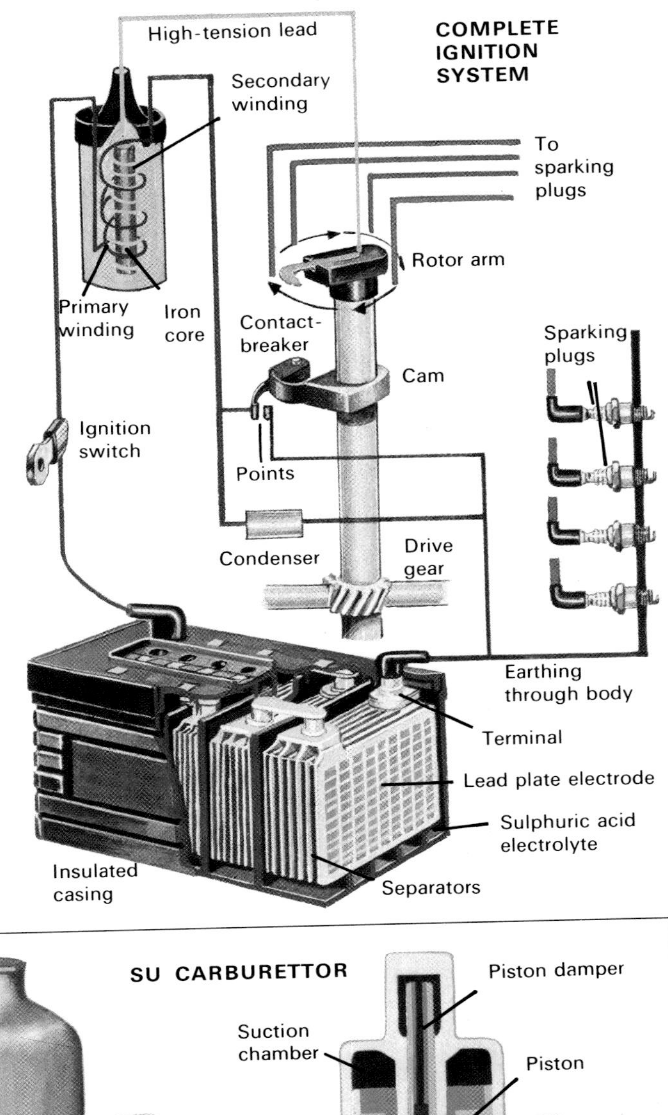

Left: Photograph of the 'O series' Princess engine, viewed from the rear and cutaway to show the integral transmission. On the right is the large flywheel and mated clutch assembly. At bottom left in the final-drive differential assembly. Between the two are the gearbox shafts carrying the gearwheels.

Top right: Diagram showing the typical ignition system of a petrol engine. The rotor arm, contact-breaker and condenser are all located in the distributor.

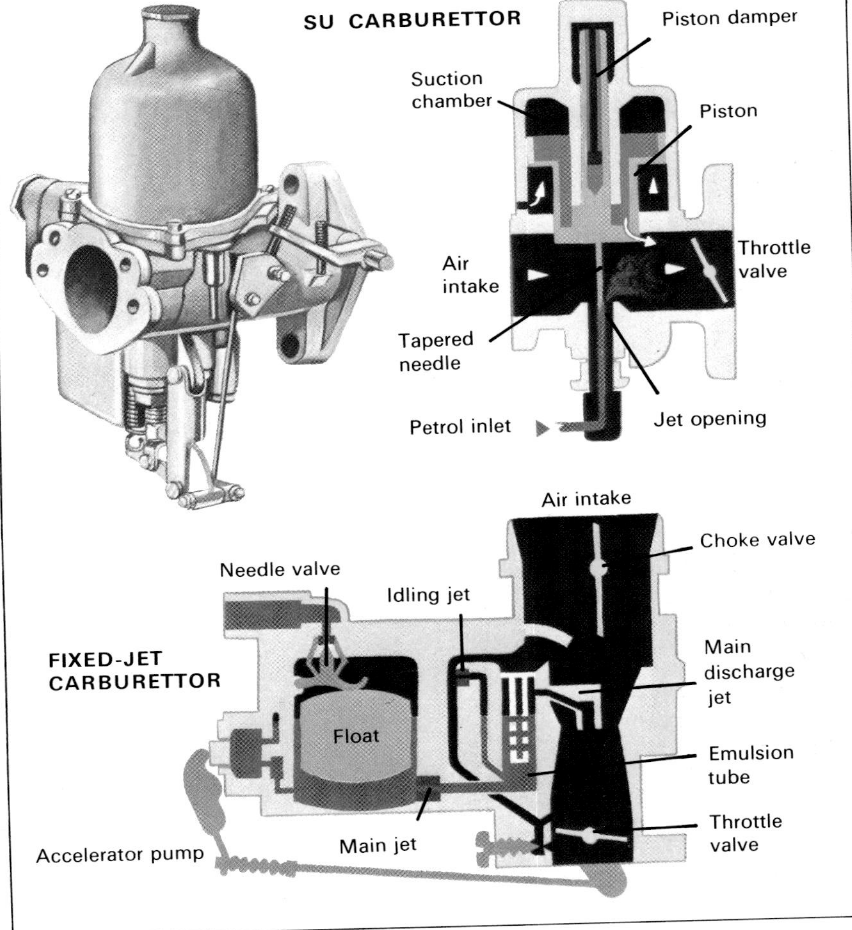

Right: Illustrations showing the two main types of carburettors, the fixed-jet and the variable-jet, of which the SU is an example. In the fixed-jet type, petrol is discharged through different jets as the engine's fuel requirements change. In the SU type, the requirements are met by the tapered needle progressively withdrawing from the jet opening.

voltage pulse of electricity in the secondary coil. This pulse is delivered by a rotating contact – the rotor arm – to each sparking plug in turn via contacts in the cap of the distributor.

To compensate for the power it continually must supply to the coil, the battery must itself be continually charged. This is done by an electricity generator, either a dynamo or an alternator (see page 60). The generator is driven by the fan-belt from a pulley on the crankshaft. The car battery is of the lead-acid type. It has plates (electrodes) of lead (negative) and lead peroxide (positive), dipping into a solution of sulphuric acid (the electrolyte). The battery produces electricity as the plates react with the acid and change into lead sulphate. When the generator feeds electricity to the battery during the charging process, the lead sulphate changes back to lead and lead peroxide.

Cooling Systems

The petrol mixture burns in the cylinders at a temperature of over 700°C. But less than a quarter of the heat produced is converted into power. The rest must therefore be rapidly removed from the engine, or it will overheat. The metal will expand and the moving parts seize-up. A good deal of the surplus heat leaves the engine with the exhaust gases, and the rest is dissipated by the cooling system.

Most modern engines are water cooled. A typical system is outlined at bottom right. The cylinder head and block contain passages (jackets) through which cooling water flows. As it does so it absorbs surplus heat from the cylinders. Now very hot, it circulates to the top of the radiator and falls through the tubes therein. The radiator tubes are embedded in a network of thin fins which offer a large surface area to the air passing over them. By the time it has fallen to the bottom of the radiator, the water is cold enough to be recirculated through the engine. This is done by a pump mounted on the same shaft as the fan, which helps draw the air through the radiator.

The water leaves the cylinder head through a thermostat, which is a temperature-operated valve. When the engine is cold, the thermostat remains closed, thereby preventing water from flowing through the radiator. This allows the engine to warm up rapidly to its working temperature. When that temperature is reached, the thermostat opens to allow normal circulation through the radiator. The thermostat valve is opened by the expansion of a volatile fluid inside a bellows or of wax inside a brass container.

In very cold weather antifreeze must be added to the water-cooling system. This is a liquid such as ethylene glycol which depresses the freezing point of water – it lowers the temperature at which the water freezes into ice. If the cooling water does freeze and expand into ice, the engine can be seriously damaged.

This drawback is not experienced by engines with the alternative, air-cooling system, such as the Volkswagen 'Beetle' engine. In such a system the cylinders have fins all around them to increase the surface area available for heat dissipation. Air is forced over the fins by a powerful fan, which is thermostatically controlled to maintain a suitable engine temperature. Air-cooled engines are not as quiet as water-cooled engines, in which the water helps absorb the noise of combustion. Most motorcycle engines are air-cooled.

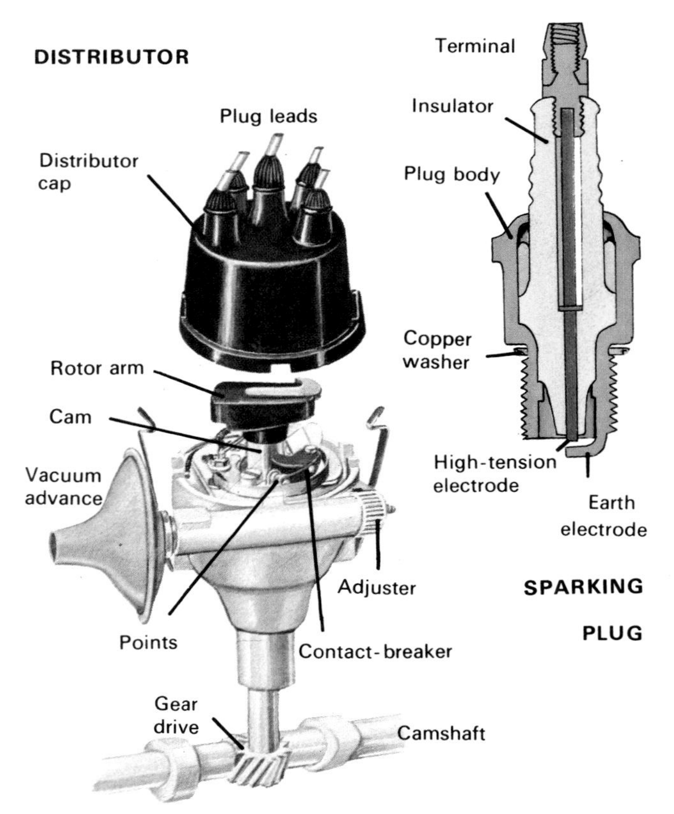

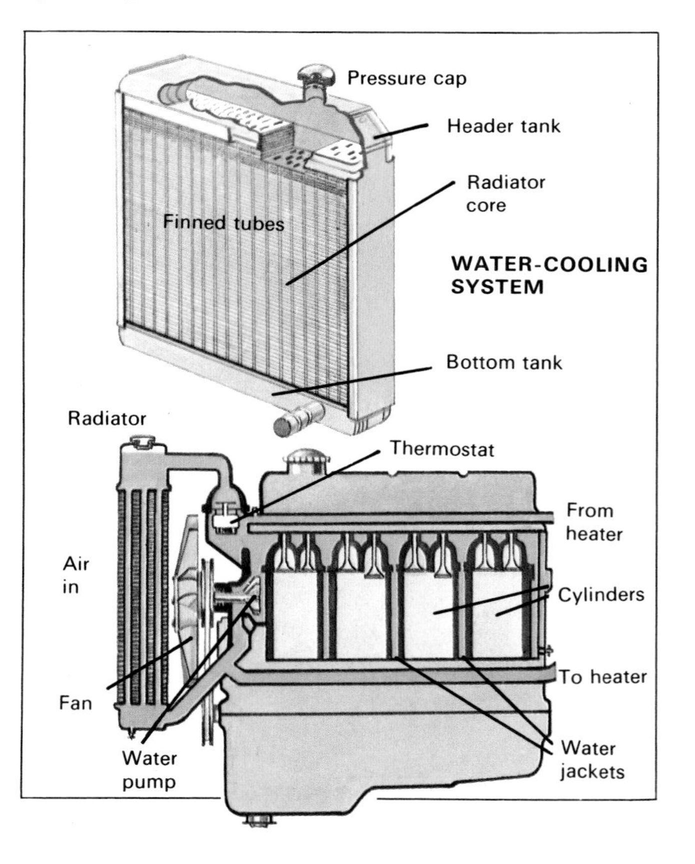

TWO-STROKES

For cars the 4-stroke petrol engine cycle is almost universal, but many smaller engines, including many of those used in motorbikes, often work on a 2-stroke cycle. They produce power every two strokes of the piston, and are designed rather differently from a 4-stroke engine, as the diagrams show. Gases enter and leave the cylinder through ports (holes) not valves. The piston itself acts like a valve, opening and closing the ports as it moves.

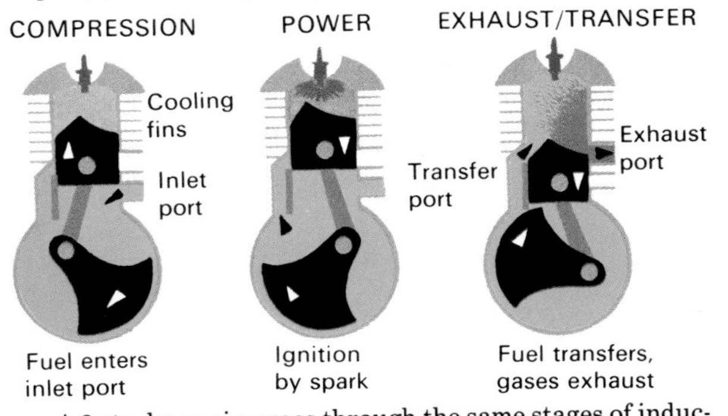

A 2-stroke engine goes through the same stages of induction, compression, power and exhaust as a 4-stroke engine does. This particular design uses the crankcase as a reservoir to hold fuel mixture before it goes into the cylinder. As the piston moves up it compresses the fuel mixture above it and also uncovers the inlet port, which allows fuel into the crankcase. So on the upward stroke induction and compression take place. The mixture is then exploded by a spark, and the piston travels downwards on its power stroke. Near the end of its stroke it uncovers the exhaust port and the transfer port. The exhaust gases escape through the one and fresh mixture is transferred into the cylinder through the other.

In the alternative fan-scavenged 2-stroke design a fan is incorporated in the fuel-intake pipe. This fan pumps fuel mixture into the cylinder at the bottom of the piston stroke, when the inlet port is uncovered.

One of the main advantages of the 2-stroke engine is its simplicity. A major disadvantage is that the incoming mixture mingles with the exhausting gases.

Air cooling is much simpler than water cooling, though not as effective in deadening the sound of combustion. Cooling is effected by surrounding the engine cylinders with tapered fins, which offer a high surface area to cooling air. Air cooling allows a more compact engine design, without radiator, fan, water pump, thermostat and the like. It is almost always used in motorcycle engines.

The Lubrication System

The oil flowing through the engine also helps to remove some of the heat of combustion, but its main function is to reduce friction between the moving parts. Lubricating oil is carried in a reservoir (the sump), which bolts on to the bottom of the crankcase. It is pumped from the sump through filters and into feed channels leading to the main bearings and to the rocker shaft in the cylinder head. The common gear-type oil pump consists of two meshing toothed wheels revolving in a figure-of-eight shaped housing. They sweep the oil around the edge of the housing as they rotate. The pump delivers the oil under a pressure of over 4 kg/sq cm (60 lb/sq in).

From the main bearings oil passes through holes in the crankshaft into the big-end bearings of the connecting rods. Oil flung off the rotating crankshaft lubricates the cylinder walls and the small-end bearings. Other bleed channels feed oil to the camshaft bearings and timing chain or gears. Eventually all the oil makes its way back to the sump to be recirculated.

Diesel Engines

In another kind of internal combustion engine, the oil or diesel engine, oil is the fuel, not petrol. It is not ordinary lubricating oil but a much lighter grade known as gas oil. The diesel engine burns this fuel in a totally different way from the petrol engine. The heat to burn the fuel comes from the compression of air in the engine cylinders. The engine is said to work by compression-ignition, as opposed to spark-ignition in the petrol engine.

The diesel is most widely used for heavy-duty applications in industry and transportation, rather

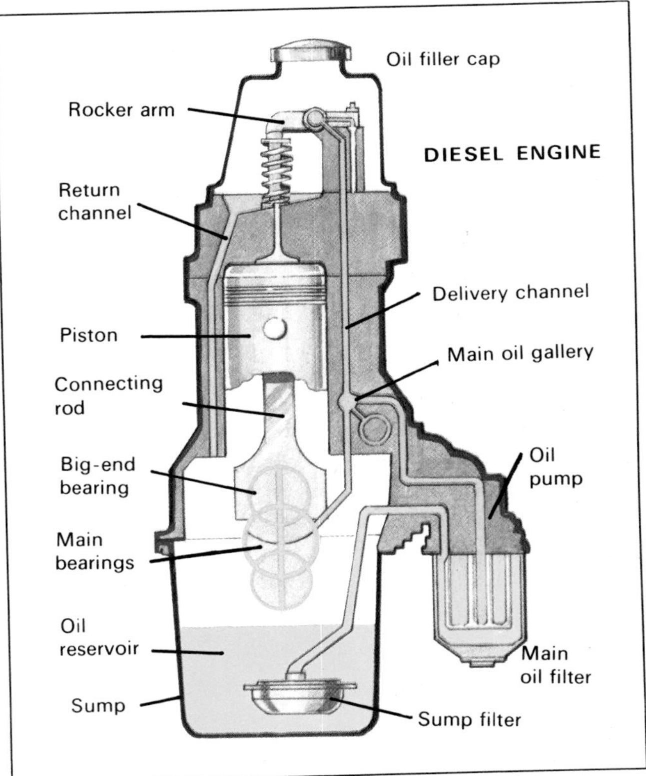

than for powering cars, though several diesel cars are now being made. Diesel engines are used primarily to drive electricity generators, ships, railway locomotives, heavy lorries and coaches. They are in general heavier and noisier than petrol engines but they use cheaper fuel, which they burn more efficiently. Whereas a petrol engine has a thermal efficiency of less than 25 per cent, the diesel's can be as high as 40 per cent. Also, the exhaust fumes from diesels are less toxic than those from petrol engines.

With the exception of the method of ignition, the diesel engine is constructed and operates in essentially the same way as the petrol engine. It is a reciprocating piston engine which delivers its power via a crankshaft. It may be water-cooled or air-cooled, and it may work either on a 4-stroke or a 2-stroke cycle. Broadly speaking the smaller engines use the 4-stroke, and the larger ones the 2-stroke cycle.

In the 4-stroke cycle of a diesel engine, air is sucked into the cylinder through the inlet valve on

Above: Cross-section of a diesel engine of 'V' design. The two banks of pistons in the arms of the 'V' drive a common crankshaft.

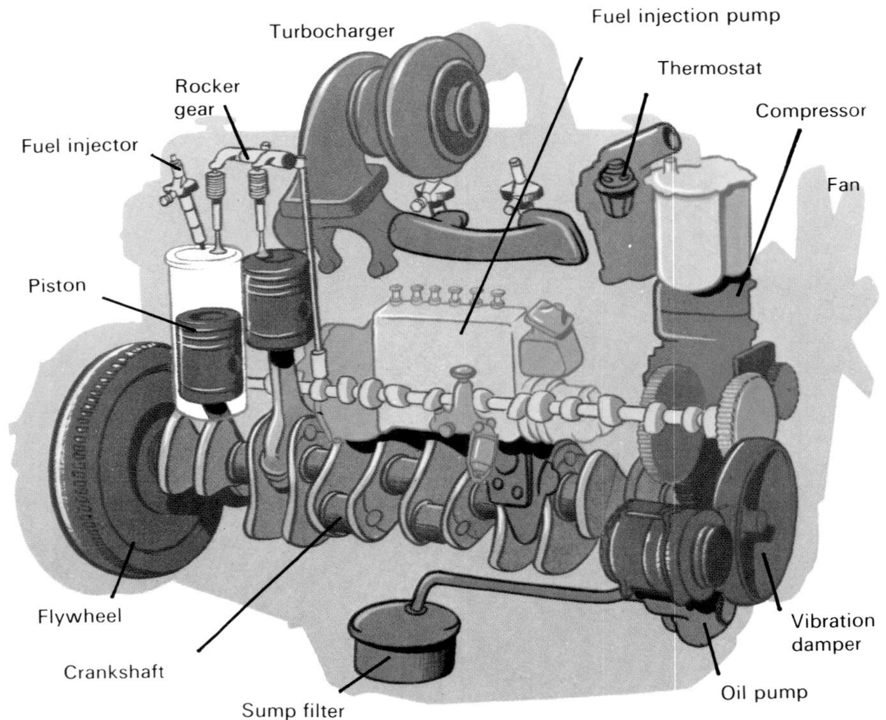

Left: The essential elements of a typical 'in-line' diesel engine design. The main differences between this and a typical petrol engine are the turbocharger to force air into the cylinders; the lack of a carburettor; and a fuel-injection system. The whole engine is more ruggedly built to withstand the greater pressures developed in the cylinders.

Right: A different kind of internal combustion engine, which often burns oil, powers this French gas-turbine locomotive.

the first, downward stroke of the piston (induction). The inlet valve closes as the piston begins to rise on its second stroke (compression) and compresses the air. When the piston nears the top of its stroke, a carefully metered amount of fuel is injected into the compressed air. The heat of compression has raised the temperature of the air to over 500°C, and the injected fuel immediately vaporizes and burns. The hot gases produced expand and drive the piston down on its third stroke (power). On the final, upward stroke (exhaust) the piston expels the burnt gases from the cylinder through the open exhaust valve. The cycle is then repeated.

The diesel engine has to be sturdier than the petrol engine because the air must be compressed perhaps 22 times. This is nearly three times as much as in the petrol engine. Particularly in the larger engines, the power output is significantly increased by supercharging, or compressing the air before it is delivered into the cylinders. A high-speed compressor or blower is used for supercharging. It is often driven by a turbine spun by the exhaust gas leaving the cylinders. This leads us fortuitously to a more powerful type of internal combustion engine.

Gas Turbines

The ordinary kind of internal combustion engine – the petrol and diesel – suffers from the disadvantage

THE WANKEL ENGINE

A novel petrol engine, invented by the German engineer Felix Wankel in the 1950s, produces rotary motion directly without the need for a crankshaft. It consists of a three-lobed rotor rotating eccentrically inside a figure-of-eight casing. Three spaces, or chambers are formed between the sides of the rotor and the casing. As the rotor turns, these chambers expand and contract in size. This is utilized in the operating cycle, which mimics the four-stage working of a conventional engine.

Fuel mixture is sucked through the inlet port into one of the segments as the rotor turns (diagram 1). As it continues to turn the mixture starts to be compressed (2). When fully compressed (3), the mixture is ignited by a spark. The hot gases produced expand and drive the rotor round. As it continues to turn, it sweeps the spent gases out through the exhaust port (4).

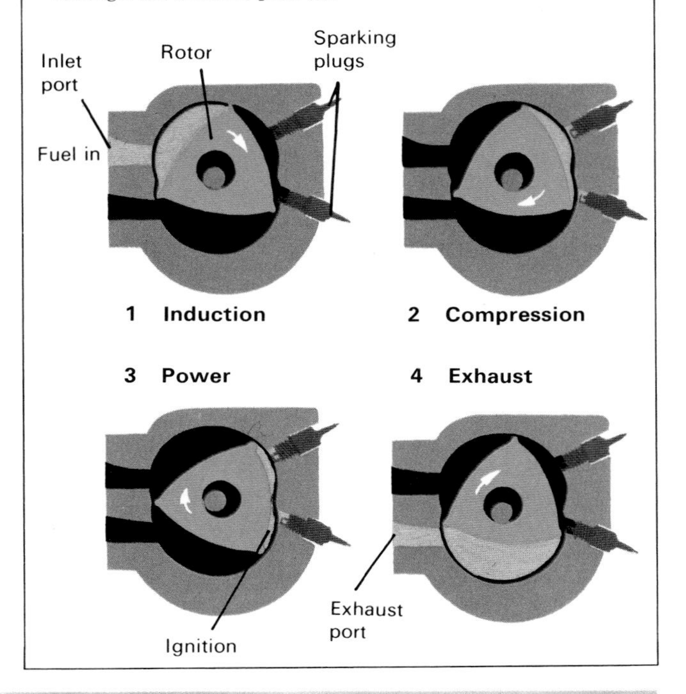

that it is reciprocating. The hot gases produced by combustion push pistons up and down. The reciprocating motion then has to be converted into rotary motion by a crankshaft before it can be used. This wastes energy and causes vibration.

Another form of internal combustion engine, the gas turbine, overcomes the problem by using hot combustion gases in an entirely different way. It uses them to spin the blades of a turbine rotor, just as a steam turbine uses high-pressure steam to spin its rotor (see page 47). The result is an efficient, smooth-running machine, free from vibration.

The gas turbine now has many applications in industry and the field of transportation. In industry gas turbines drive such things as electricity generators, pumps and compressors. Gas-turbine generators are now widely used as stand-by units in power stations because they can be quickly brought on-stream when demand is high. In the field of transportation the dominant use of the gas turbine is for aircraft propulsion, in the form of the jet engine (see page 44). Gas-turbine engines are also being increasingly used in naval ships and in locomotives. Attempts have been made to scale them down for use in cars and lorries, so far without much success.

Industrial Gas Turbines

The basic operating principles of a gas turbine are these. Air is drawn in through an inlet and compressed to several times atmospheric pressure by a compressor. The compressed air is delivered to a combustion chamber. Fuel is sprayed into the compressed air in the chamber and ignites to produce hot gas. The gas is directed through nozzles onto the turbine, which is coupled to the compressor and to the load, whatever it might be. So as the turbine spins, it drives the compressor and at the same time performs useful external work.

The spent gas then exhausts to atmosphere. This is the commonest arrangement, known as the open cycle. Alternatively, the gas operating the turbine may be circulated in a closed system, being successively compressed, heated, expanded and cooled. This is called a closed cycle.

The turbine illustrated on page 44 is a simple open-cycle unit. It is of two-shaft design. One shaft carries the compressor and its driving turbine, while the other carries the power turbine that delivers useful work. The compressor rotor carries several sets of blades. These slot in between rings of fixed (stator) blades on the compressor casing. Each pair of fixed and moving blades is called a stage. A multistage arrangement allows much greater compression than a single stage would. Flow through most of the compressor is axial – along the axis. But the final compressor stage is centrifugal – it flings the air outwards into ducts leading to the twin combustion chambers.

Fuel injected at the top of the combustion chambers forms an explosive mixture with the compressed air and burns. Initially it has to be ignited by a gas torch, but thereafter it burns continuously until shut off. Combustion in each chamber is centred around a louvred-wall flame tube in the middle and produces temperatures of about 780°C.

The hot gas from the chambers is ducted to the turbine inlet. It passes first through the two-stage, charging turbine (high pressure), which is coupled to the compressor, spinning it as it expands. It continues to expand through the two-stage power turbine (low pressure), which delivers useful output. The two turbines rotate in opposite directions, which is a good way of maintaining mechanical stability. Typical running speeds are 8000 revolutions per minute (rpm) for the compressor and 700 rpm for the power turbine.

Gas turbines run at very high temperatures and must be constructed in part of high-quality temperature-resistant materials. Careful design, however, allows more conventional materials to be used elsewhere. In the design described above air is bled from the compressor to cool the turbine casing, the rotor discs, and the bearing housings. The parts of the machine subjected to full gas temperatures such as the turbine blades, are fabricated in Nimonic alloy, an alloy containing mainly nickel, chromium, iron and cobalt.

The efficiency of a simple open-cycle unit can be increased in several ways. One is by intercooling. This means compressing the air in two or more steps and cooling the compressed air after each. This keeps the volume of air to be compressed each time to a minimum. A second way of increasing power output is to incorporate a second combustion chamber between the high-pressure and the low-pressure turbines. This is called reheating (or reheat). Overall thermal efficiency is improved by including a heat exchanger in the cycle so that the hot gases exhausting from the turbine preheat the air going to the combustion chambers.

Jet Engines

The gas-turbine engine was developed initially to power aircraft. It was designed to propel them, not by propeller as previously, but by means of a high-speed jet. And aircraft gas turbines are now always called jet engines. Jet engines operate in essentially the same way as the gas turbines described above. The difference is that it is the power in the exhausting gases which is important.

The engine is designed so that the exhaust gases leave the rear of the engine at high speed. Reaction to the gases shooting backwards creates a forwards force or thrust that provides the propulsion. It is a common misconception that the jet derives its power by pushing against the air. Nothing could be farther from the truth. The air actually inhibits jet action, and jet engines work most efficiently at high altitudes, where the air is thin. Transatlantic airliners, for example, regularly cruise at an altitude of about 10,000 metres (33,000 feet).

The rocket also works on this principle of jet propulsion. The difference between a jet and a rocket is that a jet needs to 'breathe' air from the atmosphere, whereas a rocket does not – it is completely self-contained and can thus work in the vacuum of space (see page 200).

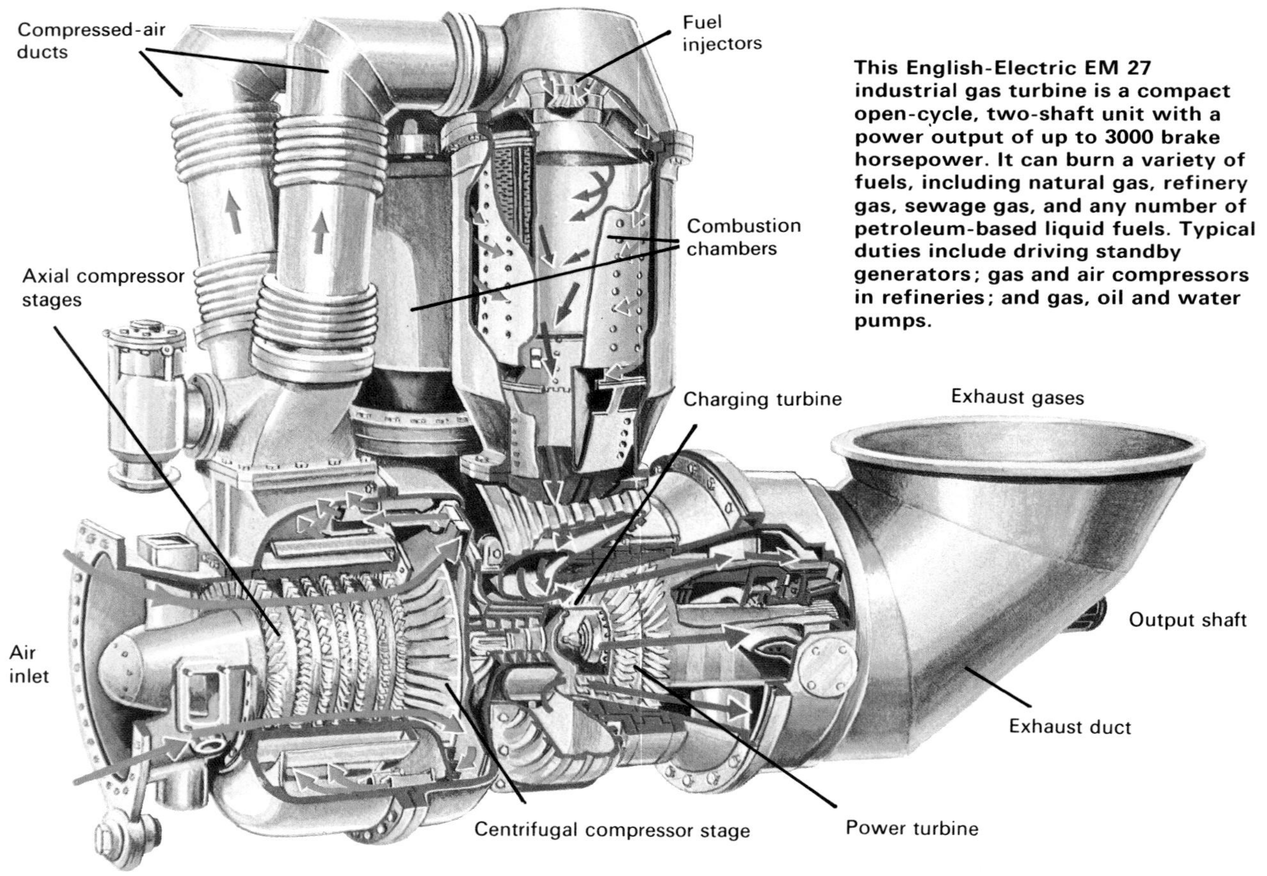

This English-Electric EM 27 industrial gas turbine is a compact open-cycle, two-shaft unit with a power output of up to 3000 brake horsepower. It can burn a variety of fuels, including natural gas, refinery gas, sewage gas, and any number of petroleum-based liquid fuels. Typical duties include driving standby generators; gas and air compressors in refineries; and gas, oil and water pumps.

Above: A fighter pilot switches on reheat to boost acceleration. 'Reheat', or 'afterburning', involves injecting more fuel into the hot jet exhaust, thereby increasing thrust but at the expense of high fuel consumption. Note the thin-section wings and overall clean aerodynamic design which characterizes planes that operate at supersonic speeds.

Right: Essential features of three main types of jet engines, all of which are in widespread aviation use.

Compared with land-based gas turbines, jet engines are very compact and have a very high power-to-weight ratio. Each of the four engines that power the Boeing 747 'jumbo jet', for example, weighs under 4000 kg (8800 lb) but has a thrust four times greater. In other engines the power/weight ratio may be 7 or more.

But you can't have something for nothing, and the fuel consumption of jet engines is very high indeed. The jumbo jet's engines each burn something like 4500 litres (1000 gallons) an hour, in about 200 tonnes of air. The airliner thus has to carry a huge amount of fuel – 190,000 litres (42,000 gallons) – for its transatlantic crossings. The universal fuel for jet engines is a form of kerosene (paraffin).

The Jet Family

There are several kinds of jet engines, which differ in the way their compressors and turbines are arranged. The main types are illustrated on the right. The simplest is the single-shaft, or single-spool turbojet. This has a small compressor driven by a single turbine in the jet exhaust. A twin-spool turbojet has two compressors, driven independently by two turbines.

In the turboprop, or prop-jet, one of the compressors is linked to a propeller, which provides most of

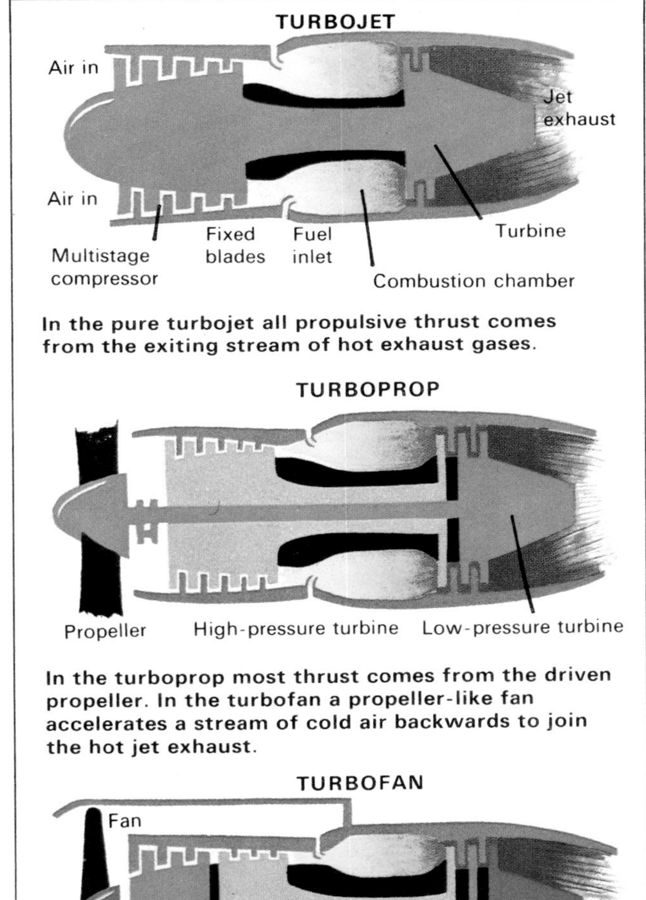

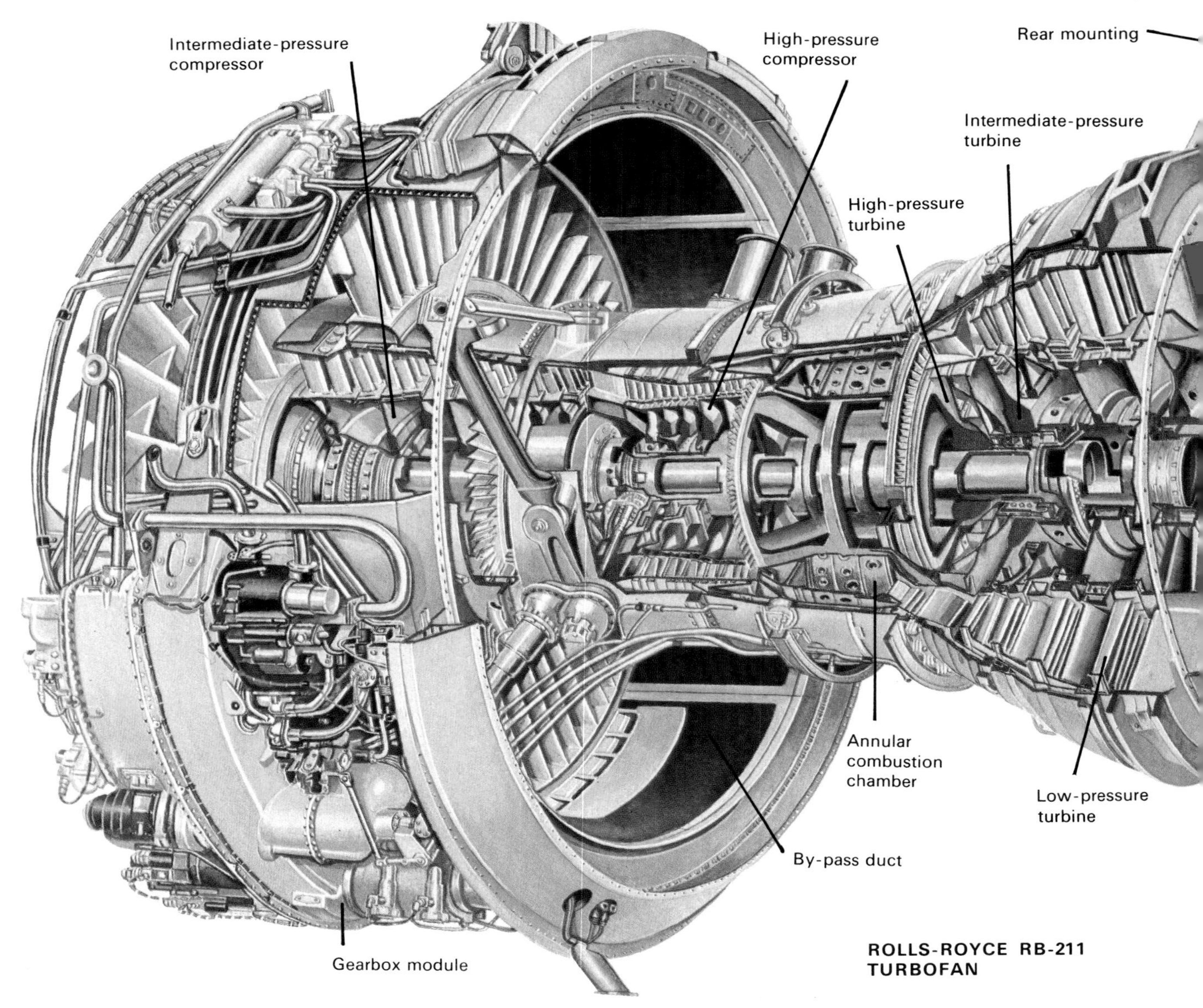

ROLLS-ROYCE RB-211 TURBOFAN

the propulsive thrust of the engine. Only a little comes from the jet exhaust. Finally there is the turbofan, also called fan-jet and by-pass turbojet. It is recognized by the large fan in the air intake which channels air not only into the compressor but also through a duct around the outside of the engine.

Most of the fastest fighter aircraft have turbojet engines. They are often equipped with a reheat, or afterburning facility. This injects fuel into the hot gas stream after it leaves the turbine and produces highly increased thrust. Because it is wasteful of fuel, reheat is only used for short periods to assist take-off or rapid climbing, for example. Turbojets can thrust fighter planes to speeds of more than three times the speed of sound (Mach 3), which at high altitudes is equivalent to about 3000 km/h (2000 mph).

The turboprop engine by contrast is suited only to relatively low-speed flight, for the optimum cruising speed of a propeller-driven aircraft is only about 800 km/h (500 mph). Turboprops are therefore most widely used in transport aircraft and also in helicopters. They are more compact than piston engines of equivalent power, and more efficient than turbojets.

Turbofans also operate more efficiently than ordinary turbojets, and they can do so at any speeds. The stream of cold air that by-passes the compressor mixes with the hot gases exhausting from the engine to form a 'fatter' and slower jet which makes for more efficient propulsion. This design reduces fuel consumption (by 20–25 per cent), increases range and is quieter, compared with a turbojet of equivalent power.

Whispering Giants

These advantages have made the turbofan the favoured engine for the modern large airliner like the Boeing 747, the Lockheed Tristar, the Douglas DC 10 and the European Airbus. The RB-211 engine (illustrated above) developed to power the Tristar, is typical of the giant turbofans now being used. It is of three-shaft design. Air is drawn into the engine intake by the fan, a large-bladed rotor over 2 metres (7 feet) in diameter. It delivers five times as much

Above: A tiny model traction engine vividly demonstrates the power of expanding steam.

Below: The mighty RB-211 turbofan installed in the wide-bodied Lockheed Tristar. The Tristar has three RB-211 engines, one slung beneath each wing, the other in the tail.

air into the by-pass duct as it does into the compressors – its by-pass ratio is 5.

The air is compressed in two steps, first by an intermediate-pressure (IP) compressor comprising seven stages, then by a six-stage high-pressure (HP) compressor. Overall the air is compressed to 27 times its original volume – the compression ratio is 27:1. The compressed air is delivered to the annular (ring-shaped) combustion chamber into which fuel is sprayed through 18 atomizing ducts.

The fuel burns at temperatures approaching 900°C, and the hot gases spin in turn the single-stage HP and IP turbines and the three-stage LP turbine. These turbines drive, through three concentric shafts, the HP and IP compressors and the fan respectively. High-strength titanium alloy is used for the fan and IP compressor rotor blades, while Nimonic alloys are used for the HP compressor and turbine rotor blades. The thrust developed by the RB-211 is over 20,000 kg (45,000 lb) for a weight of only 3300 kg (7300 lb).

Steam Turbines

Steam turbines are among the most powerful of the machines yet constructed by man. They operate at high temperatures (up to 600°C) and high pressure (up to 350 kg/sq cm, 5000 lb/sq in). They are used not only in electricity generating stations, of course, but are also the favoured power source for propelling large ships.

In a steam turbine steam spins the rotor as it expands from a high to a low pressure. Some machines exhaust the steam to atmosphere, but power-station turbines have a condenser to turn the exhaust steam back to water and thereby create a very low pressure. The pressure drop through the turbine and thus its efficiency is then much higher. In general a condensing steam turbine has a thermal efficiency of around

The rotor shaft of the low-pressure section of a 660 megawatt steam turbine being lowered into position.

40 per cent, that is, 40 per cent of its heat energy is changed into mechanical energy.

As in the gas turbine the blades of the turbine rotor slot into stationary blades on the turbine casing, each set of stationary and rotor blades being called a stage. The stationary blades act as nozzles to direct the steam onto the rotor blades. The design of the rotor blades varies from machine to machine. In an impulse turbine they are roughly cup shaped, and are forced round as steam impinges on them. A reaction turbine has simple angled blades, which spin by reaction as steam expands and accelerates through them. In most turbines the blading is of intermediate design, and the rotor spins by a mixture of impulse and reaction.

Steam expands several hundredfold as it passes through the turbine. The energy of expansion cannot be tapped adequately by a single stage, only by a series of stages – a multistage arrangement. Because the steam is expanding, each successive stage must offer a larger blade area for subsequent expansion. This means that the diameter of the rotor blading progressively increases stage by stage.

The simplest type of steam turbine has a single cylinder, but several cylinders are generally used in power-station turbines to extract the maximum amount of energy from the steam. Usually the rotors in each cylinder are joined end to end and are coupled to the generator shaft. This is called a tandem-compound design. In other designs each cylinder rotor is coupled to an independent generator, this being termed a cross-compound arrangement. Also included in the design of multi-cylinder machines is provision for reheat – the steam exhausting from one cylinder is reheated in the boilerhouse furnace before being fed to other cylinders.

The steam-turbine design shown opposite would be classed as four-cylinder, tandem-compound, double-flow with reheat. High-pressure steam from the boiler house passes into the first, high-pressure (HP) cylinder through control valves. They connect with a governor which adjusts the steam flow to give a steady turning speed. This is essential in power-station turbines which have to drive generators at exactly the right speed (see page 60).

The steam leaving the HP cylinder is then led back through the furnace to be reheated before entering the second, intermediate-pressure (IP) cylinder. Note that the rotor blades are getting larger. The exhaust steam from the IP cylinder leads into the twin low-pressure (LP) cylinders. These are of double-flow design, the steam flowing in opposite directions from a central inlet.

The LP cylinders exhaust to the condenser. This is a large vessel through which run some 20,000 tubes carrying cooling water. The steam is rapidly cooled, and condenses back into water. Since water occupies only 1/1700 of the volume of steam, a low pressure is created – about 1/15 of an atmosphere. The condensate is then recirculated back through the boiler to raise more steam. Vast quantities of cooling water are required in a large power station. A 2000-

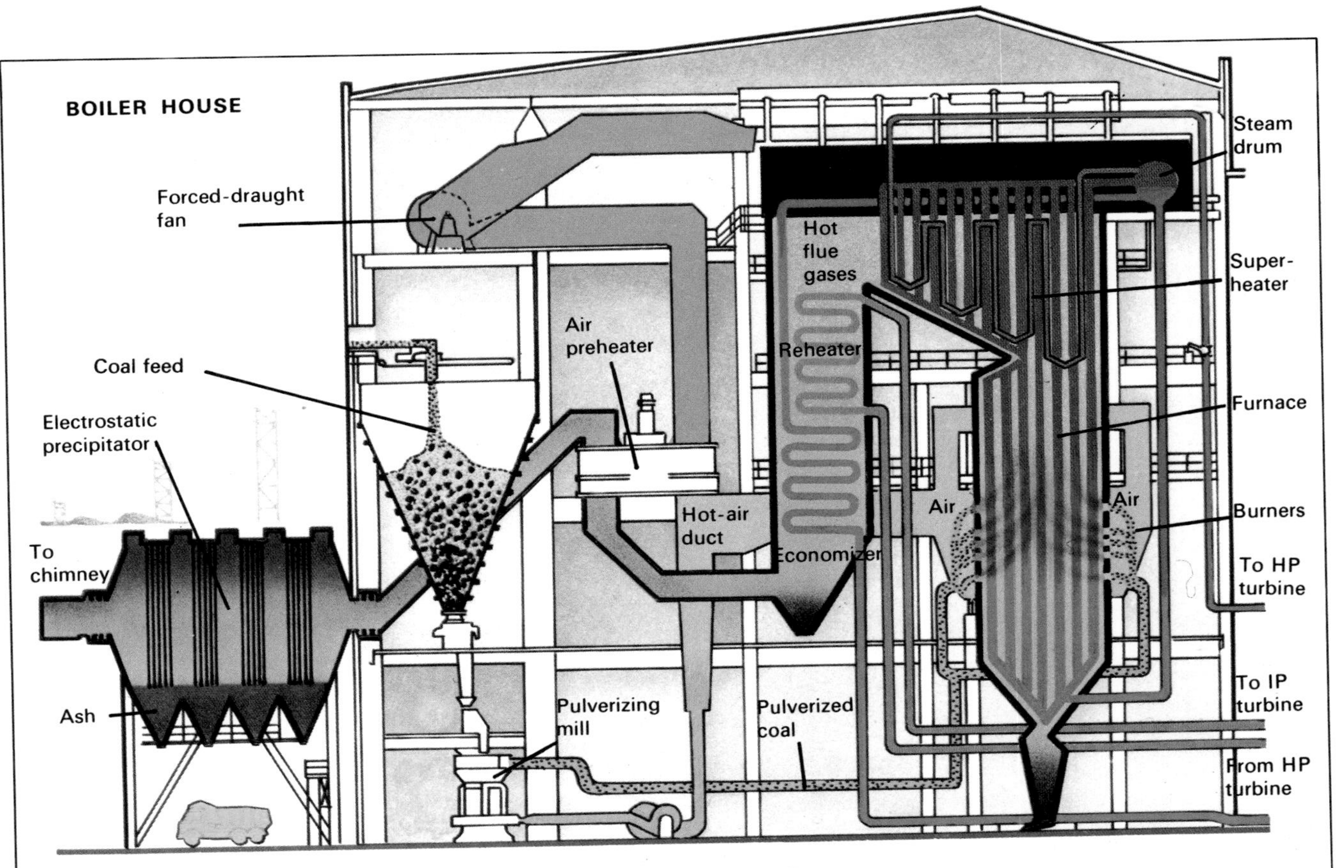

GETTING UP STEAM

In the majority of conventional power stations coal is used to fuel the steam-raising boiler, though oil, natural gas and even peat may also be used. It is often burned in a powdered, or pulverized form, in which state it burns most efficiently.

In a modern coal-fired plant, coal is carried by conveyor from the fuel store into the bunker. It then falls into the pulverizing mill, where it is crushed under rollers or moving balls. Some of the air drawn from the top of the boiler house by a fan is pumped through the pulverizing mill and blows the powdered coal through pipes to the burners of the furnace. There it mixes with preheated air and burns fiercely. The hot gases produced rise through the boiler, heating up the water in the long boiler tubes.

The water turns to steam, which enters the steam drum at very high pressure (about 170 atmospheres). The steam is then piped through the superheater tubes at the top of the boiler and becomes even hotter. It is then fed through a valve into the high-pressure cylinder of the steam turbine. It may be hot enough to make the steam pipe glow red-hot (over 550°C).

After the steam has passed through the high-pressure cylinder, it is returned to the boiler to be reheated. It is then fed back into the intermediate and low-pressure cylinders of the turbine. The condensate (condensed steam) is heated and then pumped back to the economizer, which heats it further, whereupon it passes into the steam drum.

The flue gases leaving the boiler preheat the incoming air and pass via the electrostatic precipitator into the chimney flue. The electrostatic precipitator contains electrically charged plates, which attract the fine ash particles in the flue gases. Periodically the plates are hammered so that they discharge the dust they carry. This falls into hoppers for disposal.

To prevent corrosion and lime scale being deposited in the boiler system, the feed water is pre-conditioned. It is chemically treated to remove impurities and 'hardness', and degassed.

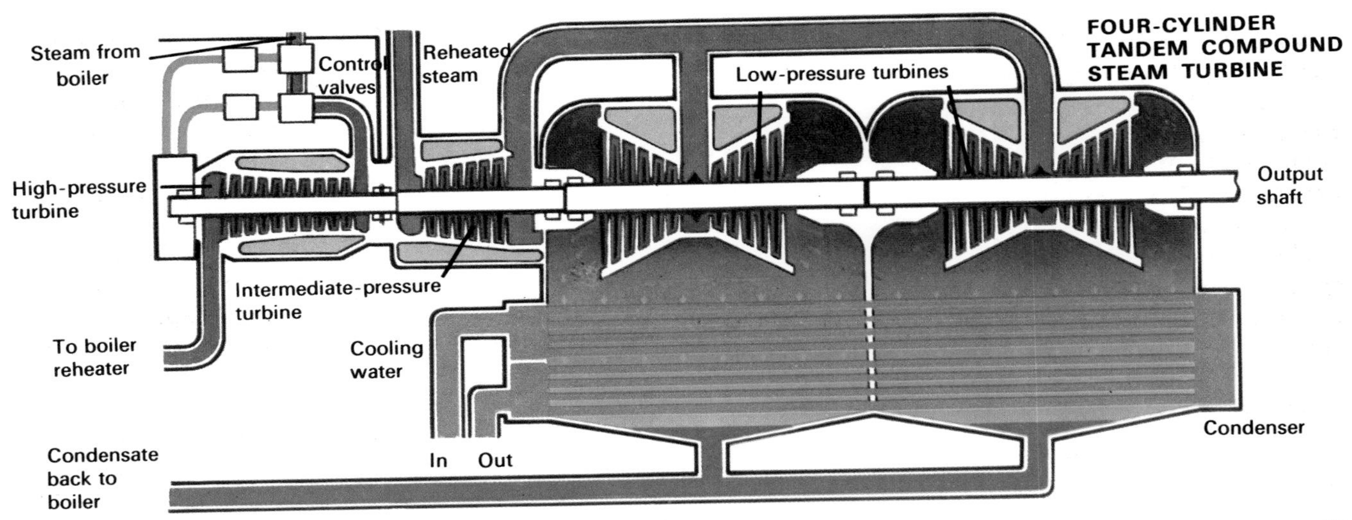

megawatt plant requires some 225 million litres (50 million gallons) of water an hour for cooling purposes.

For this reason many steam power stations are built on rivers and estuaries where abundant cooling water is available. Inland stations on the other hand have to use cooling towers to cool the water. In these towers the water is cooled as some of it evaporates into the air.

Hydroelectricity

Flowing water packs enormous power, which in suitable locations can be harnessed to drive electricity generators. Power stations that harness water power are termed hydroelectric power (HEP) plants. They use water turbines, or hydroturbines to extract the energy from the flowing water.

The siting of a HEP plant is dictated by geography and the weather. Mountainous regions with high rainfall are ideal locations. There, water stored in a reservoir at high level, can be channelled through tunnels or pipes ('penstocks') to a HEP plant at a lower level. The natural difference in level, the 'head', provides the power. The potential energy of water in the reservoir is released as kinetic energy of flow to spin the turbines.

Mountainous sites provide a high-head, low-volume flow. But an equivalent amount of power can be produced by a low-head, higher-volume flow, such as that of a deep, wide river. In fact there are more HEP plants operating at low-to-medium head say, 30–200 metres (100–660 feet), than there are at higher ones. These plants are also usually much nearer habitation than the high-level sites, so transmission problems are not as acute.

Wherever the HEP is located, a valley, stream or river must be dammed to form a storage reservoir or create a suitable head. In the mountains narrow natural gorges provide ideal sites for concrete-arch dams. They curve towards the water and brace themselves against the solid rock. Damming rivers on lower-level sites is usually more difficult. The dams are often embankment dams, built of rock rubble and soil with an impervious clay or concrete core. They are roughly triangular in cross-section and swallow enormous amounts of materials. (The 2·7 km (9000 ft) embankment dam across the River Indus, the Tarbela Dam, has a volume of 122 million cubic metres, or 160 million cubic yards!) These days the cost of constructing dams is very high indeed and most HEP projects are government-financed and may form part of a broader scheme that includes improving navigation, flood control and water storage.

Water Turbines

There are a number of different kinds of hydroturbines, each of which is suited to a particular head of water. For the highest head a Pelton wheel is most suitable. It consists of a wheel with buckets arranged around the outer edge. Water is directed at the buckets as a jet from a nozzle, and spins the wheel round. The wheel may be mounted so that it rotates vertically, in which case a twin-jet arrangement is common. When the wheel rotates horizontally, four or more jets are used.

The Pelton wheel is known as an impulse turbine. It imparts its kinetic energy to the wheel as it impinges on the buckets and changes direction. The other main types of hydroturbines work on the principle of reaction, as does the rotating garden sprinkler. Water goes into this type of turbine (and the sprinkler) at low velocity and escapes through the turbine runner (rotor) at high velocity. Reaction to the accelerating flow spins the runner.

The most widely used reaction turbine is the Francis, which is suited to a wide range of heads (3–600 metres, 10–2000 feet). The Francis runner resembles a ship's propeller, carrying several vanes set at an angle. It is known as a mixed-flow turbine because water enters it radially (from the side) and leaves axially (along its axis). The water enters the Francis and other reaction turbines from a spiral casing that imparts a whirl to it.

Another mixed-flow turbine, called the Deriaz, has movable vanes which enable it to be converted into a pump. It is used in many pumped-storage schemes.

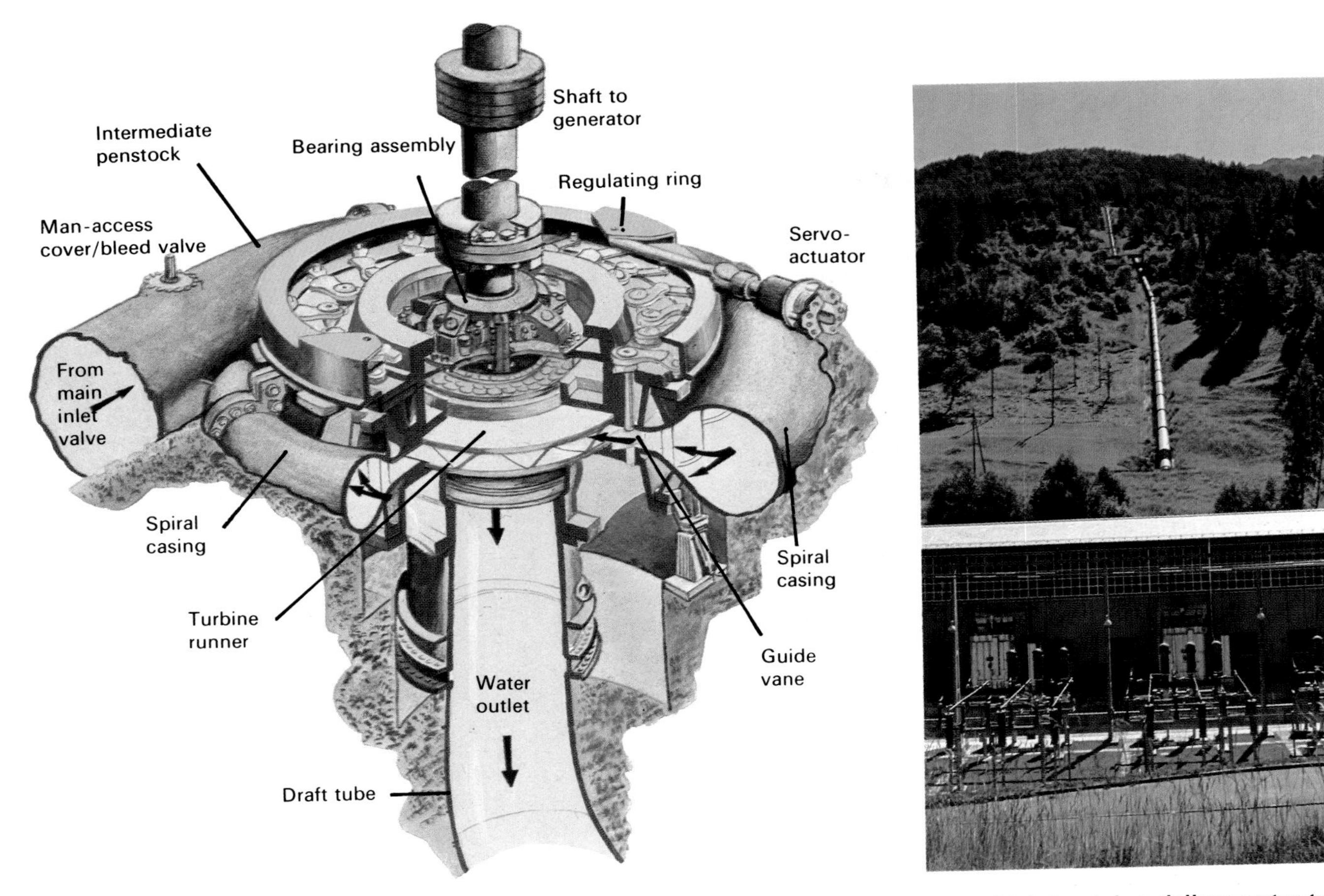

In these schemes surplus off-peak power is used to convert the turbine into a pump and pump water back to the high-level reservoir to boost the head.

The other widely used hydroturbine, the Kaplan, or feathering-propeller type, also has movable runner vanes. Unlike the Francis the water flows into and out of the Kaplan axially. It is particularly suited to low-head situations and can be adjusted to widely differing rates of flow by varying the pitch of its vanes. Simpler fixed-vane propeller turbines are also widely used for low-head, high-volume sites.

In its usual configuration the Kaplan turbine is mounted vertically. But the Kaplan turbines designed for the River Rance Tidal Power plant, in Brittany, are mounted horizontally. They are particularly suited to the very low head utilized.

The Rance tidal plant, which has been operating since 1966, could well be pointing the way ahead. Several places in the world have a sufficient tidal range (rise and fall) to make tidal-power stations an attractive proposition. They include the Bay of Fundy in Canada and the Severn estuary in the West of England, where the tidal ranges are as high as 15 metres (50 feet) and 11 metres (35 feet), respectively.

Dwindling Resources

Most electricity in the world is produced by steam-turbine generating stations. And these burn coal, oil or natural gas to produce their energy. These fossil fuels accumulated in the Earth's crust over periods of hundreds of millions of years. But they are now being used at such a rate that an energy crisis is upon us.

Supplies of oil may last for another 20–30 years if

Above: Penstocks from a high-level dam deliver water to the turbines at the Val d'Anniviers hydroelectric power station in Switzerland.

Top left: A cutaway of a pump turbine of a type used in pumped-storage hydroelectric schemes, known as a vertical reversible Francis. In the generating, turbine mode, the main inlet valve admits water into the spiral casing. It is directed by the guide vanes radially into the turbine runner, from which it exits vertically down the draft tube. The guide vanes are adjusted in unison when the servo-actuator moves the regulating ring.
Opposite top: The turbine hall of a steam-generating power station. The steam enters through the ducts on the right of the machines and exhausts to condensers below floor level.

Opposite bottom: The power station at the foot of the Hoover Dam on the Arizona/Nevada border. The turbines are housed in the massive concrete foundations of the arch-gravity dam, which is 221 metres (726 ft) high. Power output is 1350 megawatts.

Below: The mountainous terrain of Switzerland is ideal for dam-building, as here in the Val de Moiry.

we are lucky. Supplies of coal may last for a century or more. But all fossil fuels will one day run out. This will be a catastrophe not only for the power industry but also for the chemical industry, for the fossil fuels provide essential raw materials for plastics and synthetics of all kinds.

When coal, oil and gas have gone, how do we produce power? Can we build many more hydroelectric power (HEP) plants which, of course, use a renewable energy source – flowing water? Unfortunately, though more HEP plants will be built, there are not enough suitable sites available worldwide to make an appreciable impact on the energy shortfall.

One solution to the problem is already available, in the form of nuclear power. The advantage of nuclear power is that a tiny amount of nuclear fuel can provide a colossal amount of energy. Under ideal conditions, weight for weight, nuclear fuel could produce three million times as much energy as coal.

A major disadvantage of nuclear power, however, is the potential danger to the environment. The waste materials from nuclear power plants are dangerously radioactive. They give off penetrating radiation that in large doses may burn, cause cancer and leukemia, induce genetic changes, and even kill. Disposal of radioactive waste presents problems that have yet to be satisfactorily solved.

Power from the Atom

To understand the principles behind nuclear power we must first look into the very nature of matter. All the matter which makes up the universe, be it gas, liquid or solid in star or planet, rock or insect, is composed of atoms. Each chemical element – hydrogen, iron, carbon, mercury – is made up of different kinds of atoms.

Atoms are incredibly small – they measure only a few hundred-millionths of a centimetre across. And they are mostly empty space. Their mass is concentrated at their centre in a nucleus, which contains tiny particles called protons and neutrons. Protons have a positive electric charge; neutrons are electrically neutral. Around the nucleus circle clouds of even tinier particles called electrons, which have a negative electric charge. In its normal state an atom has as many circling electrons as it has protons in its nucleus, and is thus electrically neutral.

Atoms combine with one another to form molecules by means of their electrons. The number of electrons, and therefore protons, in the atom thus helps determine the atom's chemical nature. Every chemical element has different properties from its fellows because its atoms contain a different number of electrons and protons. But the presence of more or less neutrons in an atom has no effect on its chemical nature. So atoms of the same element can be different in as much as they can contain different numbers of neutrons. Such atoms are called isotopes, a term frequently met in nuclear physics.

In chemical combination an atom's electrons may

The awesome spectacle of an atomic bomb explosion, which has the explosive force of tens of thousands of tonnes of conventional high-explosive. But it is more deadly because it generates lethal radiation, and radioactive debris that can be wafted by the wind to the four corners of the globe and deposited as dangerous fallout for months on end.

Opposite: But there is another, beneficial side to the nuclear coin. The very radiation that kills can be made to cure when kept under control. Many hospitals now have radiotherapy units which use low-level radiation to treat patients with tumours, for example.

temporarily leave it, but its nucleus remains unchanged. For the majority of atoms the nucleus is perfectly stable – it remains the same for ever. But some atoms have an unstable nucleus, which will spontaneously change in one way or another and emit radiation. These atoms are termed radioactive. The radiation they emit may be gamma-rays (very short-wave electromagnetic radiation) or streams of particles such as beta-rays (electrons) and alpha-rays (helium nuclei).

The Mystery of Fission

The most interesting radioactive atom from our point of view is uranium. Naturally occurring uranium is composed mainly of the isotope uranium-238 (U-238) – the nucleus of its atoms contains 238 particles. But it also contains traces of the isotope U-235. U-235 holds the key to nuclear power because it can be made to split, or fission. When it does so, it releases an enormous amount of energy.

Where does this energy come from? It comes from

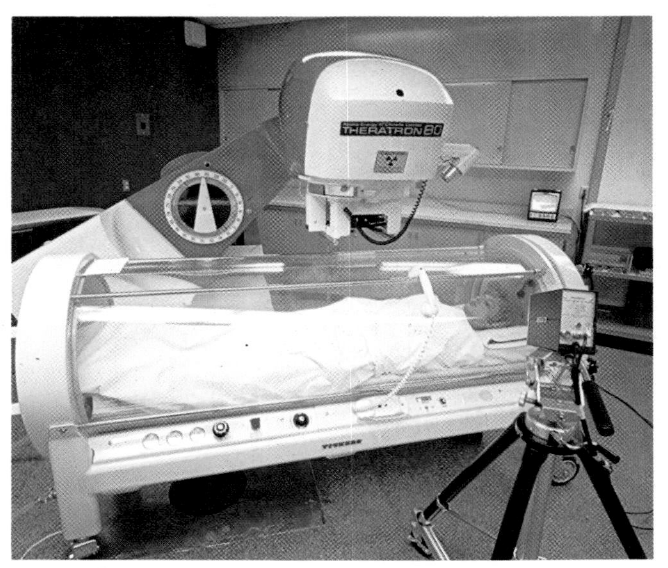

the destruction of a tiny amount of matter. If you could weigh a U-235 nucleus before fission and the fragments produced by fission, you would find that the latter weighed fractionally less. The lost mass has been converted into energy, for mass and energy are equivalent, as Einstein summed up in his famous equation $E = mc^2$. E is the energy released by the destruction of mass m, and c is the velocity of light.

We can bring about fission in U-235 by bombarding it with neutrons. When a U-235 nucleus captures a neutron, it becomes unstable and splits. As well as releasing energy, fission produces two or more neutrons. These neutrons may go on to split other U-235 atoms, which in turn will produce more neutrons to split still more atoms.

If there is a large enough lump of U-235 present (a critical mass) enough neutrons will be retained within it to split atom after atom at an ever-increasing rate. Then a chain reaction is said to occur. The energy release is colossal. An uncontrolled energy

Above: The uranium 'fuel' used in nuclear reactors is highly radioactive and must be handled remotely with equipment like this. The operator uses electro-mechanical 'hands' on flexibly jointed arms, and watches through special leaded, radiation-proof windows.

release is catastrophic; this happens in the atomic bomb. In the bomb several million million million million atoms of uranium split in a millionth of a second and release as much energy as tens of thousands of tonnes of TNT. Temperatures of several million degrees are produced, vaporizing the bomb fragments and everything over a wide area, so creating the awesome fireball, where temperatures reach several million degrees.

Taming the Bomb

To harness the mighty power of the atom we have to find some way of controlling the chain reaction so that the energy release is gradual rather than instantaneous. The nuclear reactor is the means of doing this.

The diagram below shows the main features of a typical reactor. There are several kinds of reactors, which vary in detail but function in an essentially similar way. Controlled nuclear fission takes place in the reactor core, which contains the uranium 'fuel'. The core is honeycombed with passages through which a coolant circulates to extract the heat produced by fission. The coolant transfers its heat to water in a heat exchanger, and the water is turned to steam. This is then led off to drive a steam turbine coupled with an electricity generator in the conventional way.

Most reactors include in their core a substance called a moderator. The function of the moderator is to slow down neutrons because low-speed, or 'thermal' neutrons bring about fission more readily in U-235 and are not as readily captured by U-238. Such reactors are called thermal reactors. An advanced kind of reactor, at present few in number, is designed to use fast neutrons and thus requires no moderator. It is called a fast or breeder reactor (see box, opposite).

Control over nuclear fission is exercised by means of control rods that dip down into the core. They are made of a material that readily absorbs neutrons such as boron or cadmium. They are pushed in or

Below: One kind of British-designed nuclear reactor not much favoured elsewhere.

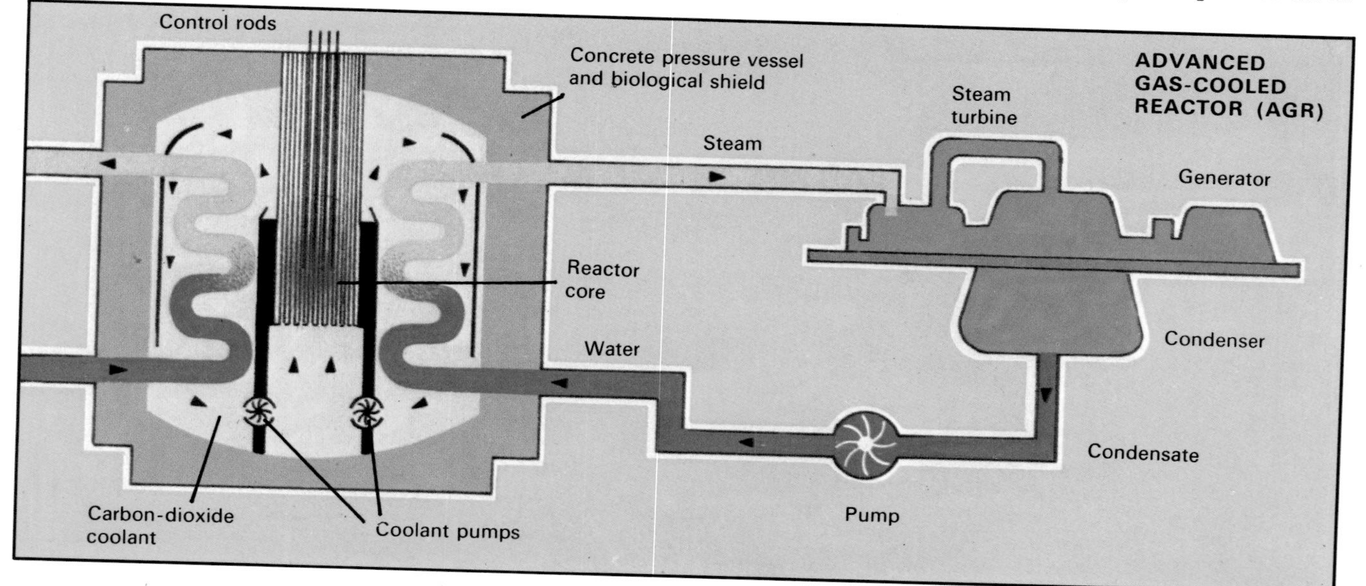

THE FAST REACTOR

Proven reserves of uranium are limited and are insufficient, when 'burned' in thermal reactors to sustain the world's demand for energy for much more than a hundred years or so. But used in another kind of reactor the available uranium can be made to last for many centuries more. This reactor is the fast, or breeder reactor. It uses fast neutrons to bring about fission and actually 'breeds' more fuel than it consumes.

For its fuel the fast reactor uses uranium enriched with plutonium-239, which is produced as a by-product in thermal reactors. Plutonium-239 is fissionable by fast neutrons, so no moderator is required. The coolant in the reactor is liquid sodium at a temperature of about 600°C and at low pressure ($\frac{1}{3}$ atmosphere). The liquid sodium extracts heat from the core and passes through an intermediate heat exchanger, where it heats up liquid sodium in a secondary circuit. That sodium passes through a second heat exchanger where it boils water into steam to feed the steam turbogenerator.

The reactor core is surrounded by a 'blanket' of uranium carbide in stainless steel cans. The spare neutrons produced during fission convert some of the non-fissionable uranium-238 into fissionable plutonium-239. This is the part of the reactor that does the 'breeding'. From time to time the cans in the blanket are removed, and the plutonium extracted.

Great controversy rages over the introduction of the fast breeder reactor because it breeds plutonium. Plutonium is one of, if not the most dangerous substances in existence. It is reputedly 20,000 times more toxic than cobra venom, and retains its lethal radioactivity for tens of thousands of years. Storage and protection of such a deadly material thus presents enormous difficulties in the short as well as the long term. But with the energy crisis deepening every year it seems that these difficulties must be overcome. No realistic alternative to fast reactors as a continuing energy source has yet come to light that can cope with anticipated future power demands when the fossil fuels have gone, though many argue otherwise (see page 218).

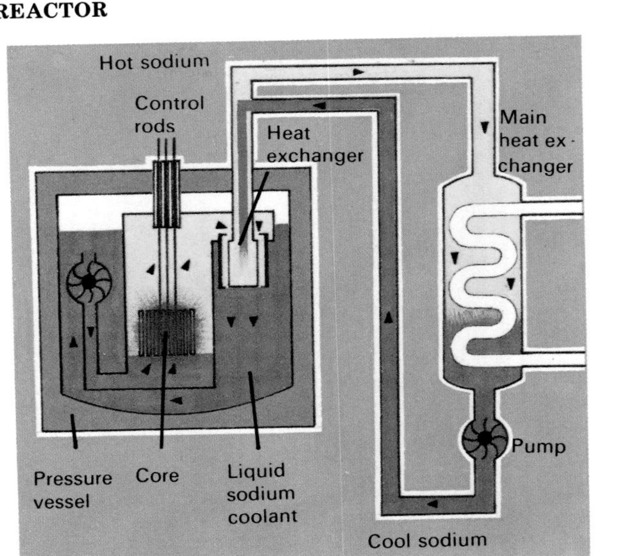

pulled out to adjust the number of neutrons available for fission. When inserted fully, they absorb so many neutrons that the chain reaction cannot get underway. Elaborate fail-safe systems ensure that whenever anything untoward happens within the reactor, the control rods automatically shut it down.

As well as producing heat, fission produces high levels of radiation. The fission fragments are also highly radioactive. To protect operating personnel and the environment in general, the reactor is built inside a containment vessel, capable of withstanding enormous pressures from within. Over 2 metres ($6\frac{1}{2}$ feet) thick, it is made of reinforced concrete and lined with steel.

Because of their design, there is no possibility that a reactor can blow up like an atomic bomb. The worst thing that could happen, if for example, the coolant was suddenly lost, would be a meltdown of the core. But so stringent are the safety measures adopted in a nuclear plant that the probability of such an accident occurring is infinitesimally small.

Practical Reactors

Worldwide by 1980 there were some 200 nuclear reactors with a power output of over 150 megawatts. The United States (68), Russia (26) and Britain (24) had the most reactors. But only in a few countries do nuclear power stations contribute significantly to the nation's electricity. In Britain and the United States, for example, they contribute more than one-eighth of the total electricity. The capital cost of building a nuclear plant is colossal – up to £1000 million. But the electricity it produces is now

Below: A cutaway drawing of what will soon become the world's first full-scale fast-breeder reactor. Known popularly as 'Super Phénix', it is being built near Creys-Malville in south-west France. It will be liquid-sodium cooled and have a power output of 1200 megawatts when it comes on stream in the early 1980s. Construction, by an international consortium, will cost an estimated 6000 million French francs.

THE WAY IT WORKS

1 Reactor building
2 Unit vent
3 Steel containment vessel
4 Polar crane
5 Personnel hatch
6 CEA change platform
7 Missile shield
8 Steam generators
9 Main steam pipes
10 Safety injection tanks
11 Pressurizer
12 Control element drive mechanism cooling equipment
13 Control element drive mechanism
14 In-core drive mechanism
15 Reactor vessel
16 Reactor vessel support columns
17 Reactor coolant pumps
18 Steam generator shield wall
19 Fuel-handling bridge
20 Refuelling canal
21 Fuel-transfer tube
22 Reactor drain tank
23 In-core instrumentation
24 Containment spray headers
25 Feedwater pipe
26 Containment ventilation cooling units
27 Component cooling heat exchanger
28 Component cooling pumps
29 Charging pump
30 Pipe chase
31 Cable chase and HVAC duct
32 Shut-down cooling heat exchanger
33 Fuel-handling building
34 Cask-handling crane
35 Fuel-handling bridge crane
36 Spent fuel pool
37 Watertight gate
38 Transfer tube gate valve control wheel
39 Fuel carriage
40 Transfer system winch
41 Cask area
42 Air-conditioning equipment room
43 Fuel shipping and receiving area
44 New fuel storage area
45 Waste management area
46 Decontamination area
47 Hold-up water storage tank
48 Reactor make-up water-storage tank
49 Refuelling water-storage tank
50 Main steam valve enclosure
51 Main steam isolation valves
52 Main steam safety valves
53 Diesel generator
54 Switchgear area
55 Control room
56 Computer room
57 Control area HVAC equipment
58 Turbine building
59 HP turbine
60 LP turbines
61 Generator
62 Alterrex (exciter)

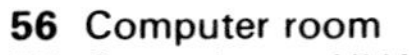

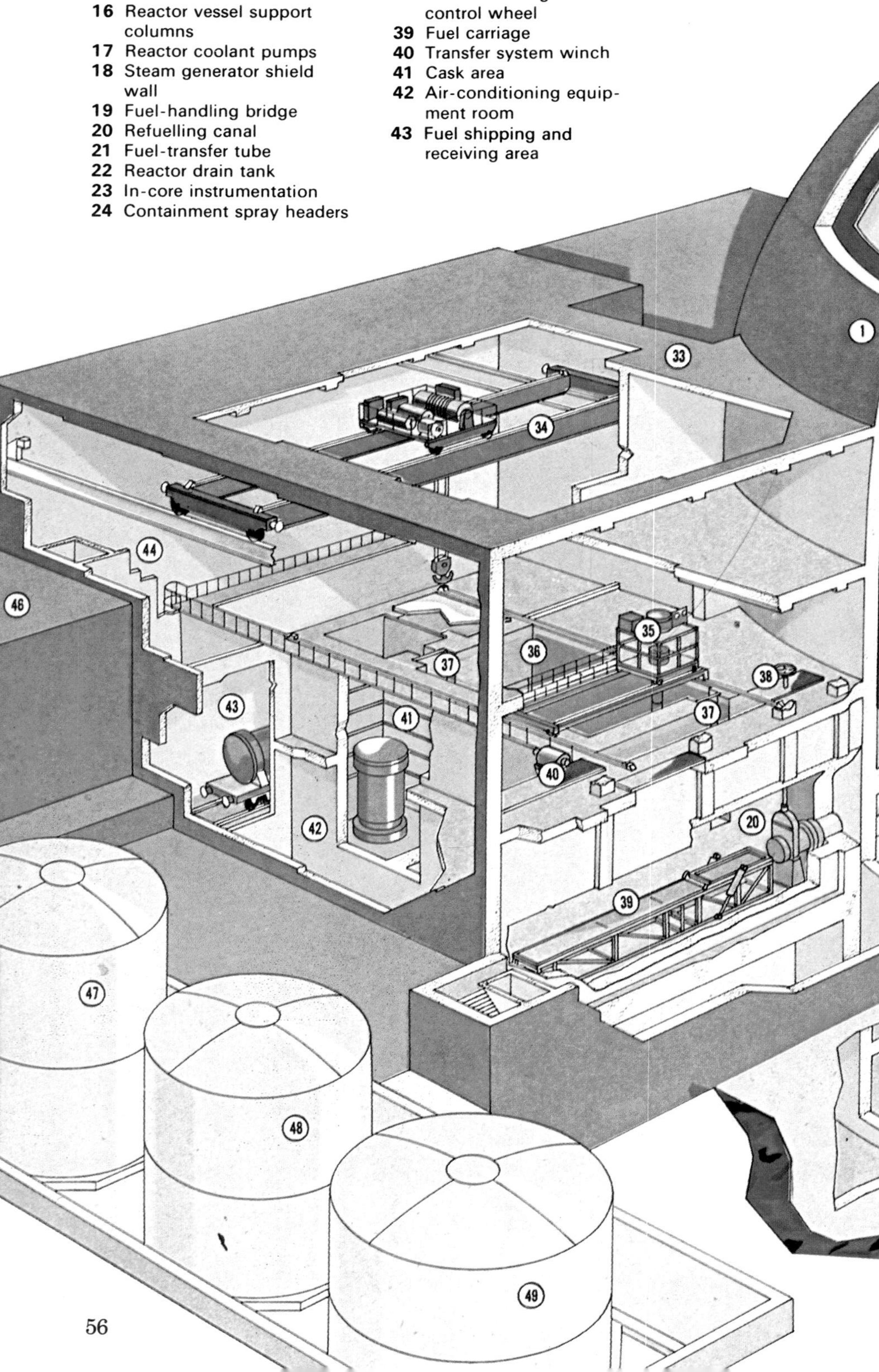

CHEROKEE/PERKINS NUCLEAR POWER STATIONS

Detailed design layout for one of the three identical nuclear reactor complexes which will form nuclear power stations in Cherokee County, South Carolina, and at Perkins, North Carolina. Each reactor unit has an effective power output of 1280 megawatts.

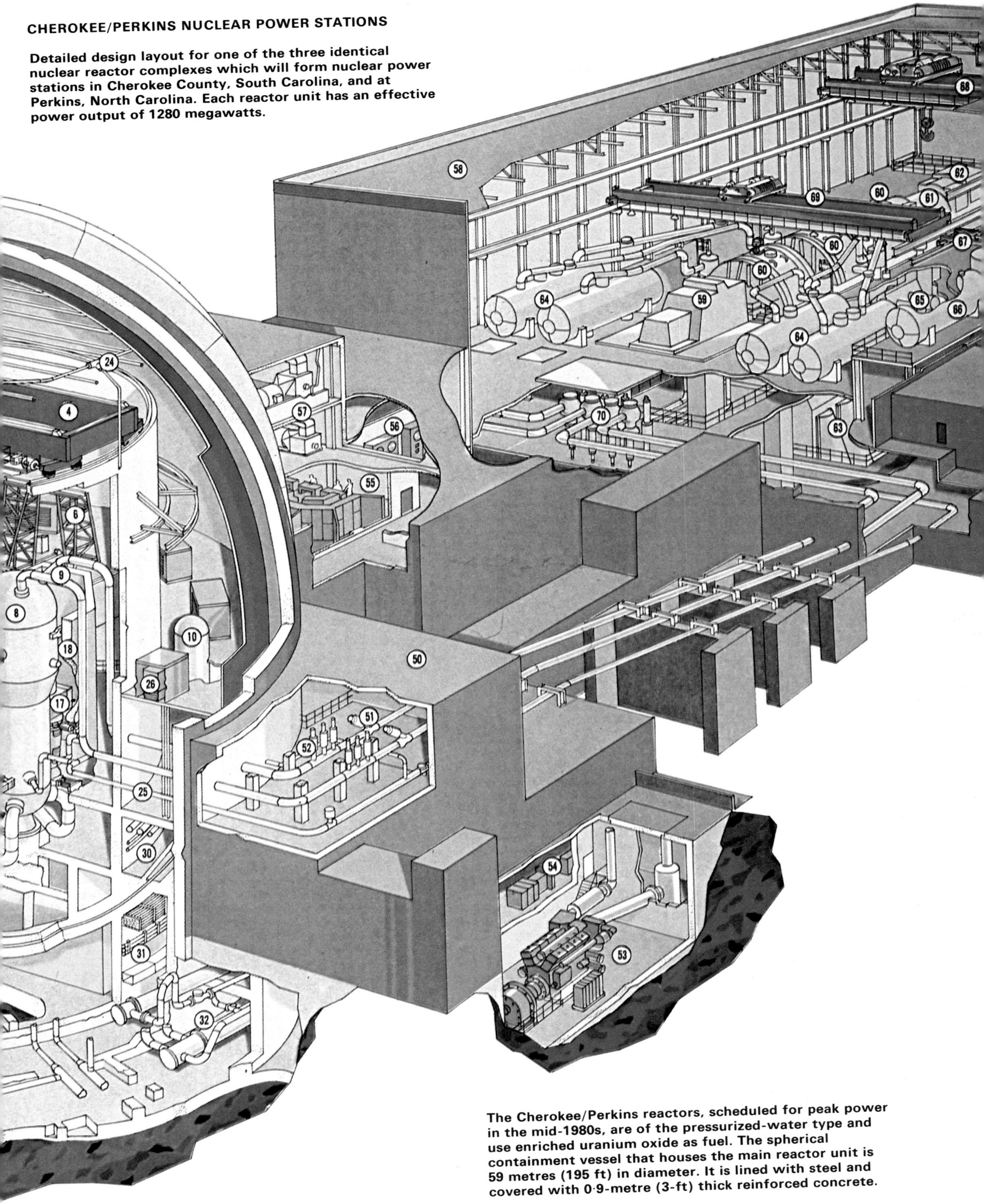

The Cherokee/Perkins reactors, scheduled for peak power in the mid-1980s, are of the pressurized-water type and use enriched uranium oxide as fuel. The spherical containment vessel that houses the main reactor unit is 59 metres (195 ft) in diameter. It is lined with steel and covered with 0·9-metre (3-ft) thick reinforced concrete.

significantly cheaper than that produced in a conventional power station burning fossil fuel.

The widely used thermal reactors differ primarily in the type of moderator and coolant they use. They also use different types of fuel. Some use natural uranium, in which the proportion of U-235 is about 0·7 per cent. Most, however, use enriched uranium in which the proportion of U-235 has been substantially increased. This allows for a more compact core design. Whereas natural uranium is generally used in metal form, enriched uranium is used in the form of uranium dioxide pellets. These pellets are contained in small stainless steel or zirconium alloy cans, forming what is called a fuel pin. Clusters of these pins form a fuel element.

The kind of reactor known as Magnox uses natural uranium as fuel, graphite as moderator, and carbon dioxide as coolant. The advanced gas-cooled reactor (AGR) uses slightly enriched uranium dioxide as fuel, but the same moderator and coolant. It operates at $1\frac{1}{2}$ times the temperature (650°C) and twice the pressure (40 atmospheres) as the Magnox design. The high-temperature reactor (HTR) also has graphite as moderator, but it has to use helium as coolant. At the operating temperature (720°C) carbon dioxide would attack graphite. The fuel is uranium dioxide enriched to 10 per cent U-235.

The other types of reactors use ordinary water or heavy water as moderator. Heavy water is water formed from an isotope of hydrogen called deuterium, which has an extra neutron in its nucleus. It is a very efficient moderator. It is used, for example, in the steam generating heavy water reactor (SGHWR) and the CANDU reactor. In the latter it also acts as the coolant. The SGHWR, however, uses ordinary light water as coolant.

Light water is employed as both moderator and coolant in pressurized water reactors (PWR) and boiling water reactors (BWR). In the PWR water is circulated at very high pressure (about 150 atmospheres) through the core and heat exchanger. This arrangement allows for a very compact design in which the core measures typically about 3·7 metres (12 feet) high and 3 metres (10 feet) in diameter. It is this kind of reactor that powers nuclear submarines. In the BWR, which operates at less than half the pressure of the PWR, water actually boils inside the pressure vessel. And the steam produced is led directly to the steam turbine, without the need for a heat exchanger.

Generating Electricity

In a power station the turbines, whether spun by expanding steam, gas or flowing water, drive generators to produce electricity. The generators may be driven indirectly through gearing, but usually they are driven directly from the turbine shaft. The majority of turbines are mounted horizontally – their rotor rotates around a horizontal axis, and they are coupled to horizontally mounted genera-

Above: An aerial view of Cottam power station which shows the typical features of an inland steam-turbine generating facility. Most prominent are the eight conical cooling towers, which cool the cooling water pumped through the condensers.

Left: A nuclear reactor refuelling machine.

tors. But the large hydroturbines are mounted vertically, and so the generators are mounted vertically above them. They are thus rather different in design.

The largest power stations in the world are hydroelectric. The Russian plant at Krasnoyarsk on the River Yenisey in Siberia has a total output of over 6000 megawatts from 12 turbogenerators. The largest steam-generating power plant is the 2500-megawatt coal-fired station at Paradise, Kentucky. It consumes over eight million tonnes of coal a year. Individual turbogenerators producing over 600 megawatts are now common, and several turbogenerators of double that output are operating.

First Principles

All electricity generators, large and small, work on the principle of electromagnetic induction. If you move a conductor in a magnetic field, you induce, or set up within it an electric current. The converse also applies. If you pass electric current through a conductor in a magnetic field, you make it move. This explains how an electric motor works. Generators and motors are in fact very similar in design, and the same machine under different conditions could function as both generator and motor.

A very simple generator is illustrated below. It consists of a single coil of wire that is rotated between the poles of a magnet. Depending on the way in which the electricity is tapped from it, it may produce either alternating (two-way) current or direct (one-way) current. The alternating-current (AC) generator is called an alternator, and the direct-current (DC) generator a dynamo.

In a relatively small alternator a stationary external magnetic field is provided by electromagnets. These are formed by winding coils of wire (field coils) around iron pole pieces, and are energized by an independent DC supply. They may have two, four or more poles. This stationary part of the

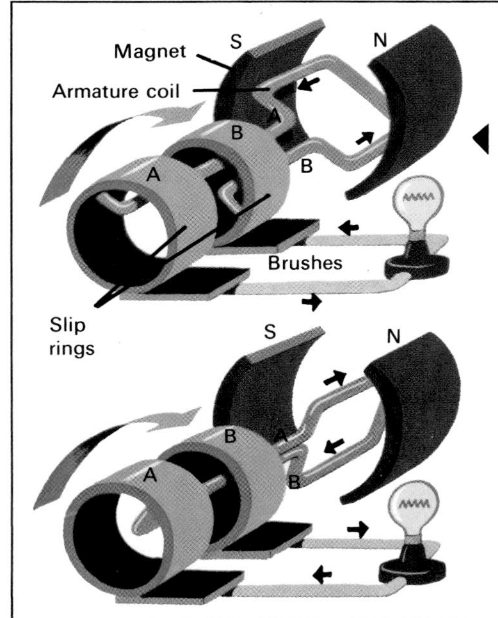

AC GENERATOR
In a simple alternator the ends of the armature coil are connected to two slip rings. From these current is taken via brushes and fed to a light bulb. As each side of the coil moves alternately up and down, the direction of the current generated in it changes.

DC GENERATOR
In a simple dynamo the ends of the armature coil are connected to the halves of a commutator. This is split in such a way that the generated current always leaves the brushes flowing in the same direction, irrespective of the direction of current in the coil.

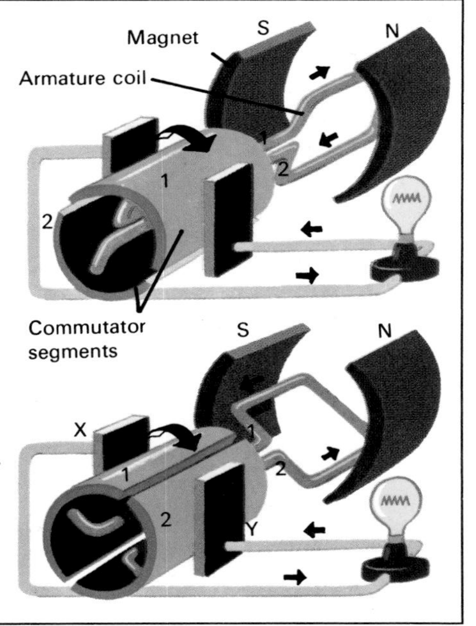

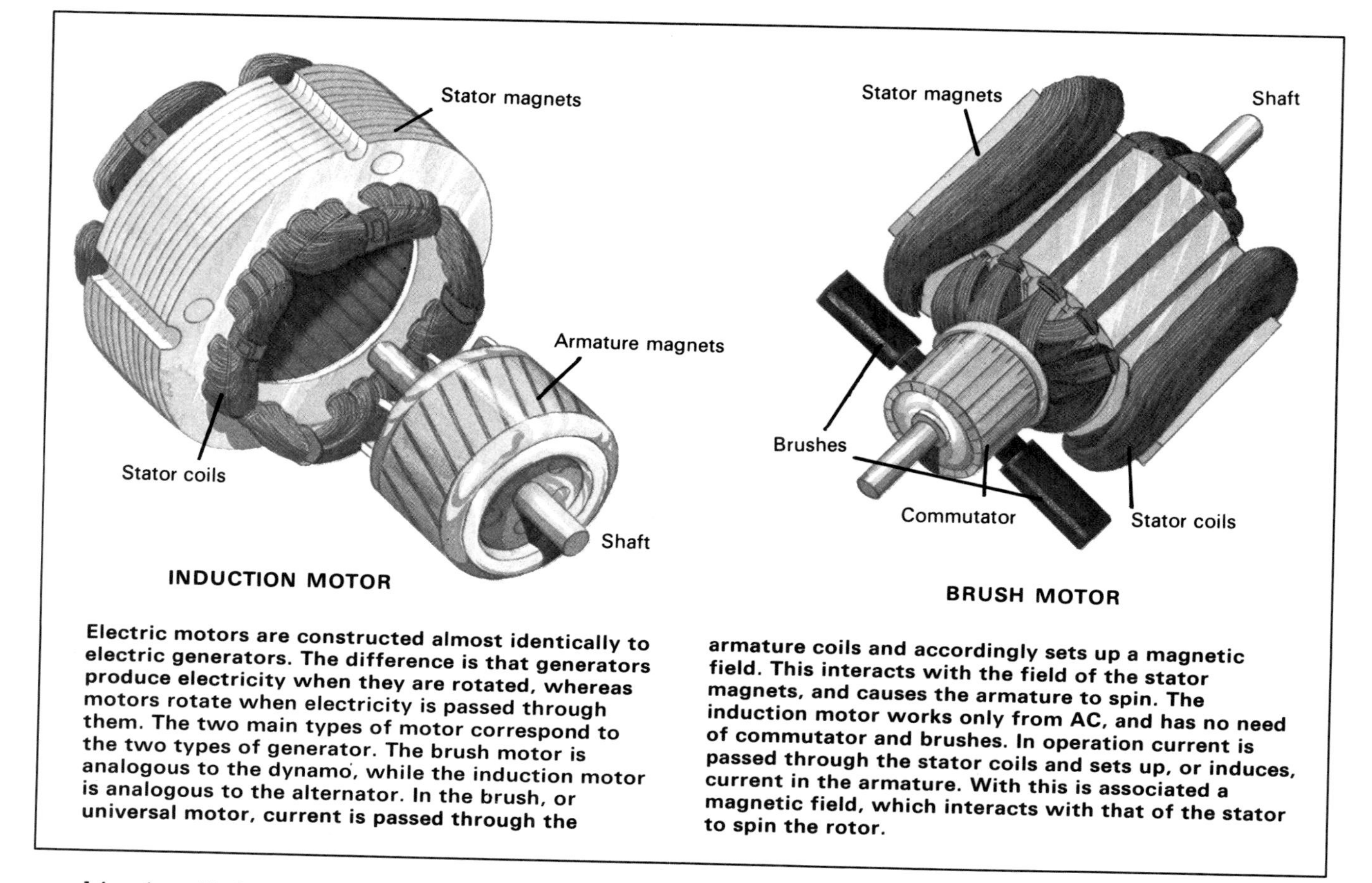

Electric motors are constructed almost identically to electric generators. The difference is that generators produce electricity when they are rotated, whereas motors rotate when electricity is passed through them. The two main types of motor correspond to the two types of generator. The brush motor is analogous to the dynamo, while the induction motor is analogous to the alternator. In the brush, or universal motor, current is passed through the armature coils and accordingly sets up a magnetic field. This interacts with the field of the stator magnets, and causes the armature to spin. The induction motor works only from AC, and has no need of commutator and brushes. In operation current is passed through the stator coils and sets up, or induces, current in the armature. With this is associated a magnetic field, which interacts with that of the stator to spin the rotor.

machine is called the stator. Many coils of wire are wound around a shaft (rotor) which is free to rotate inside the stator. This is called the armature. Alternating current is induced in the armature coils as they cut across the stationary magnetic lines of force. It is tapped by means of slip rings and carbon brushes.

The construction of a dynamo is essentially the same, the major difference being that the current is taken from the armature via a commutator not slip rings. The commutator is so designed that current always flows into and out of the same brushes, although the current in the coils changes direction every half-turn. It consists of segments of copper insulated from one another. The ends of each coil of wire on the armature are joined to opposite segments of the commutator.

The current needed to energize the field coils of dynamo may be taken from the DC output from the armature. This may be done in three main ways, termed series, shunt and compound, and gives rise to generators with slightly different characteristics.

Turboalternators

For large power stations the alternator design described above would be impractical. The turbines drive the rotors at speeds of up to 3600 revolutions per minute, and the voltage generated is typically 25,000 volts. Taking such a high voltage via slip rings and brushes would set up severe sparking which would quickly destroy the brushes.

In large alternators, therefore, the armature coils are wound on the stator, and the field coils are wound on the rotor. Alternating current is induced in the stator coils by the rotating magnetic field of electromagnets on the rotor. Direct current to energize the field coils is supplied by a small independent generator mounted at the end of the rotor shaft. It is called an exciter. The exciting current is fed to the rotor via slip rings, a commutator, or 'brushless' connexions.

The stator core of a modern high-output generator is made of low-loss iron – iron in which energy loss by eddy currents is minimal. It is fabricated, not all in one piece, but of thin laminations individually insulated. This further minimizes internal energy losses. Slots in the inner periphery of the stator take the stator windings. The conductors that form the windings consist of a number of copper laminations insulated from one another and from the core.

The design of the rotor differs from generator to generator according to the driving speed and the power output. High-output generators (say, over 20 megawatts) generally have a cylindrical rotor which is machined from a single forging of high-quality steel. Axial slots are cut in the body to take the field windings. Lower-output generators are often of the so-called salient pole type. They have typically two or four pole pieces, around which the field coils are wound. The poles may be machined with the rotor from a single forging, or they may be formed from separate laminations and dovetailed into the rotor body.

The generation of high-voltage electric current

in the stator coils and the electrical losses in the generator body result in the production of a great deal of surplus heat. So generators must be adequately cooled. Fans are therefore mounted on each end of the generator rotor to force air through holes or slots in the stator and rotor.

At higher powers this kind of ventilation is inadequate, and the air is water cooled by a heat exchanger before being passed back through the generator. For generators with an output of 100 megawatts or more, cooling has to be effected by hydrogen. Hydrogen not only has three times the heat conductivity of air; it also is much less dense and thereby reduces frictional losses. The whole generator system must be totally enclosed to prevent leakage, which could allow explosive hydrogen/air mixtures to form.

The type of alternator described is called a synchronous generator. This means that the frequency of the alternating current – the number of 'waves' of electricity generated per second – in the stator coils is related directly to the speed at which the rotor turns. Two main frequencies are used in the world – 60 hertz (cycles per second), used in the United States and Canada, and 50 hertz, used in Europe. To produce a 60-hertz current a 2-pole generator must rotate at 3600 rpm (revolutions per minute). To produce 50-hertz, it must rotate at 3000 rpm. The turbines driving the generators must be precisely governed to maintain a constant speed.

Left and right: The beginning and end of the road for hydroelectricity, generated in a power plant at the base of a dam. Transmission lines carry it out of the dammed gorge (left) and then many hundreds of kilometres to the consumer, via a substation. Its voltage is reduced first at the substation and finally by small pole transformers (right) outside the consumer's house.

Below: A geothermal power station located in Northern California, near the appropriately named town of Geyserville. It pipes the high-pressure steam of a natural geyser field directly to steam turbines. By 1980 the United States was generating about 1000 megawatts of electricity from geothermal sources. Geothermal sources are being tapped elsewhere mainly for residential district-heating schemes, as in Reykjavík (Iceland) and Paris.

Power Transmission
The power stations that produce the electricity cannot often be located near to where the power will be used. Fortunately electricity, unlike other forms of energy, can be transmitted by cable over long distances with minimal loss.

In the majority of cases the conductors, or transmission lines, are carried overground on tall steel towers, or pylons. The conductors, which are usually made of aluminium reinforced with a steel core, are between 25 and 60 mm (1 and 2·5 in) in diameter. They hang from the towers by strings of glass or porcelain insulators.

The usual transmission system is 'three-phase'. It uses three conductors in which voltages and currents are out of step by one-third of a cycle. This system, which results in a steady power flow, is three times more efficient than a two-conductor, or 'single-phase' system would be. (The domestic electrical supply is a single-phase system.) A tower usually carries two 'circuits' of three conductors.

The voltage at which electricity is usually produced in electric generators – 25,000 volts AC – is, however, not suitable for long-distance transmission. The power losses in the lines would be too great. It is much more efficient to transmit electricity at very high voltages. The cost of associated equipment rises as the transmission voltage rises, however, and there is an optimum voltage for a given distance. Over relatively short distances the voltage may be as low as 132,000 volts, but over greater distances, 275,000, 400,000 and 735,000 volts may be selected. Experiments with even higher voltages are already well advanced.

Stepping Up, Stepping Down
The generator voltage is boosted to such levels by a transformer. A transformer consists essentially of two coils of wire wound around an iron core. It operates on the principle of electromagnetic induction. A current is set up, or induced in one coil, when the current flow through the other changes.

Now, an alternating current changes continuously. When it is fed to one coil of a transformer (the primary), a continuous current is induced in the other (the secondary). The interesting thing about the transformer is that if the secondary coil has more turns than the primary, current will be induced in it at a higher voltage. If the secondary has 10 times as many turns, the output voltage will be increased, or stepped-up to 10 times the input. Conversely, if the secondary has only one-tenth as many turns as the primary, then the output voltage will be reduced, or stepped down to only one-tenth of the input. At a power station a step-up transformer increases the generated voltage for transmission. At the other end of the transmission lines, step-down transformers at substations reduce the voltage for consumer use. A major substation might reduce the transmission voltage to 30,000 volts for distribution to a town or centre of heavy industry. Intermediate substations would then reduce the voltage to, say, 11,000 volts to feed hospitals and light industries and further substations that supply 240 volts to shops and houses.

Each generator in a power station has its own transformer, which feeds the high-voltage electricity into sets of common conductors called busbars. These connect with the transmission lines. Between

FUEL CELLS

In 1959 British inventor Francis Bacon came up with an ingenious new source of electricity – the fuel cell. To date fuel cells have had their most interesting use in space. They powered the Apollo spacecraft on their triumphant journeys to the Moon (see page 212). At the same time they provided the crews with their drinking water!

The Apollo fuel cells ran off hydrogen and oxygen, plentifully available in spacecraft where in liquid form they are used as rocket propellants. In the fuel cell gaseous hydrogen and oxygen are supplied to specially designed platinum electrodes and under the catalytic influence of the platinum they combine to form water. In so doing they generate electricity by the reverse of electrolysis.

Many other fuels are currently being experimented with, including hydrazine and methanol. These fuels are electrochemically 'burned' in the cells to produce carbon dioxide and water. Although at present fuel cells are very expensive, they could become cost effective when produced in quantity and they have the advantage of producing no pollution.

Left: The nerve centre of a power station, where operating personnel monitor the performance of the generating equipment and control the switchgear that feeds the current into the supply grid.

Far left: Environmentally conscious American power companies have designed this new type of pylon.

Bottom left: A set of high-voltage circuit-breakers.

the transformers and busbars, and busbars and transmission lines are switches to disconnect the various parts of the system. The switches include circuit-breakers and isolators. Circuit-breakers are the main on/off switches. Isolators are used when needed to isolate the circuit-breakers from all outside electrical sources.

A circuit-breaker has to be very carefully designed. When the contacts inside it separate and interrupt the current, the current continues to surge for an instant, causing powerful arcing. To prevent fire and damage this arc must be quenched immediately. To do this, one kind of circuit-breaker has its contacts immersed in oil. Another uses a blast of compressed air to blow out the arc.

The Grid

The electricity demand varies widely from hour to hour, day to day and month to month. It also varies from region to region. An isolated power station could not operate efficiently nor satisfy the capricious demands of its customers by itself. To do so it must form part of a network, or grid in which a number of power stations are interconnected, feeding their electricity, as it were, into a common pool. Extra demand for power in one region can be met with surplus power produced in another.

In some countries the grid system operates nationwide. Britain has the largest grid system in the world, with over 140 power stations able to supply a simultaneous demand for 55,000 megawatts in England and Wales. The grid is also connected via submarine cable to the French electricity network, to mutual advantage.

On the Move
Road and Rail

Mobility is the hallmark of our present civilization. It dictates where and how we live and work and spend our leisure time. Giving us unprecedented mobility in our everyday lives is the ubiquitous motor-car, the latest in a line of wheeled vehicles whose ancestry can be traced to the early Sumerian civilization in the Middle East (see page 15). Born less than a century ago, the car is perhaps already doomed to extinction, at least in its present form.

For the car has a conflicting, Jekyll and Hyde personality. On the one hand it is a purveyor of convenient, speedy, comfortable and reliable travel. But on the other it is a gobbler of precious oil, an exhaler of obnoxious and polluting fumes, and a ruthless killer of human beings who happen to get in its way.

In all, cars amount for well over three-fifths of the 250 million or so vehicles on the world's roads today. The remainder are the smaller and larger relatives of the car, the motor-cycle, lorry, truck and bus. The motor-cycle is a more efficient form of individual transport than the car, while the truck and bus are more efficient transporters of people and goods in bulk. To cope with the ever-increasing volume of traffic, road networks are continually being expanded and revamped. Special superhighways, known variously as expressways, autobahns, autostrada, autoroutes and motorways, now criss-cross the continents, speeding inter-city travel.

For in-city travel, however, it is better to go by underground railway, or subway, if you can, for the streets are becoming increasingly congested. Underground railways are on the increase worldwide, reversing the trend of railways in general to reduce their track mileage. Railways have suffered great competition from the car for transporting passengers and from the lorry for hauling freight. But by streamlining their operations and improving their locomotives and rolling stock, they are beginning to hold their own again. And so they should, for the railway is potentially the most efficient form of transportation (save for the bicycle) yet devised by man.

Above: Rider and machine nearly part company in a cross-country motor-cycle trials event. Fun to ride and frugal on petrol, the motorbike has a charismatic appeal to the younger generation.

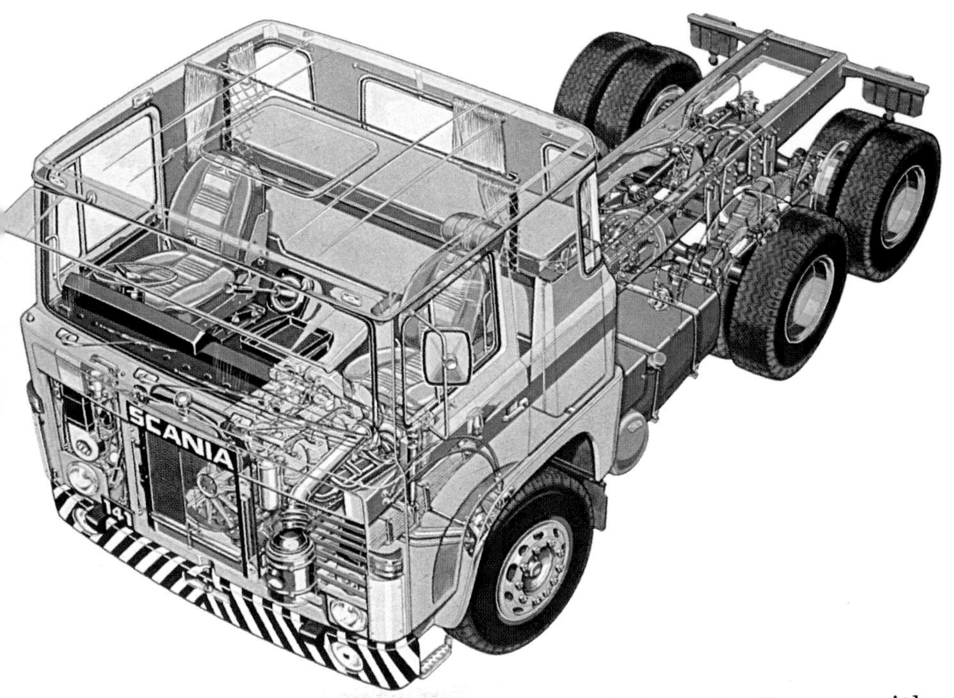

Left: Cutaway of one of the workhorses of the highways, a 60-tonne tractor truck. It is powered by a V-8, 14-litre, turbocharged diesel engine of some 370 horsepower. It has a 10-speed all-synchromesh gearbox and drives through twin axles.

Opposite: Cars negotiating the Golden Gate Bridge in San Francisco. One of the more beautiful creations of the highway engineer, this suspension bridge has a span of 1280 metres (4200 ft).

Below: British Rail's prototype Advanced Passenger Train (APT) leans into a bend, demonstrating its superior cornering capabilities.

1700s
Stagecoach
and horses
1873
Amédee Bollée's
steam carriage
L'Obéissante
1770
Nicolas Cugnot's
second steam tractor
1901
Columbia
Electric
1899 Cannstatt-Daimler
1885 Benz
three-wheeler
1893 Duryea
1903 De Dion-Bouton
1905 'Curved-dash'
Oldsmobile
1906 Renault
limousine
1913 'Prince Henry'
Vauxhall
1909 Rolls-Royce
'Silver Ghost' tourer
1914 Model T Ford
1914
Cadillac
landaulette

Stage by Stage

We are used today to travelling rapidly on the roads by car, bus and coach in comparative comfort. Our vehicles are well sprung, and the road surfaces are generally good. The development of modern road travel began when more comfortable carriages became available and roads began to improve after the Middle Ages.

The first breakthrough was the development of the passenger coach, which had its body suspended by straps from the chassis. It originated apparently in Kocs, Hungary, from where it took its name, and spread throughout Europe during the 16th century. Long distances could be travelled in a day by swapping horses at regular intervals along the route. This was done at so-called staging posts, the coaches becoming known as stagecoaches. During the 1600s coaches became lighter and used C-shaped pieces of tempered steel for springing. Glass windows were often fitted. The wheels had iron rims, which reduced wear and tear on the wheels but cut up the roads with a vengeance. Good roads had in any case been few and far between since the demise of the Roman Empire a thousand years before.

The need for better roads prompted a resurgence in road building. It began in France in the 1600s and gradually spread throughout Europe. By 1776 France had 40,000 km (25,000 miles) of good roads under construction or completed. In Britain rapid industrial expansion accelerated the demand for good communications by road. Three road builders achieved widespread fame – John Metcalf, who was blind; Thomas Telford, whose bridges were also outstanding; and John Loudon McAdam, who developed the macadam road surface of graded, crushed stones. McAdam began building his superior roads in 1815. Eleven years later he was appointed general surveyor of roads in Britain, after which time macadamizing became widespread in Britain and also in many other countries.

Horseless Carriages

But macadamizing did not come soon enough for some people. Nicolas Cugnot in France built in 1769 what is considered to be the true ancestor of the motor-car, a three-wheeled artillery tractor powered by steam. Enthusiasts in Britain, home of the steam engine, developed road vehicles of their own.

Notable among them was Cornishman Richard Trevithick. Co-inventor (with American engineer Oliver Evans) of the high-pressure steam engine, Trevithick built a steam-powered road coach he called the 'Puffing Devil'. His demonstration of it in 1801 ended prematurely when its hot boiler set fire to the coach-house of the inn in which Trevithick and his fellow passengers were lunching. He demonstrated a larger steam coach in London two years later, but decided that the rutted, pot-holed roads of the day were not the best place for steam vehicles. And in 1804 he set one of his engines on a railway track to create the first locomotive (see page 86).

Sir Goldsworthy Gurney continued the development in Britain of steam road vehicles. In 1829 he drove one of them from London to Bath and back, a distance of 300 km (200 miles), at an average speed of 25 km/h (15 mph). In the 1830s steam carriages enjoyed a vogue, but this was short-lived. The dirty, snorting monsters aroused deep hostility among railway and stagecoach operators, who feared competition from them, and the public at large, who were plain scared of them.

In 1865 Parliament passed the notorious 'Locomotives on Highways Act'. This limited the speed of steam road vehicles to 6 km/h (4 mph) in the country and half that speed in the towns. Furthermore it stipulated that they be preceded by a man with a red flag walking 20 metres (20 yards) ahead. This Act, which was not repealed until 1896, effectively inhibited the development of road transport in Britain, though inventors still continued to develop light steam carriages.

The small steam car also found favour in continental Europe and North America, where it suffered no 'Red Flag' restrictions. The most famous steam cars of all were built in the United States by the Stanley brothers Francis Edgar and Freelan O. Stanley from 1897 onwards. The steam car, however, had a rival in the electric car after the 1880s, following the introduction of an improved storage battery by Camille Faure in France. The electric car enjoyed its greatest popularity in the United States. Though cleaner and quieter than the steamer, it had the great drawbacks of lack of range and limited speed, two characteristics that dog its development even today.

Petrol Power

The new motoring fraternity were in need of a new type of engine that was safer than the steamer and had better range than the electric. This need was satisfied in the 1880s by the lightweight petrol engine developed independently by German engineers Karl Benz and Gottlieb Daimler.

Daimler was a brilliant engineer, previously employed by Nicolas Otto, who had developed the four-stroke combustion cycle of the petrol engine (see page 34). Working in Stuttgart, he and another ex-Otto employee, William Maybach, built an advanced air-cooled engine that ran at up to 900 revolutions per minute (rpm). In November 1885 they installed their engine in an old bicycle frame to produce the first motorcycle. Unbeknown to them, Karl Benz in Mannheim was experimenting on similar lines and several months previously had fitted a petrol engine to a tricycle to create the first practical motor car. Benz's engine, though generally inferior to Daimler's, nevertheless had several features of the modern engine, including mushroom-shaped valves, spark-plug ignition, and water cooling.

The early cars were built very much like the carriages of the day with provision for self-propulsion. Steering was by means of a tiller, and drive from engine to wheels was by chain or belt. In 1889 the French team of René Panhard and Émile Levassor began building vehicles which were the first true cars and not merely horseless carriages. Their 1891 design had the engine mounted at the front, establishing the shape of things to come. It also had a friction clutch and gearbox. In 1895 motoring started to become a more comfortable experience. Panhard introduced the first enclosed saloon body, while the Michelin brothers, André and Edouard, developed the pneumatic tyre, replacing the 'bone-shaking' solid rubber tyre formerly used.

The French continued to forge ahead with car design. Comte Albert de Dion and Georges Bouton combined their talents to produce sophisticated engines that were accurately machined and properly balanced. (The name 'de Dion' survives today but as a type of rear axle, which they developed.) Another French car manufacturer, Louis Renault, began by installing de Dion engines. And in his 1899 vehicle he pioneered the use of a drive shaft with flexible universal joints at each end to transmit motion from engine to rear wheels.

Frederick Lanchester in Britain, had used a drive shaft coupled with a worm drive to the rear axle, two years earlier. He was one of the few Britons who had persisted in car development, despite the inhibiting effect of the 'Red Flag' Act. Another was Percy Riley, who introduced mechanically operated valves in 1898.

Cars for the Masses – and for the Few

The foundations of the American car industry were also being laid in the 1890s. The first really successful car was built by the brothers J. Frank and Charles E. Duryea in 1893, the same year in which Henry Ford in Detroit completed his first petrol engine. In 1896 Ford built his first car; so did Ransome E. Olds, also in Detroit. By the turn of the twentieth century Ford and Olds were just two of about fifty car manufacturers in the United States. But they were poised to make an innovation that would transform the whole motor industry and make the car a workhorse of the masses, rather than a plaything for the rich.

Olds began it all with his method of assembly-line production. He employed skilled workers to make precision parts, but used only semi-skilled labour to assemble them into a car along a moving production line. His famous one-cylinder 'curved dash' model of 1901 was the first really successful American car. Henry Ford, too, saw the benefits to be accrued by assembly-line, mass production, and in 1908 produced the 'Model T', which was to become legendary. Ford concentrated on producing the one model, which meant that it could be done very cheaply. His pioneering use of thin-gauge steel for the body led to the Model T being nicknamed 'Tin Lizzie'. In 1913 Ford speeded up production by introducing the moving conveyor belt into assembly-line operations, and was soon producing 250,000 cars a year. It is a myth that they were all black! By 1927 over 15 million Model Ts had been sold.

In direct contrast to Ford's approach to car manufacture was that of the Hon. Charles Rolls and Henry Royce in Britain. They joined forces to design the perfect car, for the favoured few. In 1906 they produced an exquisite 7-litre car they called the 'Silver Ghost', arguably the finest car ever built. Perfectly engineered and as silent as a phantom, the 'Silver Ghost' earned for Rolls-Royce the reputation of being the 'best car in the world'.

The Classic Era

The mass production of the 'Model T' helped thrust the United States into the forefront of car development. In 1911 the fast-expanding Cadillac company introduced the electric self-starter and dynamo lighting. Later, direction indicators, brake stop-lights and vacuum-operated windscreen wipers came into use. In 1920 Duesenberg began fitting four-wheel hydraulic brakes. A few years later came front and rear bumpers, petrol gauges and speedometers. European car manufacturers gradually adopted these refinements, while pioneering some of their own, like servo-assisted brakes (by Hispano-Suiza in 1919) and unitary construction and independent front suspension (by Lancia in 1922).

The 'Twenties' was a glorious era for the car, producing the fast and luxurious 'Grand Touring' models like the Hispano-Suiza, and magnificent racers like the Bentley, which won the famous 'Le Mans' 24-hour race no less than five times. The 'Twenties' also saw the beginning of a new concept, in the baby car, typified by the tiny Austin 7 (1922). The even smaller Fiat 500, the 'Topolino', appeared in 1936. It was an advanced 'baby', with synchromesh on all four gears (first introduced by Cadillac in 1928), hydraulic brakes, and independent front suspension. Two years later Germany launched its version of a 'people's car', the rear-engined Volkswagen, designed by Ferdinand Porsche. The strength and reliability of the VW 'Beetle', as it came to be called, became legendary.

Where Now?

Since World War 2 the car has undergone progressive refinement into a reliable, comfortable machine with sleek lines and speedy performance. In recent years concern over energy consumption has led to cars becoming smaller and more efficient. Concern over pollution has led to the fitting of devices to limit the emission of engine fumes. And concern over safety has influenced design and led to the introduction of such things as seat belts, collapsible steering columns, and energy-absorbing front and rear ends. For the future, car designers are now taking a hard look at the past to see whether steamers and electric cars can take over when the petrol runs out, as it soon must. Other engineers are experimenting with hydrogen as fuel, which has the advantage of being non-polluting.

1927 Lancia Lambda

1929 Bugatti Type 43

1923 Austin Seven

1924 'Bullnose' Morris Cowley

1933 'Mille Miglia' Alfa-Romeo

1931 Supercharged $4\frac{1}{2}$-litre Bentley

1933 MG 'Midget'

1938 Porsche's Volkswagen 'Beetle'

1936 Citroën 'Light Fifteen'

1938 Fiat 500 'Topolino'

1947 Pinin-Farina bodied Cisitalia

1949 Jaguar XK120 open sports

1959 Issigonis's Morris Mini

1956 Citroën 'shark-nose' DS19

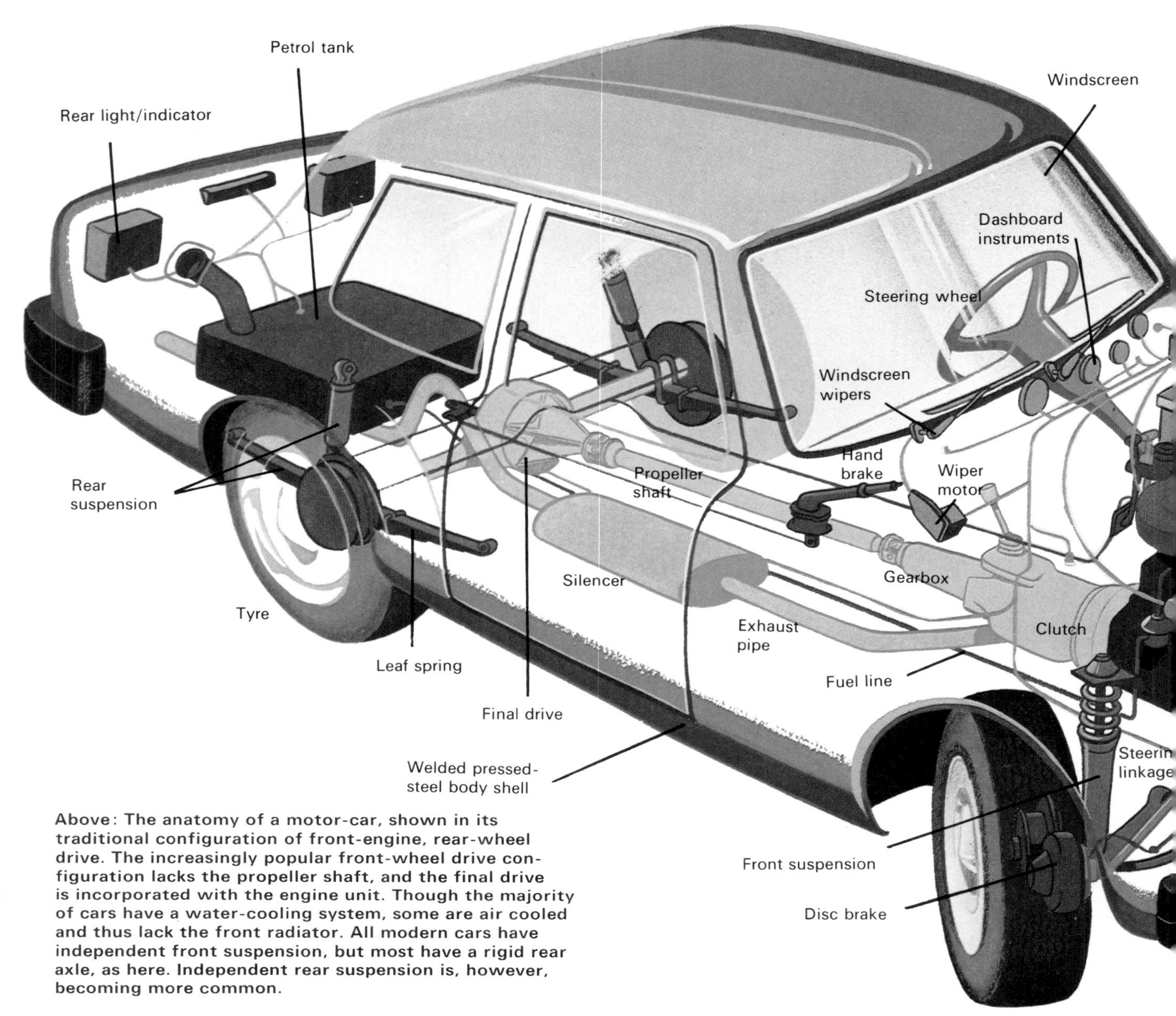

Above: The anatomy of a motor-car, shown in its traditional configuration of front-engine, rear-wheel drive. The increasingly popular front-wheel drive configuration lacks the propeller shaft, and the final drive is incorporated with the engine unit. Though the majority of cars have a water-cooling system, some are air cooled and thus lack the front radiator. All modern cars have independent front suspension, but most have a rigid rear axle, as here. Independent rear suspension is, however, becoming more common.

Systematic

The car is the most complex common machine. It is an assembly of up to 15,000 different components, varying in size from the large pressed steel shell that forms the body to tiny precision-engineered balls a few millimetres across in bearings that allow the moving parts to turn. It would be tedious and unenlightening to treat them individually; it is more feasible to consider them as part of corporate systems which have a particular function to perform.

Despite widespread differences in design, all cars are made up of similar systems. First is the engine, which provides the motive power. Most cars have petrol engines, though more and more models are now being offered with diesel engines, which work more efficiently and burn cheaper fuel. These engines are described in detail in the chapter 'Power to the People' (see page 32).

The power developed by the engine is transferred to the driving wheels by the transmission system. In many cars the transmission includes a manual clutch and gearbox. The driver selects different gears in the gearbox to make the car travel faster or slower depending on the road conditions. He uses the clutch to disconnect the gearbox from the engine while he changes gear. Other cars have automatic transmission in which gears are selected automatically.

As well as being able to go fast a car must be able to slow down and stop quickly, which is the function of the braking system. Two independent braking systems are compulsory on all cars; a foot brake operates hydraulic brakes on all four wheels, while a mechanically operated handbrake acts usually on the rear wheels only. Modern brakes are capable of stopping a car travelling at 110 km/h (70 mph) in less than 85 metres (275 ft). A quarter of this distance is the 'thinking distance', which represents the distance the car travels while the driver reacts to the braking situation.

Other major systems include the steering, which turns the front wheels at the correct angles to guide

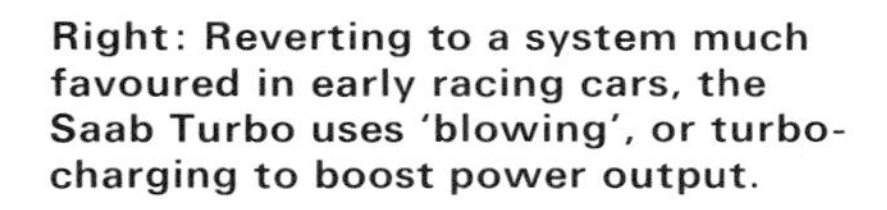

Right: Reverting to a system much favoured in early racing cars, the Saab Turbo uses 'blowing', or turbo-charging to boost power output.

Below right: The MGB roadster, whose ancestry dates back to the MG Midget of 1933. It has a cylinder capacity of 1798 cc and an 8·8:1 compression ratio.

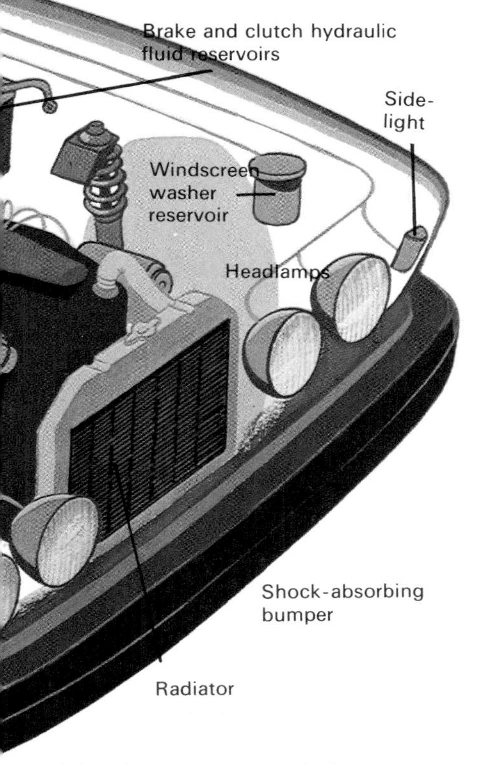

Right: Sleek aerodynamic lines and superlative performance characterize the Porsche 928.
Bottom right: The Cadillac Ventora, one of the last big family saloons to be built in the 1970s. With fuel costs soaring, small is beautiful.
Below: A Mercedes diesel racer.

Above: Car production is becoming increasingly automated. Here robot welders are spot-welding a body on Fiat's Strada assembly line.

Left: The Reliant Scimitar has an eye-catching body design in fibreglass, or GRP (glass reinforced plastic), one of the very few cars to do so. With two side doors and an opening tail-gate, it has been described as a sporting estate.

the car. It includes gearing and often power assistance to reduce driver effort. The springs and shock absorbers of the suspension system cushion the occupants of the car from irregularities in the road surface. The tyres also play a part in the suspension, but their main function is to provide firm grip, or adhesion on all types of road surface in the dry and in the wet. To most people, one of the most difficult systems to follow is the electrical, for some 60 metres (200 ft) of wiring link the electrical components in a car. The car battery and generator provide power for the ignition system and for the lights, windscreen wipers, starter motor, fan, instruments and so on (see page 39).

The Body

All the various systems of a car are supported by or mounted on the body unit. With only a few exceptions car bodies are made of pressed sheets of mild steel typically about half a millimetre (0·02 in) thick. A car body has to provide at one and the same time structural strength to support components; torsional stiffness to resist twisting forces; and beam stiffness to resist 'sag'.

Most American cars have a body formed of a superstructure built upon a separate reinforced frame, or chassis, following early bodybuilding practice. The chassis, which is usually formed of box-section members, provides practically all the strength.

Outside North America few cars are built with a separate chassis. Rather the body is designed as a single unit, or shell, which as a whole provides the requisite strength. This is known as unitary construction. The body shell is constructed of individual pressed steel panels, reinforced where necessary, which are welded together. Only the doors, bonnet and boot lid are separately hung. Some designs have a simpler body structure and have the major mechanical units, such as the engine and final drive, mounted on separate sub-frames. This arrangement helps simplify assembly during manufacture and can save time on maintenance.

In some respects unitary construction is beneficial. The shell can, for example, be readily designed with progressively collapsing and energy-absorbing front and rear ends. During a collision these ends tend to absorb the impact, thus protecting the reinforced 'safety box' that forms the passenger compartment.

The shape of a car body is very much a compromise between such things as good looks, function, ease of manufacture, passenger comfort and performance.

The performance of a car is influenced by shape because of the resistance, or drag, a body experiences when travelling through the air. Drag increases as the square of vehicle speed, which means that a car travelling at 100 km/h (60 mph) will experience *four* times the drag it does at 50 km/h (30 mph). Practically all cars now are designed with the aid of wind tunnels, imitating aviation practice (see page 126). From the aerodynamic performance of scale models of a car, that of the full-size vehicle can be predicted.

The perfect 'teardrop' streamlined shape is impractical for body design, but much can be done with fairly orthodox styling by making the body low, smooth and gently curving. Streamlining is more severe in sports and high-performance cars and reaches its zenith in racing cars, which may also utilize inverted aerofoils to prevent them flying off the road!

The Transmission System

The majority of cars have their engines located at the front, driving the rear wheels. In this conventional layout the transmission system comprises a clutch, gearbox, propeller shaft and final drive, which includes the differential.

The engine ends and the transmission begins at the flywheel, the heavy disc attached to the rear of the crankshaft. The flywheel's function is to smooth out the motion of the crankshaft, to which energy is imparted intermittently by the pistons. Making contact with the flywheel is the clutch, a kind of switch that allows the motion of the engine to be transmitted to the gearbox, or not.

The ordinary clutch is operated by foot pedal and

Below: In the past, racing cars have often proved the test-beds for innovations in car designs. But today's Grand Prix machines have little in common with the family saloon. They have scant bodywork and severe aerodynamic styling, with low ground clearance and reverse aerofoils at the rear and often at the front to create a road-hugging downforce. Massive balloon tyres with no tread are fitted to the rear wheels to improve traction. Experiments have recently been taking place in six-wheel design, two pairs of wheels taking the place of one either in the front, as here, or in the rear.

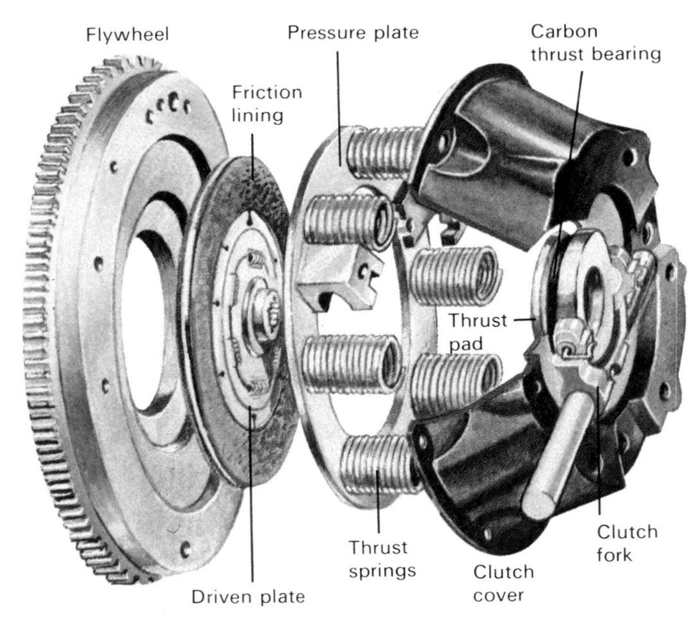

works by friction. It has a disc (driven plate) covered with a tough lining, which is attached to a shaft that goes to the gearbox. When the clutch is engaged, coil springs or a flexing metal diaphragm press another disc (pressure plate) against the driven plate and sandwich it against the flywheel. Drive is then transmitted from engine to gearbox. When the clutch pedal is depressed, the pressure plate is levered or sprung away from the driven plate, which then loses contact with the flywheel. Drive to the gearbox is thus interrupted, and gears can be changed in the gearbox without them clashing together. The pedal may operate the clutch mechanically via roads and levers, or hydraulically via master and slave cylinders, in the same way as hydraulic brakes (see page 77).

Changing Gear

A car engine operates most effectively at high speeds of revolution, up to 4000 revolutions per minute (rpm) and beyond. The car wheels need to revolve

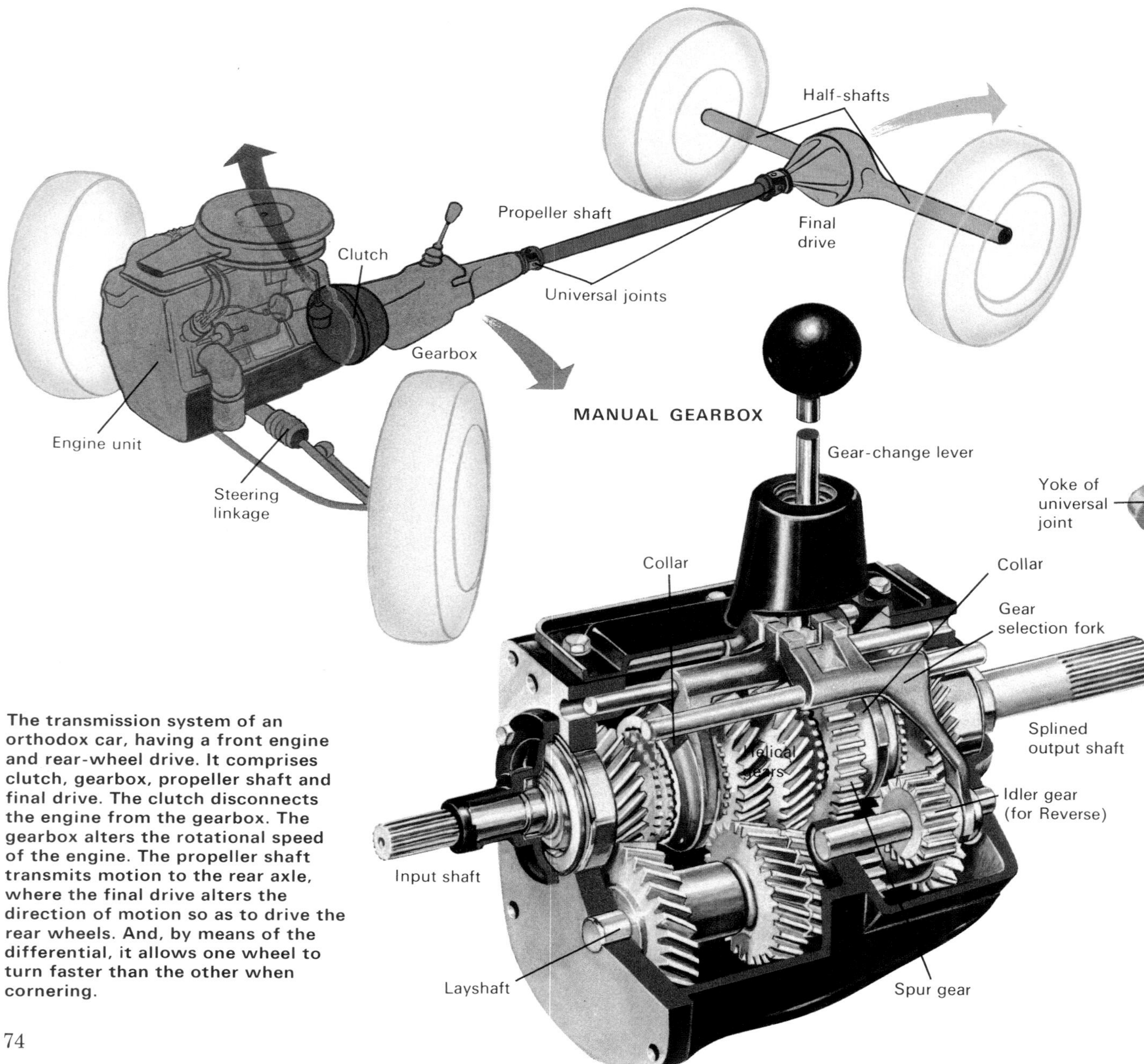

The transmission system of an orthodox car, having a front engine and rear-wheel drive. It comprises clutch, gearbox, propeller shaft and final drive. The clutch disconnects the engine from the gearbox. The gearbox alters the rotational speed of the engine. The propeller shaft transmits motion to the rear axle, where the final drive alters the direction of motion so as to drive the rear wheels. And, by means of the differential, it allows one wheel to turn faster than the other when cornering.

much more slowly even at high vehicle speeds. At a speed of 110 km/h (70 mph) the wheels make about 1000 rpm. The transmission system overall therefore incorporates reduction gearing between engine and drive wheels to allow for the difference in the speeds of rotation. In most ordinary cars the reduction gearing is about 4:1, which means that for an engine speed of 4000 rpm, the wheels revolve at 1000 rpm, and the car travels at about 110 km/h (70 mph). The reduction gearing is provided by the final drive between the propeller shaft and the rear wheels.

For travelling at steady speed on the level this gearing is satisfactory. But when the car goes uphill it tends to slow down, whereupon the engine speed drops. At the lower speed of revolution, the engine might not be able to exert enough turning effort, or torque, to propel the car up the hill. It could falter or stall. In a similar way the simply geared system would not be able to exert enough torque to move a car from a standstill. It would be too highly geared.

This is why there is need of a gearbox in the transmission. The gearbox allows the overall gearing to be reduced so that the engine can still run at high speed even when the car is travelling at low speed. It contains a number of toothed gearwheels of different sizes. They are meshed, or moved together in various combinations so as to bring about different speeds of rotation between the shaft going into the gearbox from the clutch (the input shaft) and the shaft leading from the box to the final drive (the output shaft).

The gearbox illustrated is a manual gearbox, in which the gears are changed by moving a gear-shift lever by hand into various positions. A four-speed gearbox is most common, which means that four different forward input-output shaft ratios are available. They may be something like 3:1 for first (low) gear, 2:1 for second, 1·3:1 for third and 1:1 for fourth, or top gear. The 1:1 ratio for top gear means that input and output shafts turn at the same speed. Some cars have a unit known as overdrive fitted between the gearbox and the propeller shaft to provide an even higher gear ratio. It is operated by an electrical or hydraulic switch. In addition to the forward gears there is a reverse gear to allow the car to 'back', or reverse direction.

The gearbox has its gearwheels mounted on parallel shafts. With the exception of those used for reverse, all the gearwheels remain in constant mesh. A wheel on the input shaft meshes with a wheel on the parallel layshaft and turns it round. The other wheels fixed to the layshaft likewise mesh with the wheels on the output shaft, and turn them round. When the gear lever is in the 'neutral' position, the wheels on the output shaft are free to rotate about it. No motion is transmitted from the gearbox.

When a gear is selected, the selector fork moves a collar attached to the output shaft into contact with a particular wheel, which of course is rotating. So motion is thereby transferred to the output shaft and thence to the road wheels. In top gear a collar links the input and output shafts directly, this being known as direct drive. To ensure that the collars and gearwheels mesh smoothly, they are fitted with matching conical surfaces. The cones come together first to synchronize motion before the two lock together. This is known as synchromesh.

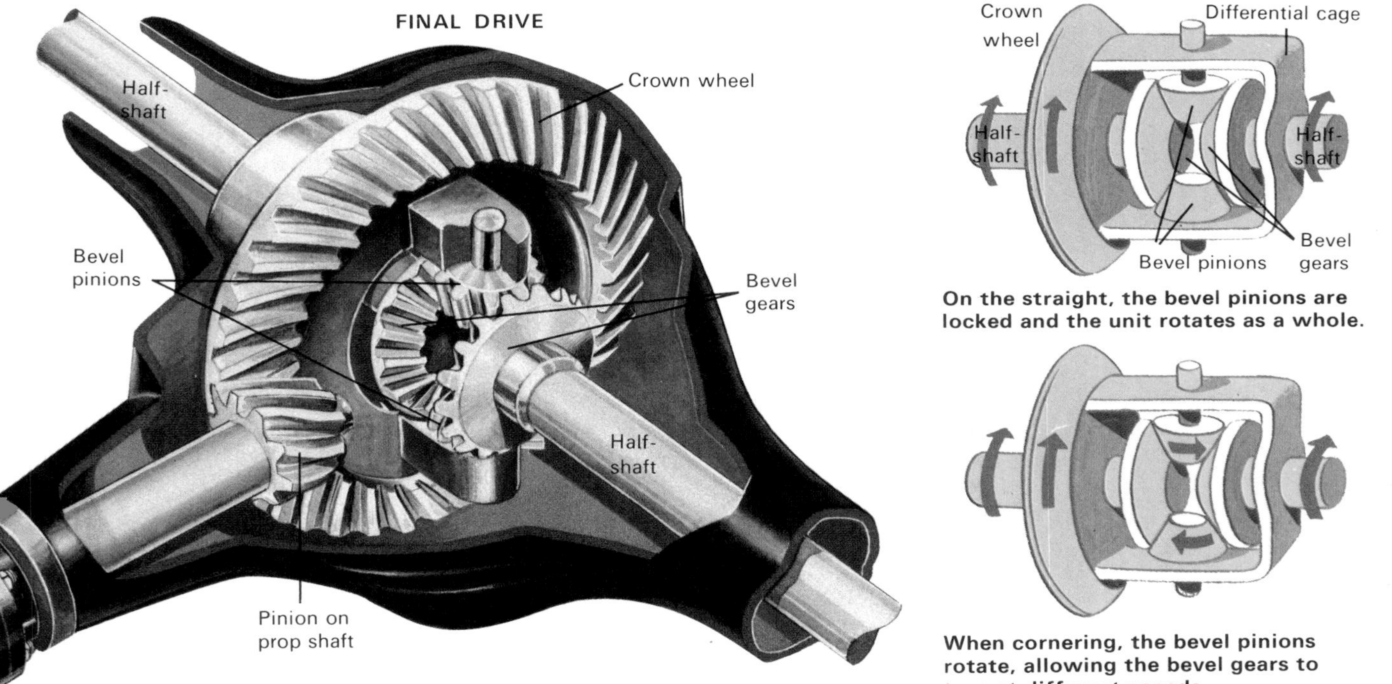

On the straight, the bevel pinions are locked and the unit rotates as a whole.

When cornering, the bevel pinions rotate, allowing the bevel gears to turn at different speeds.

Drive to the Wheels

The motion of the output shaft from the gearbox is transmitted to the drive wheels by the propeller ('prop') shaft, a strong hollow steel tube. To allow the rear axle and gearbox to move independently, as they must, the prop shaft has flexible couplings at

EPICYCLIC GEARS

In a simple epicyclic, or sun-and-planet gear, one shaft carries a gearwheel (sun). This is located at the centre of a ring (annulus), which has teeth cut on the inside. Another shaft (planet-carrier) carries two or more small gear-wheels (planets) which are located between and mesh with the sun-wheel and annulus. By holding any of these parts stationary the others can be made to rotate at different speeds. In this way the epicyclic gear acts as a miniature gearbox.

When, for example, the sun-wheel is locked, the planets revolve around it. The annulus and planet-carrier turn in the same direction, but at different speeds. When the sun-wheel is locked to the annulus, the planets are locked also, and the whole unit turns as one. When the planet-carrier is locked, the sun-wheel turns the planets, and makes the annulus revolve in the opposite direction. When the annulus is locked, the planets roll inside it and drive the sun-wheel.

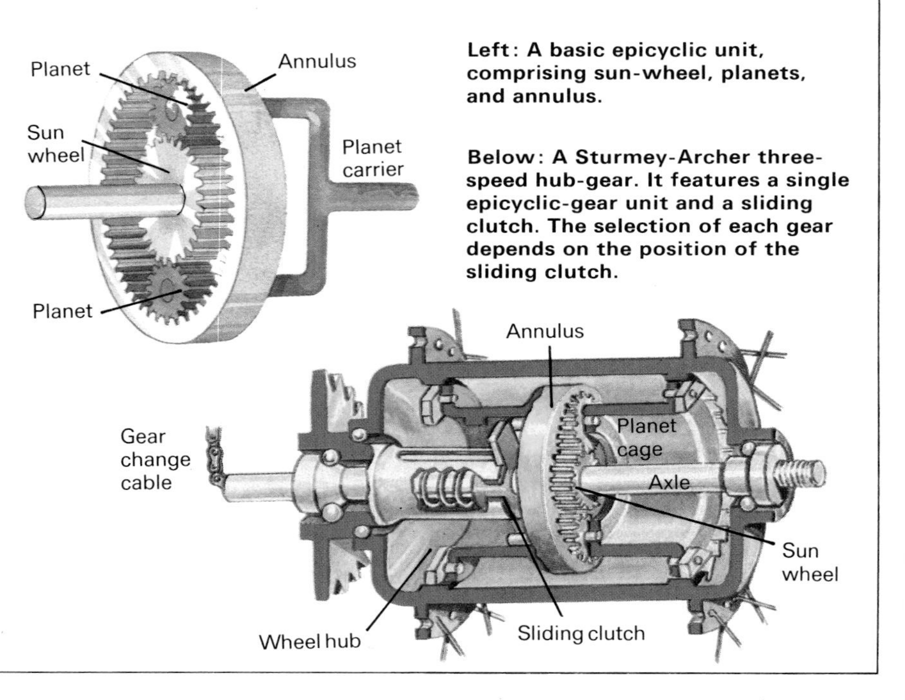

Left: A basic epicyclic unit, comprising sun-wheel, planets, and annulus.

Below: A Sturmey-Archer three-speed hub-gear. It features a single epicyclic-gear unit and a sliding clutch. The selection of each gear depends on the position of the sliding clutch.

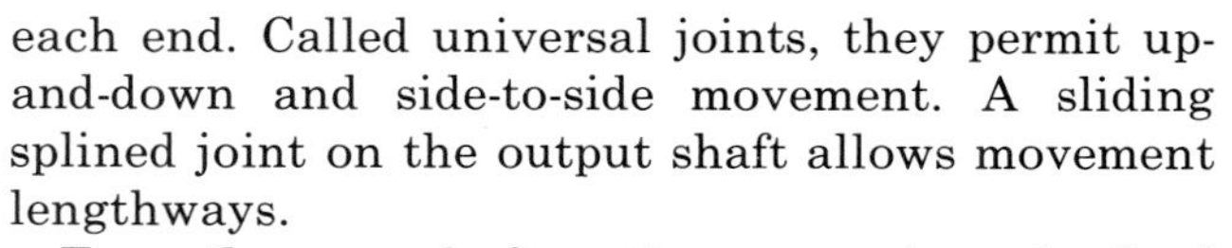

each end. Called universal joints, they permit up-and-down and side-to-side movement. A sliding splined joint on the output shaft allows movement lengthways.

From the prop shaft motion passes into the final drive. This brings about the typically 4:1 reduction gearing mentioned earlier; changes the direction of motion through 90°; and allows the drive wheels to turn at different speeds around corners. The reduction in gearing and change of direction of motion are effected by the crown wheel and pinion. The small pinion on the prop shaft turns four times (for a 4:1 reduction) to turn the crown wheel at right-angles to it once.

Drive to the half-shafts that carry the wheel hubs is accomplished by an ingenious system of right-angled bevelled gears known as the differential. The action of the differential is illustrated above. When the car is travelling straight, the small pinions held in the differential cage lock the bevel gears on the half-shafts and the whole assembly turns together. On a bend, when the inner wheel slows down, the pinions rotate and drive the outer half-shaft faster. If a car did not have a differential, but had both drive wheels driven always at the same speed by a single axle, the inner wheel would skid every time the car turned a corner.

Automatic Transmission

Practically all cars in North America and many elsewhere have an automatic gearbox, in which the gears change automatically according to the speed of the engine and the load upon it. The normal automatic gearbox is a sophisticated version of the hub gear on a bicycle.

It incorporates two or more epicyclic gear trains. Their different parts can be interconnected and locked in various permutations to bring about different speed ratios between the input and output shafts of the gearbox. This is done by means of clutches and brakebands which are brought variously into play according to the road speed (sensed by a governor) or the position of the accelerator. They work by hydraulic pressure provided by an oil pump.

With an automatic gearbox a conventional clutch is not required to facilitate gear changing. Instead a fluid coupling called a torque convertor is used. It is a turbine device in which motion is transferred from the flywheel to the input shaft of the gearbox by whirling liquid. It is so called because it can multiply the turning effort, or torque, of the engine at low road speeds.

The gear control of an automatic car is usually mounted on the steering wheel and has five positions – P (Park), R (Reverse), N (Neutral), D (Drive), L (Low gear).

The Braking System

A car has brakes on every wheel, which are applied by means of a brake pedal. Pressure on the pedal causes a brake lining to be forced against a drum or disc attached to each wheel. Friction between the lining and the drum or disc makes the wheel slow down. Heat is produced as this happens so the lining is tough and heat resisting. It contains asbestos and often metal fibres. A car also has a hand-operated brake that is used mainly for parking. It incorporates a ratchet mechanism which allows it to be left 'on'. It works mechanically by means of cables or levers and acts only on the rear wheels.

The main four-wheel, foot-operated brakes, however, work by means of hydraulic pressure. The foot pedal is linked by lever to a piston in a 'master' cylinder containing fluid. Pipes carry the fluid to

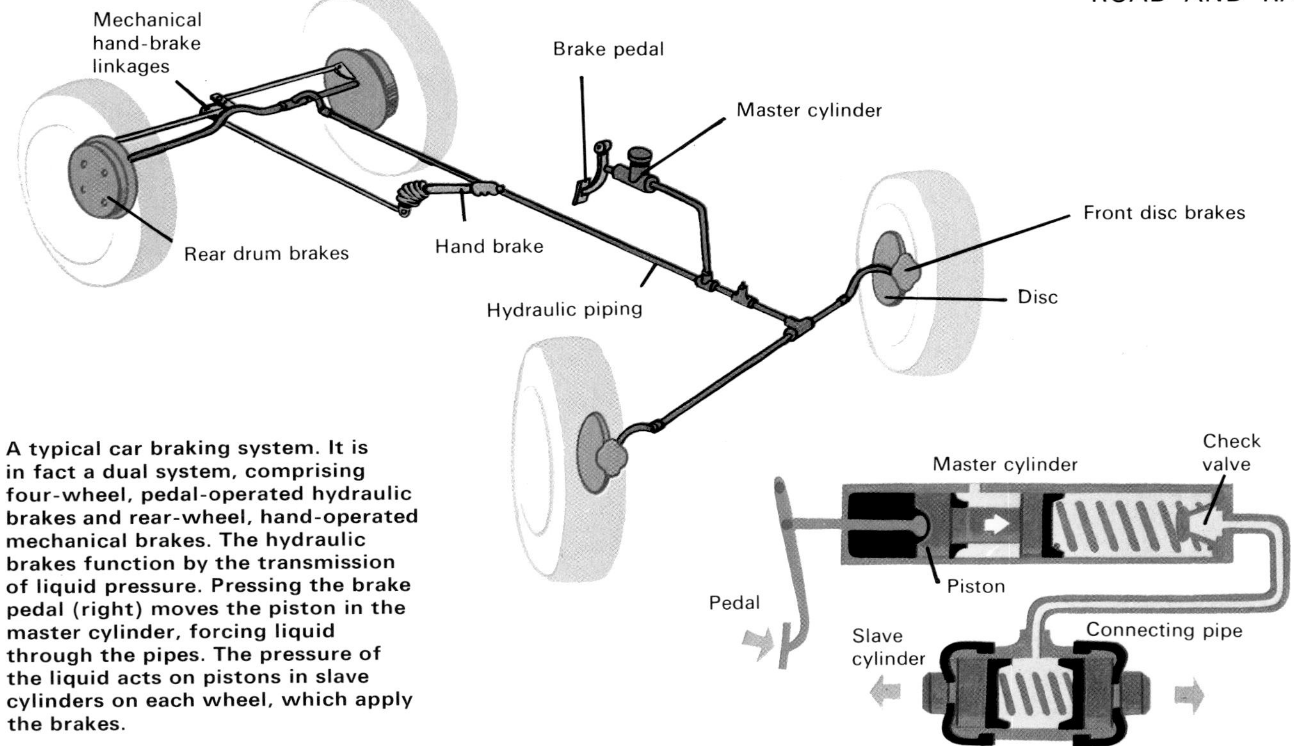

A typical car braking system. It is in fact a dual system, comprising four-wheel, pedal-operated hydraulic brakes and rear-wheel, hand-operated mechanical brakes. The hydraulic brakes function by the transmission of liquid pressure. Pressing the brake pedal (right) moves the piston in the master cylinder, forcing liquid through the pipes. The pressure of the liquid acts on pistons in slave cylinders on each wheel, which apply the brakes.

pistons in 'slave' cylinders on each wheel. These pistons are in contact with shoes or pads that carry the brake linings. When the foot pedal is depressed, the piston moves along the master cylinder and forces fluid through the pipes. The hydraulic pressure created acts against the pistons in the slave cylinders, which force the brake linings against the wheel.

The system works because liquids are virtually incompressible, and pressure applied to any part of a liquid is transmitted instantly and equally to every other part. This means that all the brakes work simultaneously, which is essential for safety. By making the area of cross-section of each slave piston larger than that of the master piston, extra braking force is achieved. (Force = pressure × area. The pressure is the same throughout the system, so force is proportional to area.)

Most modern cars have two kinds of brakes. Drum brakes are usually fitted to the rear wheels and disc brakes to the front wheels, which take more of the braking load. In a drum brake a cast iron brake drum is bolted to the wheel. The drum revolves around a stationary backplate to which are attached twin curved brake shoes covered with brake lining. The shoes are forced against the drum by movement of the pistons in one or more slave cylinders.

In a disc brake a thick cast-iron disc is bolted to the road wheel. Flat brake pads are located on either side of the disc and are pushed into contact with it

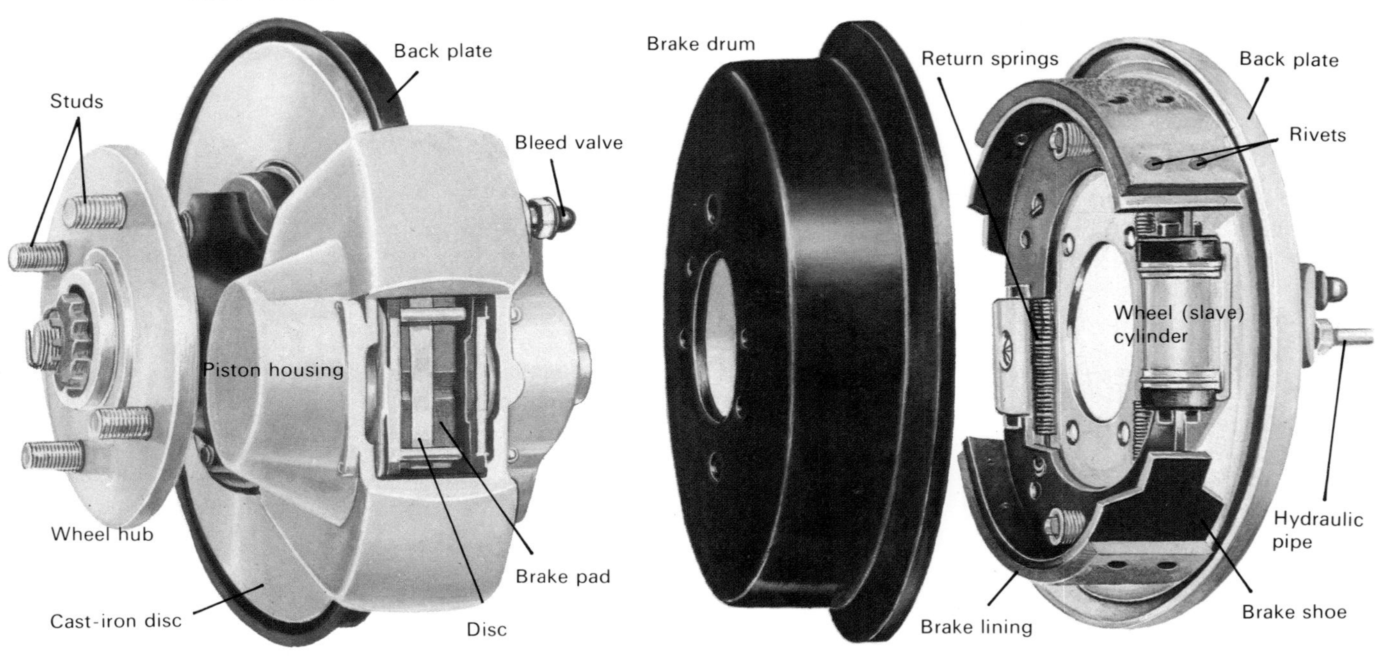

by hydraulic pistons. Braking is a caliper (pinching) action, as it is in most bicycles. Since the brake pads cover only a small area of the disc, the disc is able to dissipate the heat of braking readily. Drum brakes, on the other hand, are more enclosed and cannot dissipate heat so well. They begin to 'fade' because of overheating when they are repeatedly applied.

High-performance cars and heavy saloons need more force for braking than can readily be supplied by the driver. So they are fitted with a servo unit. This is a mechanism that uses the partial vacuum created in the inlet manifold of the engine to operate a piston that increases the hydraulic pressure in the brake pipes. Some cars are also fitted with a dual braking system for safety. They have an alternative hydraulic system acting on at least the front wheels which can take over if the main one springs a leak or otherwise fails.

The Steering System

The most important function of steering a car is performed by a number of rods and levers coupled to the steering wheel. The geometry of the steering is such that when the car goes round a bend, the inner wheel turns more sharply than the outer. It has to because it lies closer to the centre of curvature of the bend. If both wheels turned by the same amount, sliding and tyre wear would result.

One of the simplest and most widely used steering systems is the rack-and-pinion, illustrated above. The steering column ends in a small pinion that is meshed with a toothed horizontal rack. When the

Above: Most trucks have two sets of leaf-springs for rear suspension. At bottom is the main spring; above is an auxiliary spring which deflects under heavy loads. This picture also shows details of the drum-brake unit.

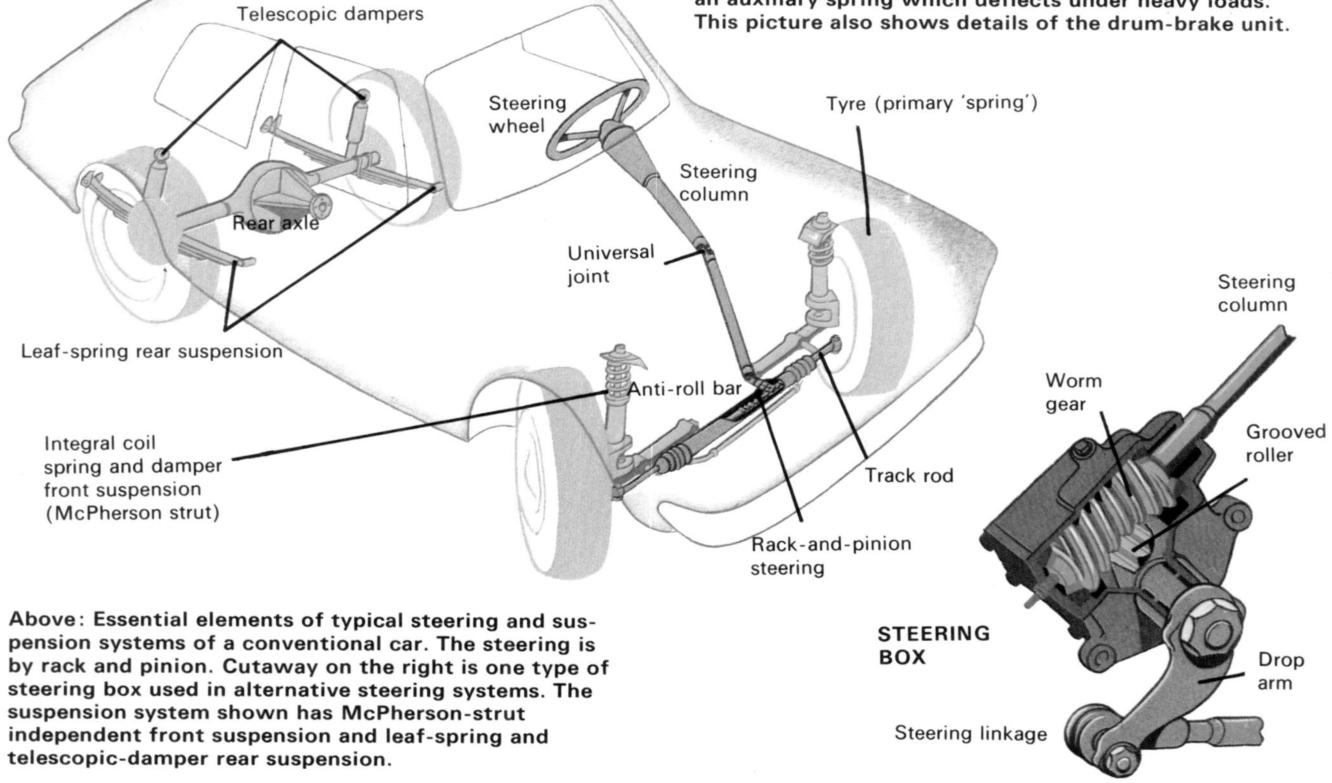

Above: Essential elements of typical steering and suspension systems of a conventional car. The steering is by rack and pinion. Cutaway on the right is one type of steering box used in alternative steering systems. The suspension system shown has McPherson-strut independent front suspension and leaf-spring and telescopic-damper rear suspension.

pinion turns, the rack moves back and forth. The movement of the rack is transmitted through flexible ball joints to arms that swivel the wheel hubs. In alternative steering systems a steering box is used to convert the circular motion of the pinion into side-to-side motion to operate the steering. It contains a kind of worm gear which operates a so-called drop arm that moves the steering linkages.

The steering-box and rack-and-pinion systems have built-in reduction gearing so that a driver need exert less effort to steer the car. But in heavy cars the steering must be power-assisted. This is particularly necessary when manoeuvring at low speeds, for example, when parking. In a power-assisted system movement of the steering column opens a valve which allows high-pressure oil to move a ram that operates the steering.

The Suspension System

There are a number of different suspension systems that help cushion the occupants of a car from the bumpiness of uneven roads. They usually incorporate springs. By their very nature springs absorb energy, in this case the energy of the up-and-down movement of the wheels. This movement compresses or extends the springs from their equilibrium position, which they immediately try to regain. Left to themselves the springs would continue to oscillate up and down for ages. And to prevent this they have to be damped.

In orthodox front-suspension systems coil springs are used in combination with telescopic dampers (also known as shock-absorbers). Dampers work

NOVEL SUSPENSIONS

Some cars do not include conventional coil springs and dampers in their suspension systems. Instead they use the cushioning effect of compressed air or rubber for springing and the flow of fluid through a restriction for damping. Best known is the hydropneumatic (fluid-and-air) suspension fitted to many Citroëns, which was introduced in the 'shark' front DS19 in 1959.

In this system each wheel is mounted at the end of an arm which is flexibly joined to the underbody of the car. A piston attached to the middle of the arm fits into a cylinder fixed to the body which broadens out to form a spherical chamber. Hydraulic fluid above the piston can pass through a valve into and out of the chamber, in the top of which nitrogen gas is trapped. When a wheel goes over a bump the piston moves upwards and compresses the gas. When the bump is passed, the compressed gas expands and forces the piston and wheel down again. The trapped gas thus behaves like a coil spring. Restricted passage of the fluid through the valve provides damping. This basic set-up is extended to provide a self-levelling capability. Fluid flows into the cylinder to correct the tendency of the body to sink under heavier passenger load. It flows out again when the load lightens.

Citroën also pioneered on their incomparable 2CV ('Deux-chevaux' – two horsepower) the concept of linked front and rear suspension units. In the 2CV system a coil spring links front and rear wheels. When the front wheel rises over a bump, the front of the car tends to rise too. The linked spring makes the rear suspension move the rear of the car upwards to match the front, helping to keep the body level. A more sophisticated linked suspension system, known as hydrolastic, has been fitted to several British Leyland cars. Their front and rear suspensions are linked hydraulically, the passage of fluid to and fro causing the front and rear ends to lift in unison.

The suspension system illustrated below is the so-called Hydrogas suspension fitted to the British Leyland Princess. It has hydraulically linked units which incorporate a gas damper.

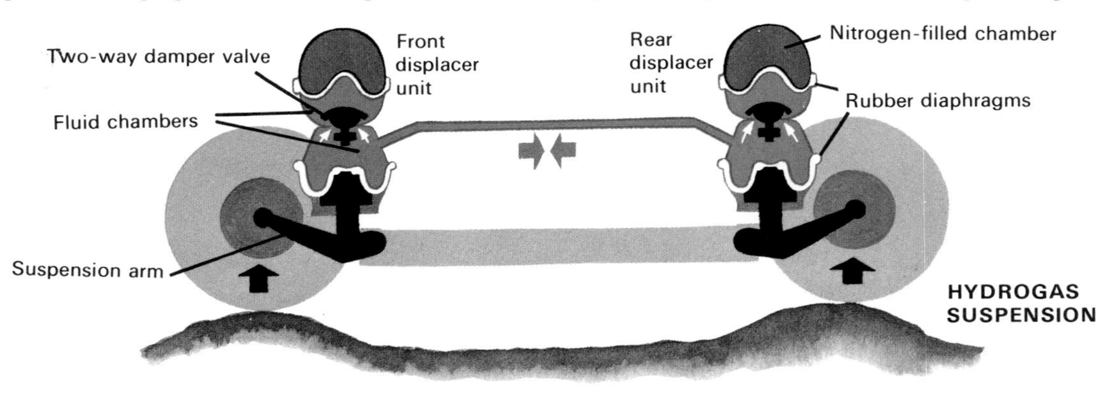

HYDROGAS SUSPENSION

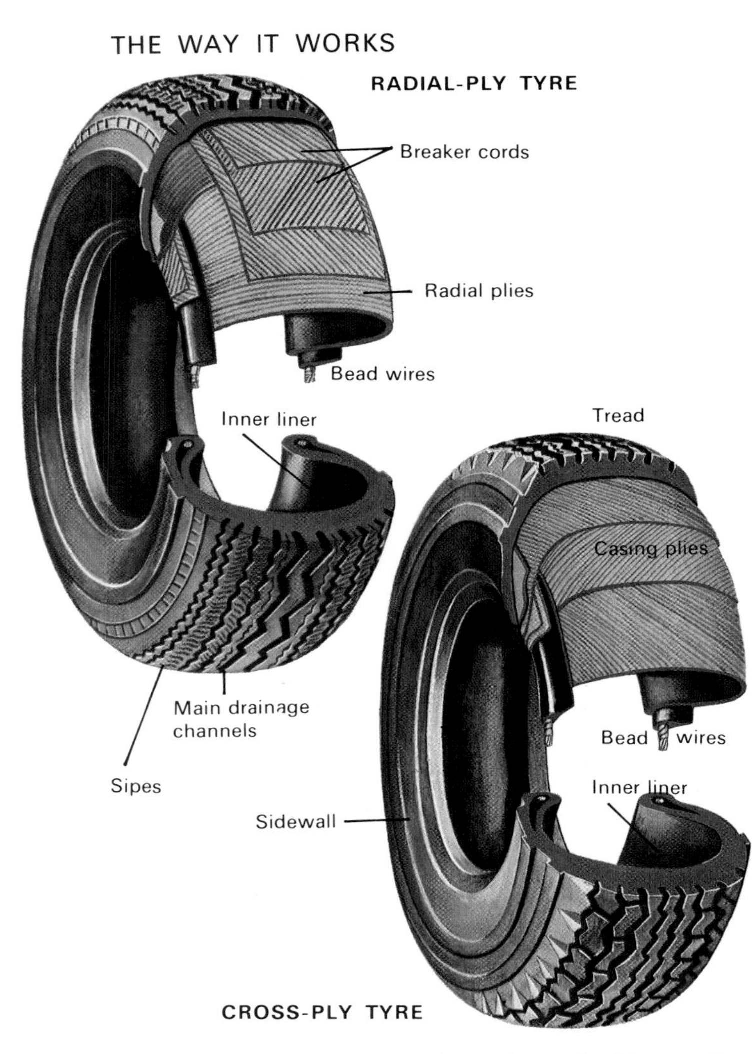

hydraulically. They contain a piston inside a cylinder filled with fluid. Small holes in the piston permit fluid to flow through them, but only slowly. This produces the damping effect.

The arrangement in the diagram shows a coil spring and damper combined in a single strut assembly. Another common system has what is called a double wishbone. Two wishbone-shaped links, are hinged at their broad end to the vehicle body and at their narrow end to swivelling members connected to the wheel stub axle. A coil spring and damper are incorporated between the wishbones and the body. In both the strut and the wishbone system the front wheels are separately suspended from the body so that the movement of one does not affect the other. This independent suspension makes for a much more comfortable ride.

In most cars, however, the rear wheels are not independently suspended. The rear-axle housing is rigid and is tilted by disturbance of either wheel. It is usually linked to the body by leaf springs at each end. Telescopic or other types of shock absorbers provide damping. Some cars do have independent rear suspension, the rear axle being made more flexible by incorporating universal joints on either side of the differential and often at the wheel hubs. A trailing arm, or semi-trailing arm design is widely used, in which each wheel is supported by an arm hinged to the body so that it can move up and down.

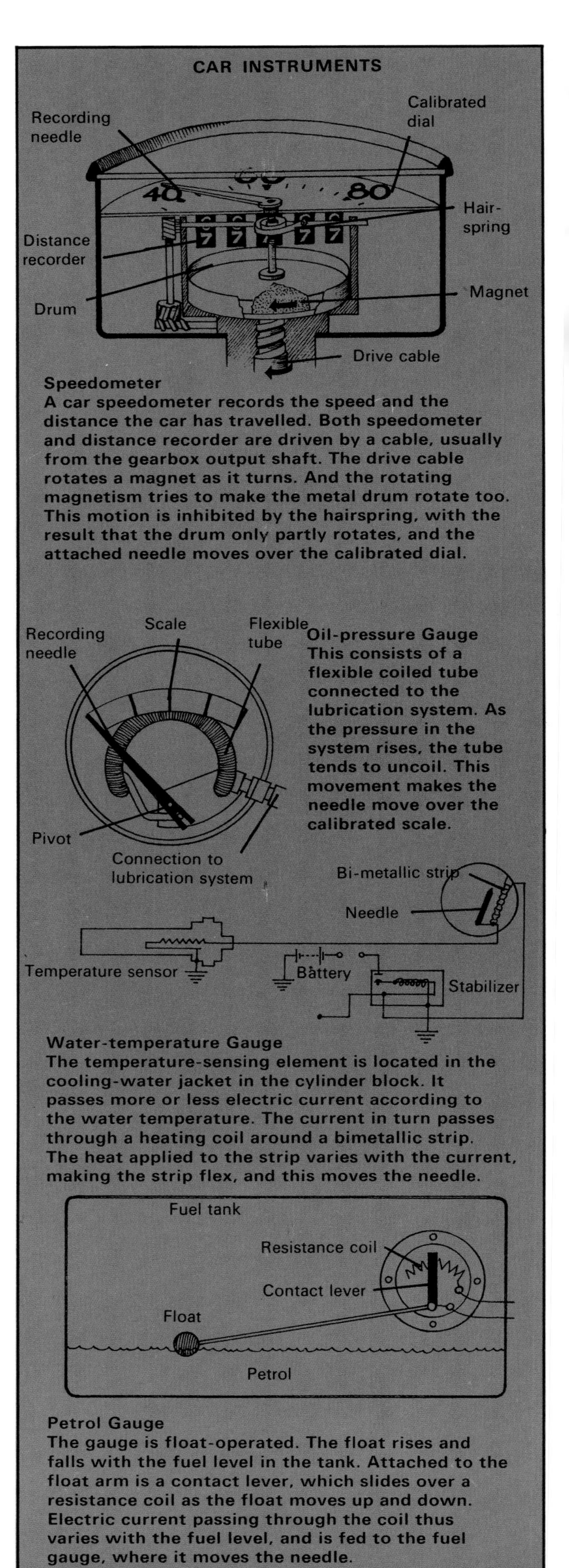

Speedometer
A car speedometer records the speed and the distance the car has travelled. Both speedometer and distance recorder are driven by a cable, usually from the gearbox output shaft. The drive cable rotates a magnet as it turns. And the rotating magnetism tries to make the metal drum rotate too. This motion is inhibited by the hairspring, with the result that the drum only partly rotates, and the attached needle moves over the calibrated dial.

Oil-pressure Gauge
This consists of a flexible coiled tube connected to the lubrication system. As the pressure in the system rises, the tube tends to uncoil. This movement makes the needle move over the calibrated scale.

Water-temperature Gauge
The temperature-sensing element is located in the cooling-water jacket in the cylinder block. It passes more or less electric current according to the water temperature. The current in turn passes through a heating coil around a bimetallic strip. The heat applied to the strip varies with the current, making the strip flex, and this moves the needle.

Petrol Gauge
The gauge is float-operated. The float rises and falls with the fuel level in the tank. Attached to the float arm is a contact lever, which slides over a resistance coil as the float moves up and down. Electric current passing through the coil thus varies with the fuel level, and is fed to the fuel gauge, where it moves the needle.

On Two Wheels

Many people begin their motoring career not on four wheels but on two, with a motorcycle, or as it is more commonly known, the motorbike. Motorbikes have always appealed to the young for they are simple to operate and fun to ride. They can manoeuvre more easily and in general accelerate more rapidly than a car. They also cost less and are cheaper to run. German engineer Gottlieb Daimler demonstrated the first practical two-wheeled motorbike in 1885, though English inventor Edward Butler had built a motor tricycle a year earlier.

Several types of motorbikes are produced, which are tailored for different uses. There are powerful streamlined racing bikes capable of speeds of 250 km/h (150 mph) or more, docile 'runabouts', and stripped-down trials or 'scrambling' machines. In the forefront of motorbike development today are Japanese manufacturers such as Honda and Suzuki. Since World War 2 they have come to dominate the motorbike scene, ousting the more traditional European producers such as Motoguzzi (Italy), Ducati (Germany) and Triumph (Britain).

Variants of the motorcycle include the moped and motor scooter. The moped was originally conceived as a light motorized bicycle with a small engine and pedals to start the engine or assist it when going uphill. Many modern mopeds, though they have pedals, closely resemble ordinary motorbikes. Motor scooters are small compact machines with small wheels, enclosed mechanics, leg protectors and

Above: The Italian Lambretta 'Grand Prix', a near-identical descendant of the original motor scooter of the 1950s. It is available in two models with 150 cc and 200 cc two-stroke, single-cylinder, air-cooled engines. With a top speed of 100 km/h (62 mph), the 150 cc version has a petrol consumption of about 30 km/litre (85 miles/gallon).

Below: The most energy-efficient travelling machine ever invented – the bicycle, thought to have been originally conceived by master-inventor Leonardo da Vinci in the 1490s. This model typifies the modern racing bicycle. It has a strong, lightweight frame of brazed steel tubing and alloy wheels. Its 10 gears are obtained by a five-speed Derailleur gear on the rear hub, coupled with a double chainwheel. The Derailleur mechanism moves the chain between the different-sized rear sprockets to achieve different gear ratios. The two different-sized chainwheels double the ratios.

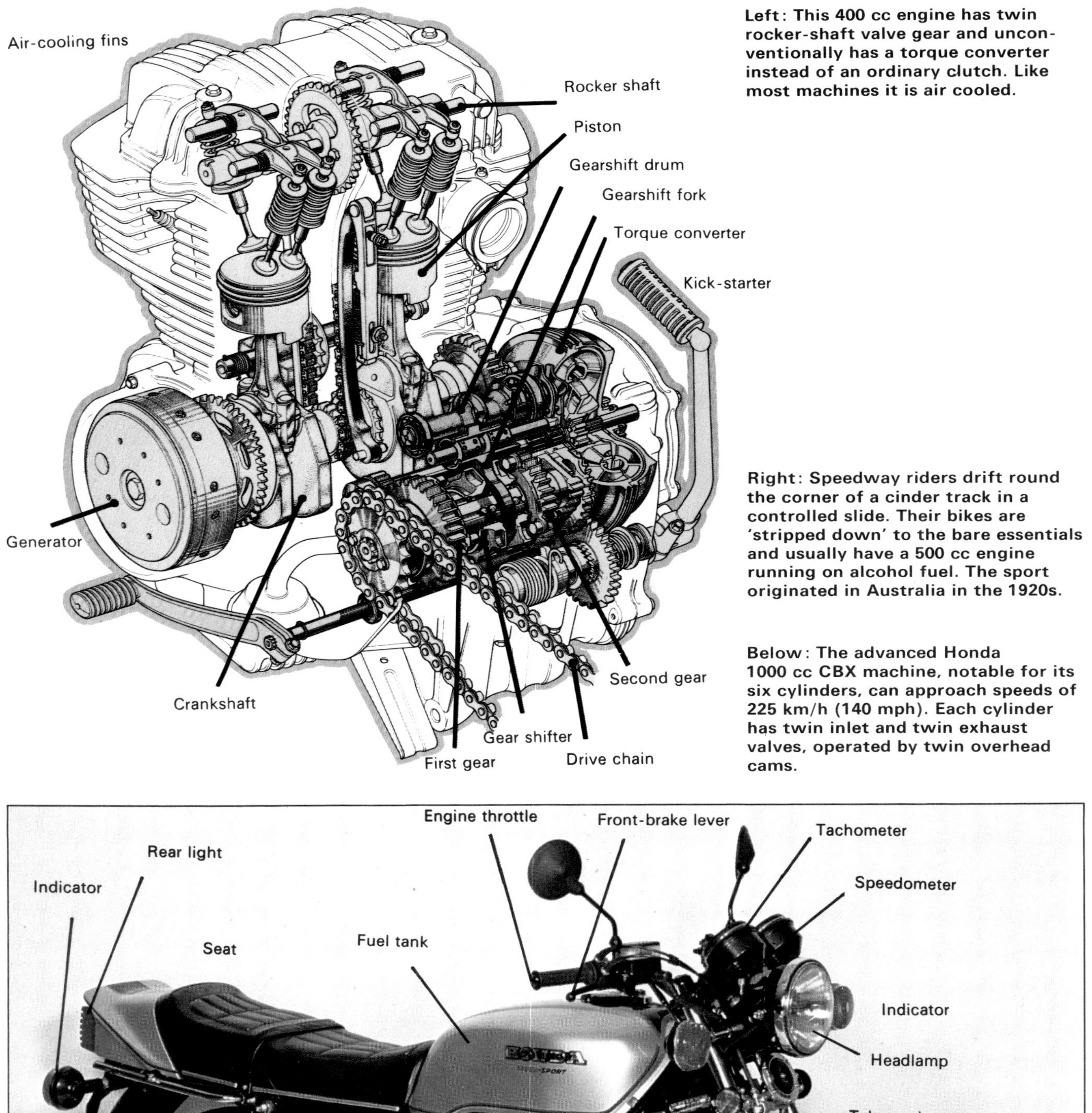

Left: This 400 cc engine has twin rocker-shaft valve gear and unconventionally has a torque converter instead of an ordinary clutch. Like most machines it is air cooled.

Right: Speedway riders drift round the corner of a cinder track in a controlled slide. Their bikes are 'stripped down' to the bare essentials and usually have a 500 cc engine running on alcohol fuel. The sport originated in Australia in the 1920s.

Below: The advanced Honda 1000 cc CBX machine, notable for its six cylinders, can approach speeds of 225 km/h (140 mph). Each cylinder has twin inlet and twin exhaust valves, operated by twin overhead cams.

often a windscreen. They are designed with ease of operation and comfort very much in mind. The Italian designers Vespa and Lambretta introduced scooters in the 1950s and still dominate the field.

The standard motorbike is built very differently from a car, but it nevertheless works in a similar way and is made up of similar systems. The engine uses petrol as fuel and has one or more cylinders. Its cubic capacity may be up to 1000 cc, with 50 cc, 125 cc, 250 cc, 500 cc and 750 cc machines being most common. The smaller machines generally have two-stroke engines (see page 41), while the larger have four-stroke, like cars (see page 36).

Each cylinder is supplied with fuel/air mixture through a carburettor, and ignition is by sparking plug. On larger machines a coil-ignition system (page 39) is used, but on others the simpler magneto-ignition suffices. A magneto is a type of electric generator. The engine may be started by electric self-starter or by kick-starter. Both turn over the engine until it fires. The engine is almost always air-cooled, the cylinders being surrounded by tiers of heat-dissipating fins. A few large models, however, do have water-cooling and have a small radiator fitted.

The transmission system of a motorbike includes a clutch, 2–6 speed gearbox and usually a drive chain to the rear wheel. The clutch is operated by a lever on the handlebars; the gearbox by a foot pedal. In some recent designs drive from the gearbox to the rear wheel is by drive shaft rather than chain.

Brakes are fitted to both wheels, the front brake being worked by a handlebar lever, the rear brake by a foot pedal. Formerly, cable-operated drum brakes were universal but in recent years hydraulic disc brakes have come into widespread use. Suspension is provided at the front by internally sprung and dampened telescopic forks and at the rear by integral coil springs and dampers (shock absorbers).

Going Commercial

A variety of other vehicles use our roads besides cars and motorbikes, carrying passengers and goods in bulk and performing other useful functions. They include not only lorries and buses, but such vehicles as cement mixer trucks, fire engines, mobile cranes, gulley emptiers, dustcarts, and road sweepers. Broadly speaking these commercial vehicles, despite their disparate roles, are mechanically very similar. A range of lorries, for example, can be derived by building different bodies on to a common chassis.

The majority of commercial vehicles are powered by diesel engine (see page 42). A 6-cylinder, in-line engine is perhaps the most common unit, with a power output of between about 150–250 horsepower (HP). But 4-cylinder in-line, V-8 and V-12 diesels are also used. Some massive dumper trucks even have V-16 engines, with a power output of 700 HP upwards.

Above: A container lorry being off-loaded at a container terminal. More and more goods are now being transported in standard-sized containers, which can be rapidly transferred between lorry, railway wagon and ship by suitable container-handling equipment, shown above. This is essentially a kind of Goliath crane. It consists of massive cross-girders supported by legs. A travelling carriage, or crab, carrying the hoisting gear straddles and runs along the cross-girders. All hoisting and movement is controlled from the operator's cage, which is attached to and angled beneath the crab.

Below: A tractor-truck of the so-called cabover design. Access to the engine is provided by a fully tilting cab. Note the vertical exhaust pipes and silencers favoured in American truck designs, which deliver exhaust fumes overhead rather than near ground level.

For greater efficiency many lorry diesel engines are supercharged – they use a compressor to force extra air into the engine cylinders.

The diesel engine is not generally as flexible as the petrol engine and operates best within a restricted speed range. Most lorries therefore have a larger number of gears than a car. Some have as many as 30 forward speeds and three reverse speeds! This is achieved by means of a main gearbox of 10 speeds, coupled with an auxiliary gearbox of three speeds. The majority of gearboxes are manually operated, though automatic and semi-automatic ('preselector') transmissions are gaining ground for some applications. As with cars, automatic transmission involves the use of a torque converter coupled with an epicyclic gearbox. To give extra traction the transmission may drive more than one axle. Two- and three-axle drive are common in larger vehicles. The lorry 'cut away' on page 65 has two-axle drive.

Twin-axle steering is often employed to provide a tighter turning circle, particularly in truck tractors. A truck tractor is the power unit of an articulated lorry, which is coupled to a two-wheeled semi-trailer. It has what is called a 'fifth wheel', located on the part which slips beneath the semi-trailer. The fifth wheel is a latching and pivoting assembly which allows the tractor and semi-trailer to mate and swivel.

The weight of a laden lorry can run into a hundred tonnes or more, so an effective braking system is vital. Smaller vehicles may have power-assisted hydraulic brakes like some cars (page 78). The larger ones have air brakes. The engine drives a compressor that charges a reservoir with compressed air. When the brakes are applied, valves open to allow compressed air into a brake chamber on each wheel. There it pushes against a diaphragm which moves linkages that force brake shoes against the wheel drum.

Left: An American 'Trackways' coach designed for comfortable long-distance travel. It is fully air conditioned and has reclining seats. The seats are set high for better vision, the luggage being stowed beneath. It is also equipped with a toilet for use in transit.

Right: The dustcart, or refuse collector, a specialist municipal vehicle essential in our consumer, 'disposable' society to rid our homes and cities of the $2\frac{1}{2}$ kg (5 lb) of rubbish we each discard on average each day. It has powerful hydraulic rams to compress the rubbish as it is tipped in.

Below right: Essentially British in origin and use – the double-decker bus with a carrying capacity of about 70 persons; it makes for efficient in-town travel.

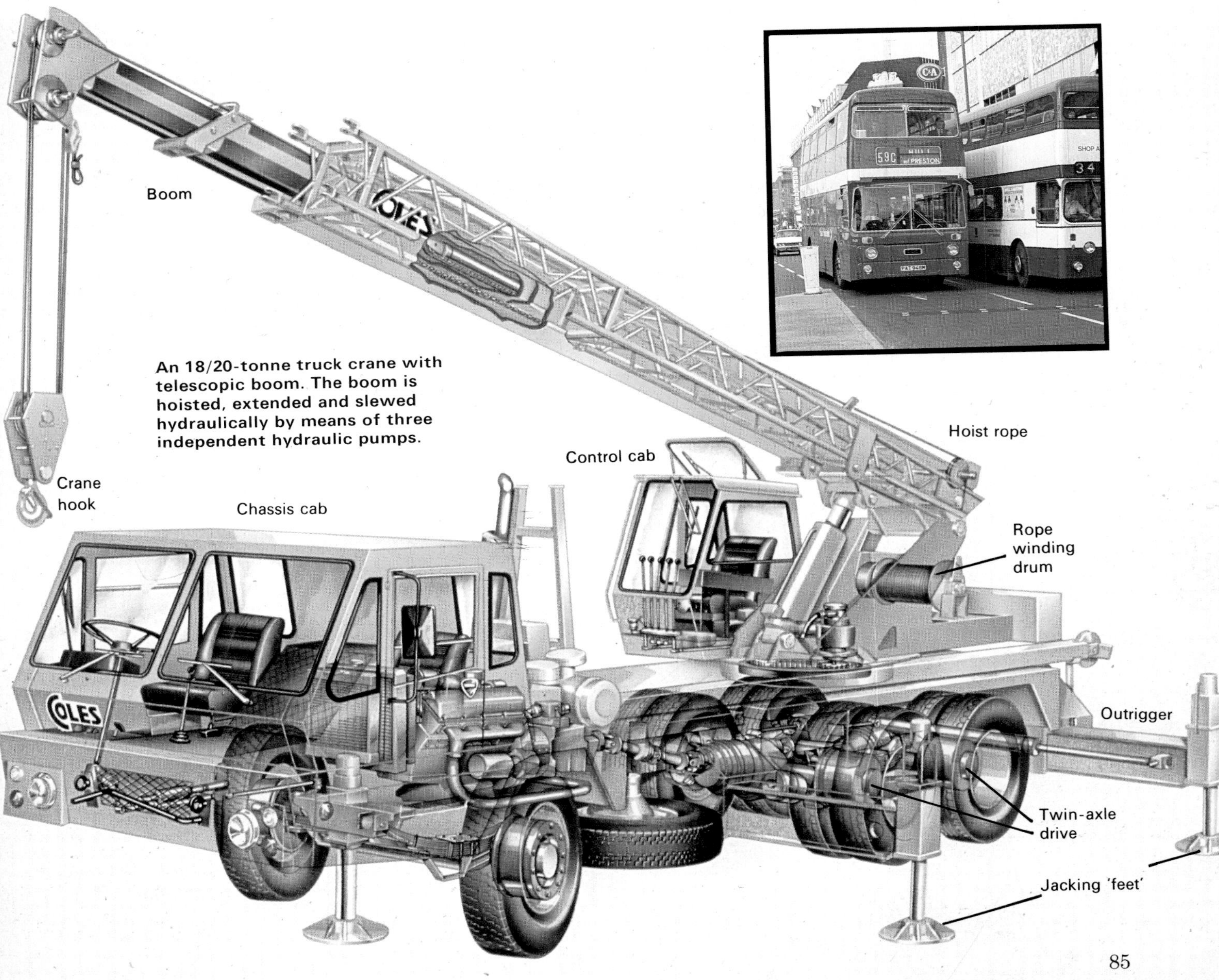

An 18/20-tonne truck crane with telescopic boom. The boom is hoisted, extended and slewed hydraulically by means of three independent hydraulic pumps.

The Iron Road

Richard Trevithick's failure to build a steam road carriage (page 67) at the turn of the eighteenth century was due largely to the appalling state of the roads at the time. So he designed a steam engine that would run on a prepared 'road' – a railway track – on which it would have a smooth ride. In so doing he pioneered a form of transport that was to revolutionize society, speed the growth of industry, and influence the settlement of the continents.

Railways had existed in Europe since the 1500s. They were used in mines to facilitate the transport of coal and ore wagons. It was much easier to roll wagons along a track of twin rails than along uneven ground. In Trevithick's day the track was known as a plateway or tramway. The rails were long pieces of angle iron of L-shaped cross-section. The flat bottoms of the rails formed the track, while the verticals acted as flanges to guide the smooth wheels of the wagons.

It was on such a track at Penydarran in South Wales that Trevithick ran his protolocomotive 'New Castle' in 1804. It was a clumsy machine with a huge flywheel and gearwheels driving both axles. It was far too heavy for the track, whose plates soon buckled and broke. John Blenkinsop achieved more success in 1812 with a locomotive which had twin vertical cylinders. They drove through gearing a cogwheel that engaged a toothed rail, or rack.

British mining engineer George Stephenson also became fascinated with steam traction, building his first locomotive, 'Blücher', in 1814. It was Stephenson who literally put railways on the map. He built and equipped the first two public railways designed for steam locomotion – the Stockton and Darlington (1825) and the Liverpool and Manchester (1830). To run on the latter Stephenson built the famous 'Rocket', which trounced its rivals in proving trials at Rainhill in 1829. It owed its superiority to its multitube boiler, which soon became standard.

Railways Go Global

Railway mania gripped not only Britain, but also Continental Europe and North America. In 1825 76-year-old John Stevens ran the first steam locomotive in the United States. Five years later the first successful American-built locomotive 'Best Friend of Charleston' inaugurated a regular service on the South Carolina Railroad. In Europe France pioneered steam railways in 1832.

By 1860 the United States had laid the foundations of what was to become the biggest railway network in the world. It had in excess of 50,500 km (30,000 miles) of track, more than double that of Britain. In 1863 work began on the first of the great transcontinental railways. Two teams of navvies started work simultaneously at Sacramento, California, driving east and at Omaha, Nebraska, driving west. They met at Promontory, Utah, six years later, after which time it was possible to travel from coast to coast. Transcontinental railways spanned Canada in 1885, and work started on the world's longest line, the Trans-Siberian, six years later. This line, which runs from Moscow to Vladivostok, is now over 9300 km (5800 miles) long.

While the spiders' webs of track spread worldwide, the railways became safer and more comfortable. As early as 1841 semaphore signalling came into use, which has only recently been superseded by the 'traffic-light' type signals. In the mid-1850s a system of interlocking points and signals was introduced to prevent a signalman switching trains onto conflicting routes. In 1872 the track-circuit system of signal setting went into use, by which a train automatically set the signals behind it at 'danger'. By then the compressed-air brake invented by George Westinghouse (1869) was rapidly becoming standard in the United States, though in Britain the vacuum brake was preferred.

Passenger comfort on long-distance routes was immeasurably improved by the adoption of the sleeping car. George Pullman in the United States pioneered this form of travel in 1859, though it took some years to catch on. The Belgian Georges Nagelmackers introduced the sleeping car to European railways in 1876 when he founded the Compagnie Internationale des Wagon-Lits.

Steam Bows Out

In the 1860s congestion in the streets of London forced railway engineers to go underground, and in 1863 they completed the first section of what was to become the world's most extensive underground railway system. It connected the mainline railway stations of Paddington, Euston and King's Cross. Initially steam locomotives were used, fitted with special, though not particularly effective, equipment to suppress smoke and fumes. London's first deep-level underground, or 'tube' system opened in 1890. It used electric traction, which had been pioneered principally by the Siemens brothers in Germany, starting in 1879. Mainline railways did not adopt electric traction until much later. Italy was first (1902) in this respect.

Railway engineers began experimenting with diesel traction soon after Rudolf Diesel announced his improved oil engine (1893). But the first practical diesel locomotive did not go into service, for shunting duties, until 1925. Seven years later a streamlined two-car diesel-electric train, the 'Flying Hamburger', began running between Berlin and Hamburg in Germany. Its success led to the widespread adoption of the diesel-electric locomotive, especially in the United States, where it still predominates. The United States also has the longest experience in operating gas-turbine locomotives, a form of traction pioneered by the Swiss in 1941. In the 1960s hover-trains, such as France's Aérotrains, which run along a monorail track on a cushion of air, showed the way ahead. It is in such trains, which are potentially capable of speeds of 500 km/h (300 mph), that the future of railways must lie.

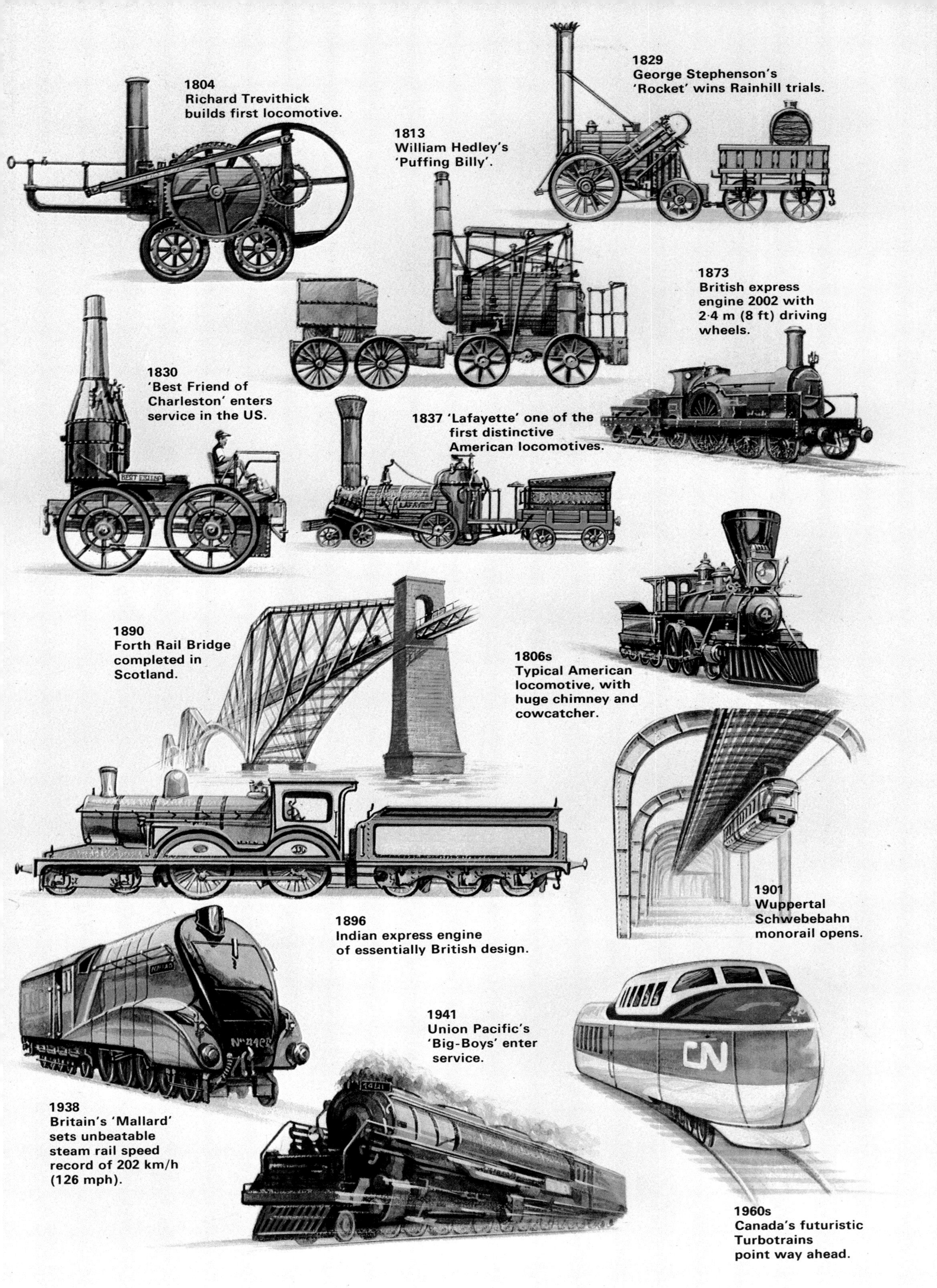

1804
Richard Trevithick builds first locomotive.

1829
George Stephenson's 'Rocket' wins Rainhill trials.

1813
William Hedley's 'Puffing Billy'.

1873
British express engine 2002 with 2·4 m (8 ft) driving wheels.

1830
'Best Friend of Charleston' enters service in the US.

1837 'Lafayette' one of the first distinctive American locomotives.

1890
Forth Rail Bridge completed in Scotland.

1806s
Typical American locomotive, with huge chimney and cowcatcher.

1901
Wuppertal Schwebebahn monorail opens.

1896
Indian express engine of essentially British design.

1941
Union Pacific's 'Big-Boys' enter service.

1938
Britain's 'Mallard' sets unbeatable steam rail speed record of 202 km/h (126 mph).

1960s
Canada's futuristic Turbotrains point way ahead.

Left: A 'Big-Boy', one of the biggest and most powerful steam locomotives ever built. Of articulated design, these monster engines developed over 6000 horsepower and weighed over 550 tonnes. Mechanical dinosaurs that they were, they dropped out of regular freight service in 1959, 18 years after they had been introduced.

Bottom left: In total contrast to the 'Big-Boys' are the swift and streamlined High-Speed Trains (Inter-City 125) of British Rail. Constructed in lightweight aluminium and powered by diesel engine, the High-Speed Trains can cruise at speeds up to 200 km/h (125 mph).

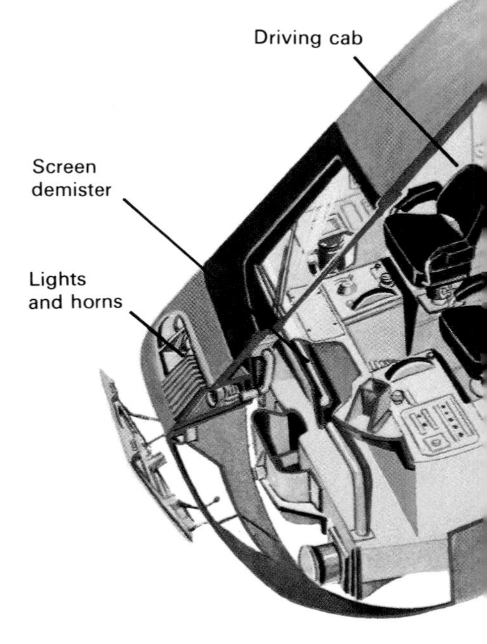

The Demise of Steam

The steam locomotive dominated the railways of the world for a century or more. But after World War 2 country after country began switching to diesel and electric traction. Today only a handful of steam locomotives are in regular use on world railways, notably in South America and Africa. Many things contributed to the downfall of steam locomotives, but essentially it was a question of economics. Intense competition from other means of transport forced the railways to improve their efficiency and streamline their operation.

Steam locomotives require a great deal of preparation before they can even start. Tonnes of coal must be loaded into the tender, together with thousands of litres of feed water for the boiler. The furnace must be lit and a good head of steam raised. Even when the locomotive is stationary for long periods during its working day, it must be kept under steam, which is very wasteful of fuel. Also it burns its fuel very inefficiently – only about five per cent of the heat energy in the coal is converted into mechanical energy to drive the wheels. This alone, in our energy-deficient age, would ensure the quick death of steam. Steam locomotives also require extensive maintenance even after a few hours' service. And their great weight, up to several hundred tonnes, gives the track they run on a terrible pounding so that this also requires continual and expensive maintenance.

The diesel and electric locomotives that have superseded the steamers are light in weight and are many times more efficient in energy utilization. They can be started almost instantly and switched off when temporarily not in use. They have rapid and smooth acceleration and can maintain high speeds for long periods. Refuelling and maintenance are easy. Last but not least they do not produce the prodigious quantities of polluting fumes and smuts that steam locomotives do.

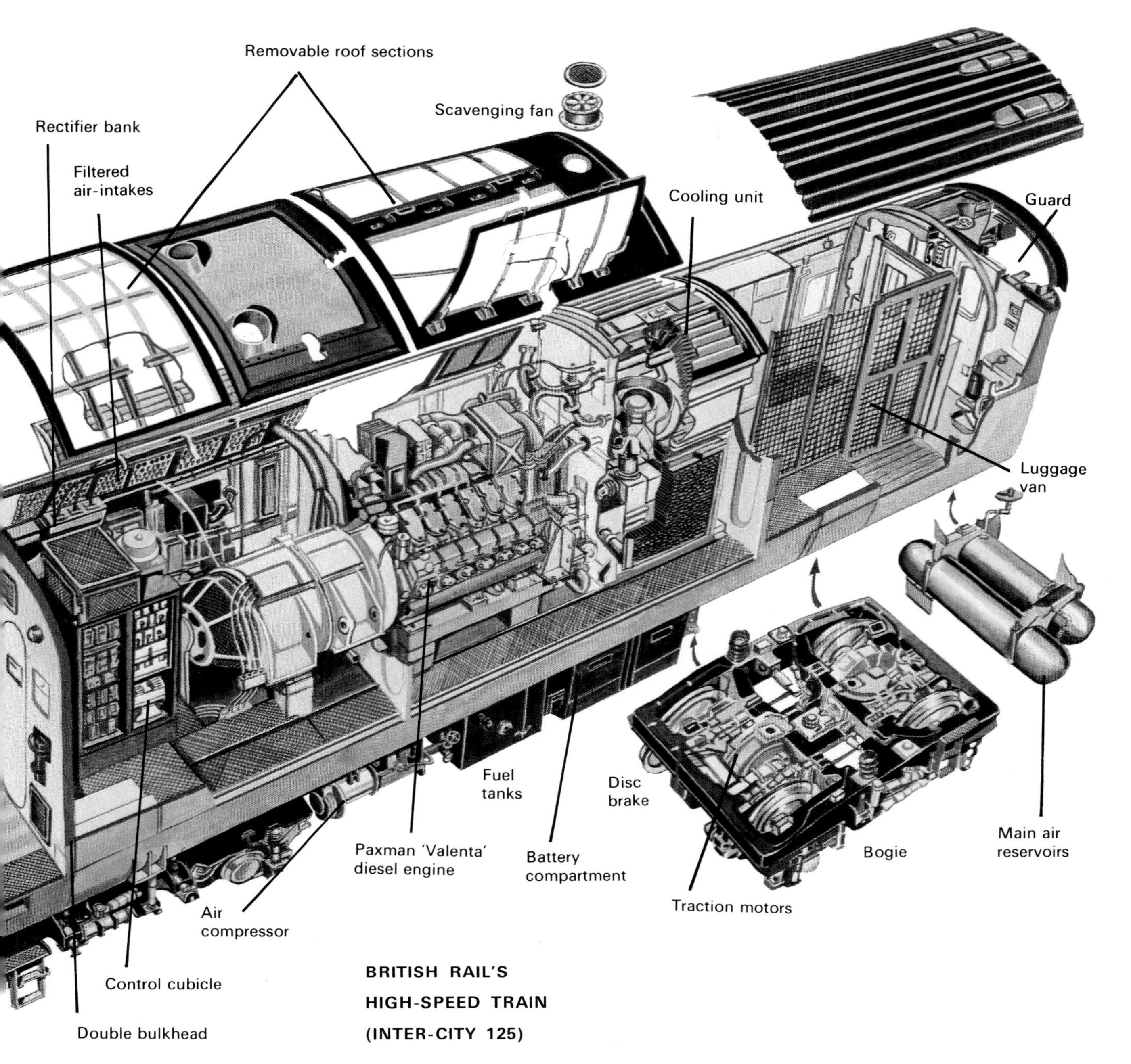

BRITISH RAIL'S HIGH-SPEED TRAIN (INTER-CITY 125)

Diesel Locomotives

Diesel locomotives use the same kind of engine as many lorries and buses. They burn oil by means of compression-ignition (see page 41). But locomotive diesels are larger and much more powerful. The large units that power mainline passenger trains have an output exceeding 2500 kilowatts (kW), or 3300 horsepower (HP). Diesel freight locomotives may have more than twice this power.

Several engine designs are employed in locomotives. Many have their cylinders in-line, like a conventional car engine, but others have them arranged in a V-configuration, driving a common crankshaft. An unusual British design has the cylinders arranged in a triangular shape; like the Greek letter capital delta (Δ). Three pairs of pistons drive three crankshafts located at each corner of this so-called deltic engine.

The majority of locomotive diesel engines work on the four-stroke cycle (page 36), rather than on the two-stroke cycle (page 41). And invariably they are supercharged, or turbocharged: that is, the ingoing air is compressed before it is delivered to the cylinders. This is done by means of a powerful centrifugal fan (blower), driven by a turbine which is spun by the exhaust gases leaving the cylinders. Compressing the air causes it to heat up, which is not desirable. So the air is passed over water-cooled pipes before it enters the cylinders.

Water is also invariably used to cool the engine itself. It is pumped through jackets surrounding the cylinders and into a radiator, through which outside air is sucked by a powerful fan. The cool water returning to the engine block from the radiator is led through a heat exchanger, where it cools the engine lubricating oil.

The world's fastest diesel-powered train – British Rail's High-Speed Train – has its two 12-cylinder, four-stroke engines in separate power cars, one of which is shown 'cut away' above.

Transmitting Power
There are several ways in which diesel locomotives can transmit power to the driving wheels. By far the commonest is electric transmission. The diesel engine is coupled directly to an electricity generator. The electricity produced is fed to electric traction motors which turn, through gearing, the axles carrying the driving wheels. The wheels of a diesel (and electric) locomotive are grouped beneath it in twin frames, or bogies, which commonly carry two or three axles. Usually all the axles are driven independently by separate traction motors. But in some three-axle designs the central axle is an idler.

Direct-current (DC) motors are almost always used for traction. They are of the series-wound type, this meaning that their field coils are linked in series with the armature coils (see page 59). This arrangement gives the very high starting torque, or turning effort, required to shift a heavy train. Many locomotives have a DC generator, whose output can be fed directly to the traction motors. Others have an alternating-current (AC) generator, or alternator, which can produce more power and is cheaper to maintain than the DC type. When an alternator is used, however, the AC output must then be converted by a rectifier into DC for the motors.

The traction motors get hot during operation and have to be cooled by air blown through them by fans (blowers). Power for the blowers is taken from a separate, auxiliary generator usually mounted on the same shaft as the traction generator. The

auxiliary generator also supplies power to drive the air compressors that operate the brakes, the lighting, the air conditioning, and so on.

Diesel railcars and small shunting engines may have mechanical transmission, which is virtually the same as that used in lorries and buses. Drive is taken to the wheels via a clutch and gearbox. In some units the clutch is replaced by a fluid coupling, and the gearbox works automatically.

Hydraulic transmission has also been used successfully in diesel locomotives, particularly in Germany and Britain. The function of the gearbox in the mechanical type of transmission is taken over by an ingenious hydraulic device called a torque converter. This consists of a centrifugal pump (impeller) and a turbine, which are housed in an oil-filled chamber. The diesel engine drives the impeller, while the turbine is coupled to the driving wheels. Between impeller and turbine is a fixed ring of guide vanes to circulate the oil from one to the other. When the impeller rotates, oil is forced against the turbine blades, causing them, and the coupled driving wheels, to turn.

Electric Locomotives

Diesels are the most common form of locomotive in North America, but in Europe and elsewhere electric locomotives have approached, and in some instances eclipsed them in importance. And the move is towards greater electrification in the future. Electric locomotives are the quietest, cleanest and potentially the most powerful locomotives of all.

Since they take their power from an external source, electric locomotives are simpler in construction than other types, being merely power converters. They also require little maintenance and are long-lasting. But they have the disadvantage that they can run only on lines that have been electrified, and electrification is very costly. It requires the provision of conductors along the track to pass on the electricity to the locomotive. Electricity sub-stations must be built at regular intervals to tap the electricity from the power-supply grid and feed it into the conductors.

The majority of electric locomotives pick up their power from an overhead conductor, or trolley wire, slung above the tracks. A sprung arm, or pantograph on top of the locomotive makes sliding contact with it. In the alternative system the conductor takes the form of a third, or live rail, which is carried on insulators and located at the side of the track. The current is collected via iron shoes which slide along the rail.

The current may be supplied to an electric locomotive in the form of DC or AC. DC is simplest to use because it can be fed directly to the traction motors, which are invariably of the series-wound, DC type. Britain and France, for example, have

Right: One of the electric TGVs (Trains à Très Grande Vitesse) now speeding on French Railways, or SNCF (Société Nationale des Chemins de Fers). It is similar in external design to the gas-turbine TGV.

Left: Essentially a bus on rails, the diesel railcar has proved popular on local, low-volume traffic routes. Illustrated is the so-called 'autorail panoramique', which operates on scenic routes in the South of France.

Top left: Twin diesels of the Canadian Pacific Railroad haul a train through Shooting Horse Pass in the Canadian Rockies. Since the 1950s diesel-electric locomotives have predominated in North America.

Below: Another French electric locomotive, of the BB 15000 class, cut away to show essential features.

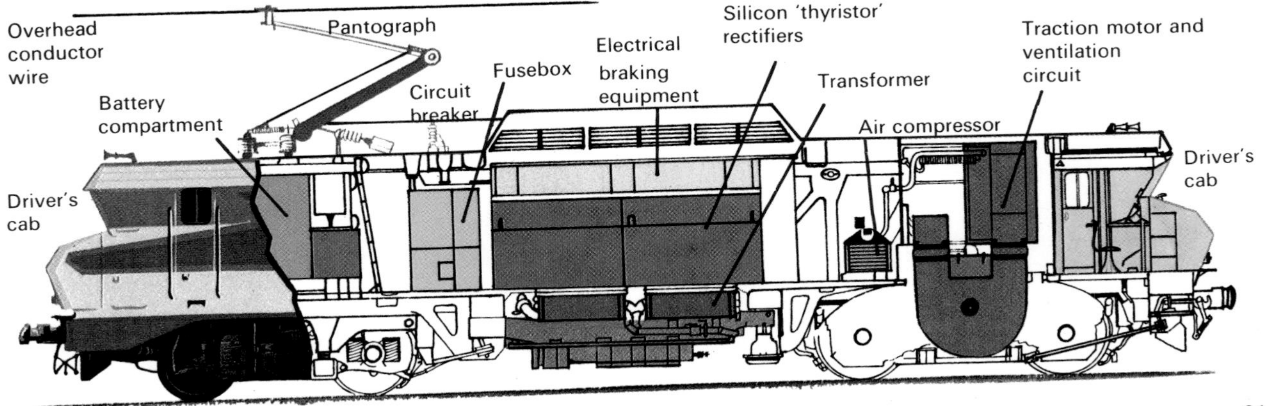

extensive DC lines. In Britain the lines work at 600 volts from a third rail, while in France they work at 1500 volts (the more usual) from overhead wires. The two French locomotives which jointly set the world rail speed record in 1955 work on this system. Designated Bo-Bo 9004 and Co-Co 7107, they achieved a speed of 330·8 km/h (205·6 mph). (The Bo-Bo designation, incidentally, means that the locomotive has its wheels, all driven, in twin, two-axle bogies. Co-Co means that the wheels, again all driven, are in twin, three-axle bogies.)

DC transmission, however, is costly because electricity substations have to be located at frequent intervals along the track, and they must incorporate expensive rectifying equipment to change the supply grid current, which is AC, into DC. It is for this reason that AC line systems are now increasingly used. In modern systems the supply voltage is 20,000 or 25,000 volts, alternating at the appropriate mains frequency of 50 or 60 hertz (cycles per second). In some systems, however, different frequencies are used, commonly $16\frac{2}{3}$ hertz in Europe and 25 hertz in the United States.

Before it can be fed to the DC traction motors the alternating current picked up from the conductor must have its voltage reduced and be converted into DC. The voltage reduction is carried out by a transformer (page 62). A rotary converter or rectifier then changes the lower-voltage AC into DC.

As with diesels the electric locomotive has an auxiliary power system. This drives the compressor and exhauster needed for braking; electric fans to force-ventilate the traction motors; and a pump to circulate cooling oil through the transformer, which becomes very hot during operation.

Speed Kings

Japan has operated the world's fastest railway route since 1972, when the New Tokaido service began. It is operated by severely streamlined 'bullet' trains, powered by overhead-wire electric traction. They cover the 160 km (100 miles) between the cities of Osaka and Okayama in 58 minutes at an average speed of 166 km/h (103 mph). They may approach their maximum speed of 255 km/h (160 mph) at times.

In 1980 British Rail's Advanced Passenger Train (APT) went into regular service. Although the prototype APT had gas-turbine engines, the production units are electric and take their power from overhead wires. The maximum speed of the APT is close to that of the Japanese bullet trains, and it has advanced features that enable it to achieve very high average speeds on existing track.

The journey time of conventional trains is dictated largely by the speed restrictions around curves. On all but the straightest routes, only modest time savings can be achieved by increasing speeds much above 160 km/h (100 mph). About half of Britain's major railway routes are curved, and 50 per cent of the curves are relatively sharp and require speed restrictions to be imposed on conventional trains.

The APT is designed to negotiate these curves at up to 40 per cent higher speeds than normal. This greatly increases average speeds and conserves energy, for the train does not need to brake or accelerate so much before or after cornering.

To increase passenger comfort when cornering fast the APT has a novel kind of suspension. It allows the coaches to 'lean' inwards by up to 9 degrees when going round curves. Each coach is tilted individually by hydraulic jacks mounted on the bogies. The tilting is controlled electronically according to the lateral (sideways) acceleration of the bolster that carries the coach. To measure the acceleration a spirit-level accelerometer is used, in which the position of the bubble is sensed electrically.

At high speeds the main source of train resistance is aerodynamic drag. This increases in proportion to the speed squared – in other words if you double the speed, you increase the drag four times. The APT is thus carefully streamlined.

Power Packed

The power units of the APT are not integral with the driving cabs as they are in conventional train layouts. They form separate power cars in the middle of the train (one or two may be used), which push and pull the other trailer cars. To keep weight to a minimum the body shells of the trailer cars are fabricated in aluminium.

The power car takes alternating current from a 25,000-volt overhead wire via a pantograph and feeds it to four 750 kW (1000 HP) traction motors. These are unusual in that they are mounted on the body of the power car rather than in the bogies. They drive the axles, not directly, but through a mechanical transmission. The power is transmitted via a body-mounted gearbox and connecting shaft, to a reduction gearbox mounted on the drive axle.

To cope with the high speed levels reached, the APT has a powerful and novel braking system. It uses a hydrokinetic, or water-turbine brake. Essentially this consists of a multivaned turbine wheel which is normally free to rotate inside a housing. It is driven round by the transmission. On the power car the brake unit is fitted to the body-mounted gearbox. On the ordinary trailer cars the brake unit is bolted between the two halves of the hollow axles. When braking is required, water is introduced into the housing where it quickly slows down the turbine wheel. In the process the water gets very hot and so is circulated through a cooling radiator. Each axle has hydrokinetic brakes.

Left: With sacred Mount Fuji as a backdrop, a 'bullet' train speeds along Japan's New Tokaido line at a speed approaching 200 km/h (125 mph). The bullet trains have operated the world's fastest rail service since 1972, but are now being rivalled speedwise by British Rail's Advanced Passenger Train (APT).

Top left: The APT during trials, when it was powered by gas-turbine. The APTs entering service have electric traction.

Below: Today's trains travel so fast that air resistance, or aerodynamic drag, becomes a major problem. To reduce drag to a minimum trains have to be well streamlined, hence the 'bullet' front of the New Tokaido trains and the 'wedge' nose of the APT. France's TGVs are similarly well streamlined. Their shape, like that of their rival speedsters, was chosen after extensive wind-tunnel testing (below).

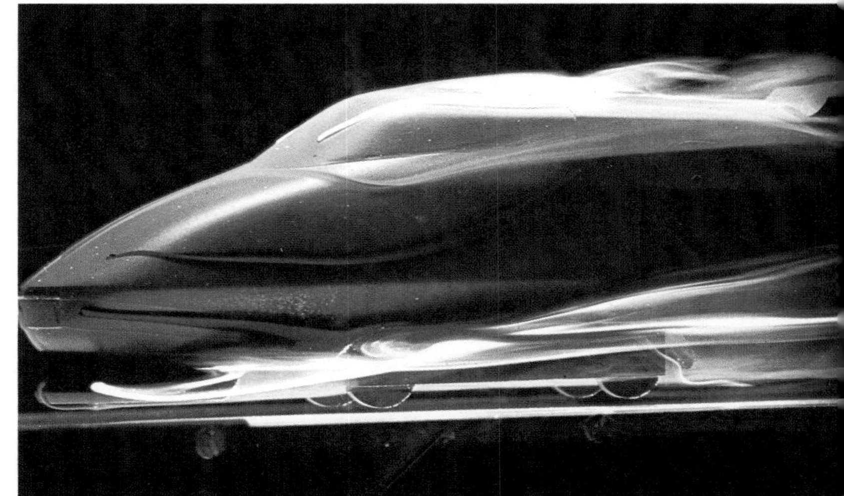

One of French Railways' turbo-TGVs travelling at high speed during trials. It is capable of speeds in excess of 300 km/h (200 mph).

JET SETS

For several decades now some of the long-distance freight routes of the Union Pacific Railroad in North America have been worked by locomotives powered by gas turbine. They are massive machines with a power output approaching 4500 kW (6000 HP). Gas-turbine locomotives are also to be found in France and Russia, and they have been used at times in Canada and Britain. Their most recent use in Britain was for testing the prototype Advanced Passenger Train in the late 1970s (see page 92). The current French turbo-TGVs 'Trains à Très Grande Vitesse' are perhaps the most successful gas-turbine locomotives to date. They have compact power units which operate efficiently even on relatively short-distance routes. Normally, gas turbines are best suited to long-haul, non-stop working.

A locomotive power unit functions in exactly the same way as any other gas turbine. It burns fuel in compressed air; the hot gases produced spin a turbine, which drives the compressor and delivers useful power (see page 43). In most gas-turbine locomotives the turbine shaft is coupled to a generator, which provides electricity to drive traction motors. Mechanical drive via gearing has been attempted, but has not been successful. For fuel, fuel oil and gas are most widely used. The gas is usually propane, stored in liquid form in pressurized tanks.

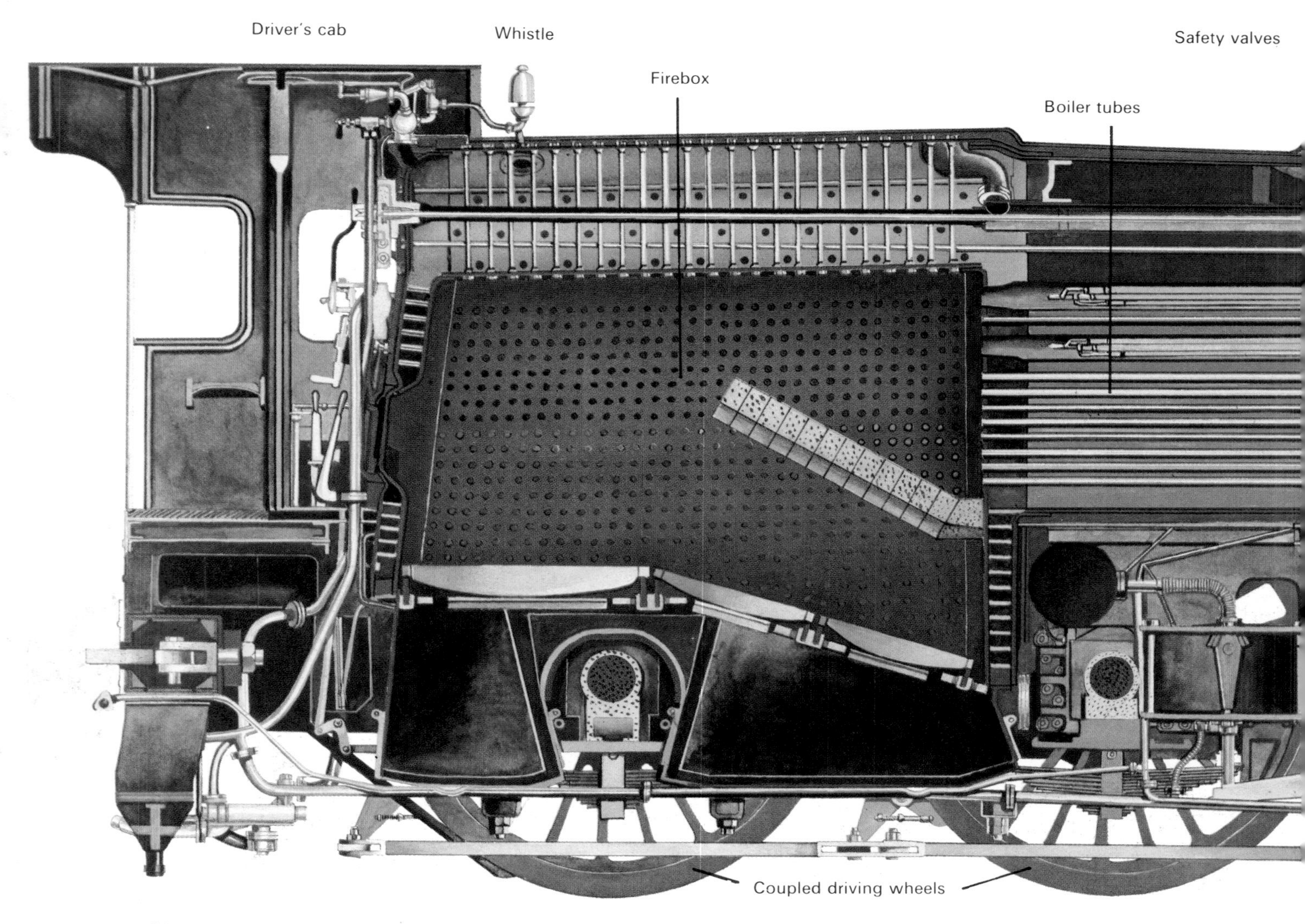

The Romance of Steam

Though diesel and electric locomotives are super-efficient, clean and quiet, they do not have the majesty and charisma of the steamers they have replaced. Belching steam and smoke and snorting like the dragons of myth and legend, steamers impress by their sheer elemental power. But they are mechanical dinosaurs, doomed to extinction by a change in climate, in this case economic climate.

Steam locomotives today survive only in a few regions of the world, particularly in underdeveloped countries. The relative simplicity of their construction and operation allow them to be handled by unskilled labour. Many lines in India, Africa, South America and South-East Asia are still steam worked, though probably not for much longer.

The steam locomotive has as its power source a reciprocating steam engine. Steam is generated in a boiler and is fed to a cylinder containing a piston. It drives the piston back and forth, and this motion is then transmitted to the driving wheels. The picture below shows the mechanical means of achieving this sequence of events in the most efficient way.

Fuel (usually coal) is burned in a furnace, generating flames, smoke and hot gases, which rise into the firebox. They are drawn into tubes running through the huge boiler that forms the main part of the locomotive. The water surrounding the tubes heats up rapidly and boils into steam. The steam collects in the steam dome at the top of the boiler and is then circulated through U-tubes ('superheaters') set inside some of the upper boiler tubes. In them it becomes superheated, reaching temperatures of 300°C or more. Its pressure at this time may be as high as 20 atmospheres.

The superheated, high-pressure steam is then led to the cylinders, which are located on each side and to the front of the locomotive. Steam passes into each cylinder through a slide valve, which admits steam to one side of the piston. As it expands, the steam drives the piston along the cylinder. When the piston is near the end of its stroke, the slide valve moves over and admits steam to the other side of the piston, driving it back. As it comes back, it forces the exhaust steam from the other end of the cylinder. This is channelled through the blast-pipe, which discharges beneath the chimney. The blast of steam up the chimney reduces pressure in the smoke box and creates a draught that draws the flue gases more strongly through the boiler tubes. The characteristic 'puff' of a steam locomotive is made by the blast-pipe discharging.

Cutaway drawing of one of Great Western Railway's (GWR) famous 'Castle' class locomotives, 'Caerphilly Castle'. Built in 1923, this 4–6–0 steamer was retired in the mid-1950s, when the days of steam were numbered.

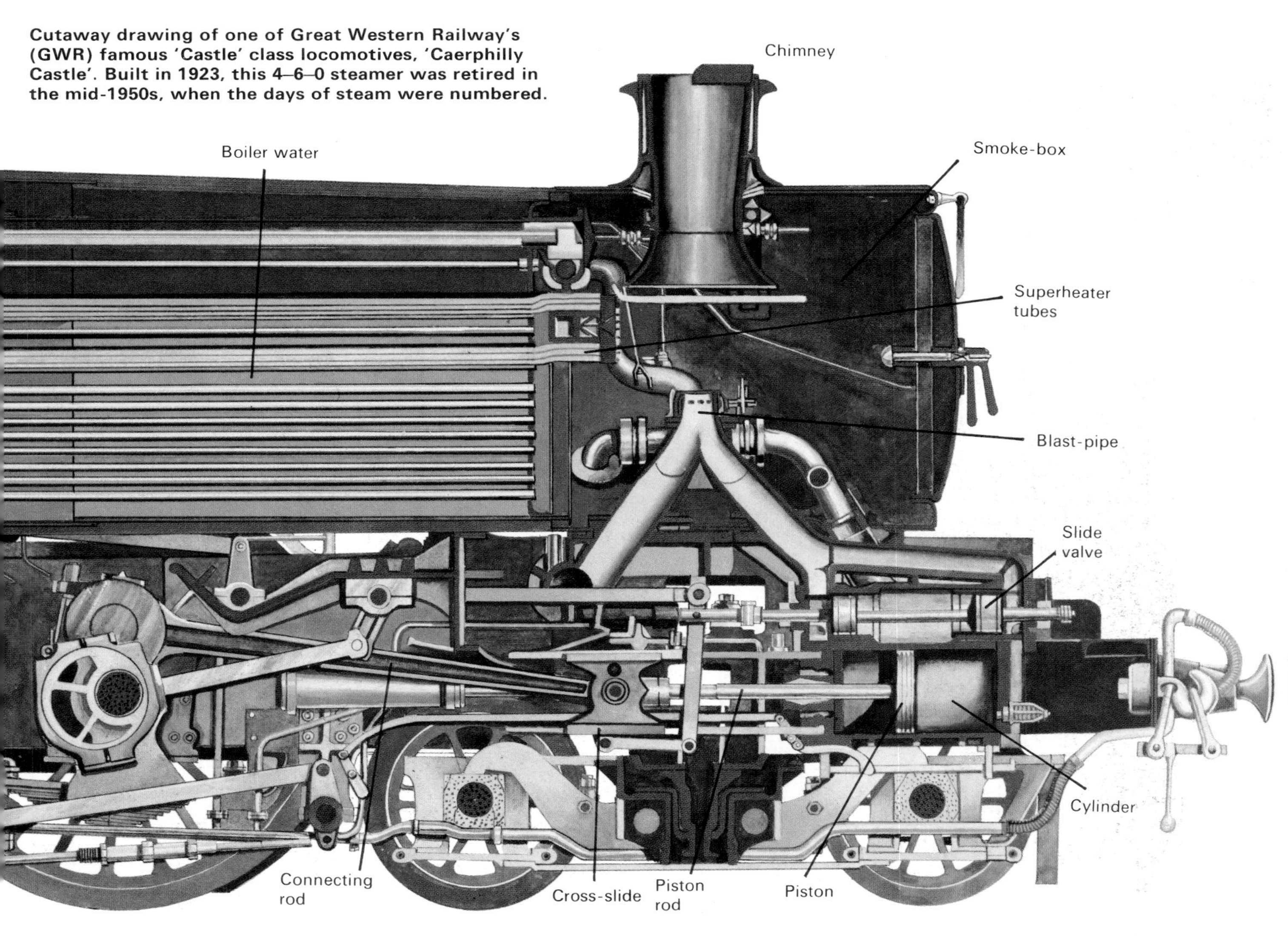

The piston is attached by a rod to the cross-head, which moves back and forth in a guide. A connecting, or driving rod transmits the motion of the cross-head to a crank on one of the driving wheels, which it thus turns round. Invariably other wheels are joined by coupling rods to the driven wheel to improve traction. Eccentric cranks on the drive wheel move rods that operate the slide valves. The locomotive driver can also move the slide valves manually by means of a hand-wheel in the cab. By so doing he can switch the steam flow from one side of the piston to the other and hence reverse the direction of travel.

Though the majority of steam locomotives have used two cylinders (one each side) some, particularly in Britain, have used three or four. This results in a smoother transfer of power and reduces the rhythmic pounding of the rails, or 'hammer blow', which can become marked with two-cylinder operation. Many French locomotives adopted another type of multicylinder arrangement. The steam is delivered first to two small high-pressure cylinders, which then discharge to two larger low-pressure cylinders before exhausting to the atmosphere. This so-called compound expansion results in higher efficiency.

Some of the most interesting locomotives are of articulated design. They have two engine units, each with its own cylinders and driving wheels, fed with steam from a common boiler. The biggest locomotives ever built, the 'Big Boys' of the Union Pacific Railway, were articulated. The two engine units were joined by an articulated coupling beneath the common boiler. With this arrangement the locomotive could negotiate sharper curves. The 'Big Boys' weighed a prodigious 550 tonnes and delivered a power output of some 4500 kW (600 HP). More widely used, however, were the articulated locomotives of the Beyer-Garratt design. They found widespread application in eastern and southern Africa, and many are still in service there. They differ from the Big Boys in having their two engine units mounted on separate frames.

FAIL-SAFE

The momentum of a train, weighing in all several thousand tonnes and travelling at speeds up to 160 km/h (100 mph), is colossal. And from the early days of the railways trains have been blessed with highly effective braking systems. The majority of the world's railways use the compressed-air brake invented by George Westinghouse in 1869. But in Britain the vacuum brake is generally used. In both systems a train-pipe, or brake-pipe, runs the whole length of the train and connects with brake cylinders on each wheel. Air is fed into or withdrawn from the cylinders to apply or release the brakes. The brake normally consists of an iron shoe which presses against the outer wheel rim. But some high-speed trains now use disc brakes similar to those used on cars (see page 77).

In the compressed-air brake the train-pipe is connected through a three-way valve to an air reservoir and the brake cylinder. When the brakes are 'off', air at a higher pressure than normal is maintained in the train-pipe. It holds down a piston in the valve in a position that allows compressed air into the reservoir but connects the brake cylinder to the atmosphere. A spring forces the piston backwards to release the shoe from the wheel. To apply the brake, the pressure in the train-pipe is reduced. The piston in the valve moves into a new position which allows compressed air from the reservoir into the brake cylinder. There it forces the piston forwards against its spring and causes the shoe to be pressed against the wheel.

In the alternative vacuum brake a partial vacuum is maintained in the train-pipe to keep the brakes 'off'. This is done by means of a suction pump, or exhauster, in the locomotive. When in the 'off' position, a vacuum is maintained inside the brake cylinder on both sides of the piston that applies the brake. To release the brake, air is allowed into the train-pipe to destroy the vacuum. It is admitted to one side of the piston and, since the other side is still under vacuum, forces the piston down the cylinder and the brake shoe against the wheel.

Both the compressed-air and the vacuum brake systems are 'fail-safe'. If the train-pipe is accidentally ruptured, the brakes will automatically be applied. In the one the pressure will be lost, and in the other the vacuum.

Right: The traditional railway track, formed of twin steel flat-bottomed 'T' rails. They rest in iron baseplates on heavy wooden sleepers, to which they are firmly fixed by spikes. The sleepers are located in a bed of stone-chip ballast, which allows rapid drainage. There are several alternative methods of anchoring the rails. The sleepers may be made of reinforced concrete to which the rails are fixed by clips and nuts to embedded bolts. The rails are now often welded together to give much longer lengths of track, which makes for smoother running.

The standard gauge for the rails – the distance they are apart – is 143·5 cm (4 ft 8½ in), the gauge adopted by George Stephenson. Some railways use narrow-gauge track, of which the metre gauge (3·3 ft) is the most common. Russian railways are among a small minority that use a broad-gauge, of 1·5 metres (5 ft). In the early days of the railways a broad gauge of 2 metres (6·6 ft) was favoured by master-engineer Isambard Kingdom Brunel, who built many railway lines in Britain's West Country to that gauge. But they were eventually rebuilt to standard gauge.

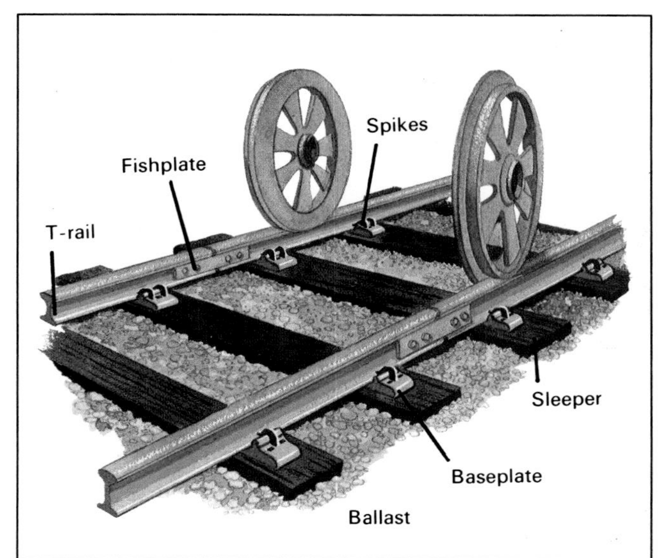

Left: Traffic control on the railways is now highly centralized, with hundreds of kilometres of track being supervised from an elaborately equipped control centre. These have largely replaced individual signal boxes spaced at regular intervals along the track. And signal setting is done by remote, push-button electrical control rather than, as previously, by means of hand levers. The signals too have been modernized and consist now of coloured lights rather than semaphore arms. Displayed in the modern control centre is an illuminated board carrying a map of the railway tracks under control. By means of electrical 'track-circuiting' the map lights up where a train is travelling along a track. The primary function of the track-circuit, however, is to set the signals immediately behind a train to 'Stop'.

Left: One of the few steam locomotives still in regular use – a Beyer-Garratt in East Africa. The best known of the articulated locomotives, it has two separate driving units coupled together. This arrangement permits tighter turns on the metre-gauge track on which the Beyer-Garratts travel.

Top left: Over most of the world, however, steam lives on in only a few locations where enthusiasts have bought and renovated small branch lines and rolling stock to run on it.

Right: Railways come into their own in transporting freight, particularly in bulk. Freight trains are in general made up of wagons bound for different destinations. Periodically they have to be sorted and re-routed. This usually takes place in huge depots called marshalling, or classification yards. This one is operated on French Railways. Each wagon in a train to be sorted is fed into the yard along a single track, which progressively branches into more and more tracks. By setting the appropriate points, the yard operator can switch wagons going to different destinations into separate trains. In modern yards closed-circuit television or electronic scanning is used to assist wagon identification.

Left: Because of their gradients, mountain railways demand a more positive form of traction than ordinary lines. This is provided by rack-and-pinion drive, a pinion on the locomotive engaging and moving along a toothed rail, or rack, laid between the ordinary rails. The rack is clearly visible in this picture of the Rigi Railway in Switzerland. It is of the 'ladder' type, introduced by Nicolas Riggenbach in 1863.

Right: The winding machinery that drives the cables which haul the cable-cars up San Francisco's hilly streets.

Top right: In San Francisco, street-cars, or trams, dovetail into an efficient public-transport service, which also includes buses, subway (the Bay Area Rapid Transit System) and the century-old but still flourishing cable-cars.

Bottom right: A cable-car on the scenic Powell and Mason line to Bay and Taylor, one of the three remaining cable lines in San Francisco.

On the Rack

Ordinary locomotives have smooth wheels and run on a smooth track. They rely simply on the adhesion, or friction, between the two for traction. On the level adhesion is good, but it falls off rapidly as the track begins to slope. For slopes greater than about one in ten (that is, one metre rise in ten metres travelled) smooth-wheeled traction is useless. In mountainous country it may be difficult to avoid steep slopes, and railways resort to the rack positive-drive system first demonstrated by John Blenkinsop as early as 1812.

The locomotive engine drives a toothed wheel, or pinion, which engages with a toothed rack fitted between the rails of the track. Three types of rack are used. Simplest is the ladder rack, whose teeth go from side to side, like the rungs of a ladder. The Swiss Vitznau-Rigi Railway uses this system. The steepest rack railway of all, the Mount Pilatus, also in Switzerland, has an alternative system designed by Edward Locher (1889). The rack has teeth cut in both sides, which engage with twin drive pinions. This system enables it to climb slopes up to one in two. But the most common rack system, the Abt (named after its inventor, Roman Abt), uses two or more stepped racks fitted side by side so that the teeth of one are opposite the troughs of the others. The highest railway in Europe, the Jungfrau, uses the Abt system. It reaches a height of over 3450 metres (11,300 feet).

Rivalling the rack railway in its ability to climb steep slopes is the cable railway. This has no locomotive; its passenger cars are hauled on the end of a cable by a stationary engine. This recalls the early days of the railways when stationary steam engines were used to haul wagons up steep slopes before steam locomotives came into widespread use. Cable railways are to be found in the Alps and in mountainous regions elsewhere. In Brazil part of the main Santos-to-Jundiai Railway is cable-hauled, as is Haifa's (Israel) underground railway.

Cable railways usually work on the funicular system. The hauling cable links two cars which travel up and down parallel tracks – one going down as the other goes up. It is driven by a powerful winch in an engine house at the upper level. Because they are linked, the descending car helps to haul up the ascending one. And the engine does not have to supply as much power as it would if it were hauling up the ascending car alone.

San Francisco has a unique type of cable railway, on which run its famous cable-cars. Conceived by inventor Andrew Hallidie in 1873, the cable-car still provides the cheapest and most attractive way of seeing this beautiful American city. They run on rails up and down the steep city streets, where

slopes of one in five are not uncommon. Between the rails is a slot, under which an endless cable runs, driven continuously by engines in a winding house (now also a cable-car museum). Each car has a clutch mechanism worked by a lever which grips the cable to go and releases it to stop.

Another form of street rail transport has survived in San Francisco – the tram, which Americans call the streetcar. Street trams are most often found, however, in European countries, including Scandinavia, Germany, the Netherlands, Belgium, Switzerland and Russia. They are invariably powered by electricity, which they usually take from overhead wires. Trams were ousted from most cities because their inflexibility on fixed rails made them ill-suited to modern highway planning and maintenance. But a move back to the tram may not be far off for trams are much cheaper to run than buses and, moreover, emit no polluting fumes. Trolley buses, too, could make a comback for similar reasons. They are also electrically powered and take their current from overhead wires, but unlike the tram run on ordinary tyred wheels.

Rise of the Monorails

Designers with an eye to the future reckon that conventional twin-rail railway tracks have had their day. The future, they say, belongs to the monorail, and indications are that they will be proved right. A monorail 'hover-train' currently holds the world rail speed record of 378 km/h (235 mph).

But the idea of the monorail is far from new, dating

back at least to 1883. In that year Charles Lartique built a 14-km (9-mile) monorail line in County Kerry, Ireland. It was of straddle design, the carriages fitting like panniers on either side of the rail. This design is found updated in the German Alweg system ('Alweg' is an acronym of the name of its inventor Axel L. Wenner Gren). The passenger car is mounted on bogies carrying two rubber-tyred wheels, which run along the top of a concrete monorail. It is driven and guided by pairs of wheels running along the sides of the beam. Japan has in operation several monorails of the Alweg design. The monorail at Disneyland near Los Angeles is also of the straddle type.

An alternative, but less common type of monorail, has the passenger cars slung beneath the rail. An outstanding example of this design is the Wupperthal Schwebebahn, which has been in commercial service almost continuously since 1901. It carries passengers between Bramen and Elberfeld, in Germany, following the River Wupper for much of the time. A more modern suspension monorail operates at Ueno Park Zoological Gardens in the suburbs of Tokyo. The passenger cars hang by curved arms from wheeled bogies that run along the top of the monorail. Each bogie has guide wheels to keep it on the beam.

In the most advanced monorail systems wheels have disappeared. The vehicle is supported a few millimetres above the track by an air cushion or a magnetic field. It is propelled by a novel kind of electric motor invented by British electrical engineer Eric Laithwaite (see Box).

Some town planners see the monorail as a means of alleviating the chronic congestion that now afflicts the streets of our large cities. Raising the track to roof-top level still leaves the roads below free to carry other traffic. One major drawback of this idea is that monorails are unsightly, to say the least. Also, those built so far have not proved themselves reliable or able to carry the necessary volume of passenger traffic.

LOW-FLYING TRAINS

The tracked air-cushion vehicle, or hover-train, has been demonstrated successfully by Jean Bertin in France, among others. Bertin calls his vehicles Aérotrains, and it was one of his vehicles that set the world rail speed record.

A hover-train uses a powerful fan to force jets of air between its body and the monorail track. A thin 'cushion' of air is produced that causes the body to lift. In this position friction between body and track is eliminated. Hover-trains can thus be propelled efficiently and up to speeds wheeled vehicles are incapable of reaching. Wheels lose their traction at about 400 km/h (250 mph).

If hover-trains are not in contact with the track, how can they be propelled? One way is to use an aeroplane propeller driven by gas turbine to push the train along. But this is far too noisy for commercial use. And the method of propulsion now favoured is the linear induction motor.

In an ordinary electric motor a rotor is made to move when electric current is passed through its coils, which are situated in a magnetic field. In the linear motor the coils have been transformed into a single straight conductor. Electromagnets are mounted around the conductor to provide the necessary magnetic field. When current is passed through the conductor, the magnetic field produced interacts with that of the electromagnets. And conductor and electromagnets move apart. In the hover-train the electromagnets are mounted on the vehicle, while the conductor is fitted to the track. Passing a hefty current through the track conductor produces a very powerful magnetic reaction that thrusts the vehicle forwards. The linear motor also features in an even more revolutionary design, which raises itself above the track by magnetic levitation.

'Maglev' (magnetic levitation) works on the principle of induced magnetic repulsion. If you pass a magnet over a conductor, you induce in the conductor an electric current, with which is associated a magnetic field. The direction of this field opposes that of the magnet, resulting in a repulsive force between magnet and conductor. A maglev train must have on board a very powerful magnet to induce the required force of repulsion needed to lift it above the track.

Only recently have magnets powerful enough become available. They are electromagnets whose coils are made of 'super-conducting' metals such as tantalum, niobium and bismuth. These metals are so called because they have the unusual property of losing virtually all their electrical resistance when cooled to the temperature of liquid helium (−268°C). When you feed electric current into them, it keeps circulating almost indefinitely, and its magnetism is also retained. An electromagnet with superconducting coils requires comparatively little power to maintain a powerful magnetic field.

Right: One of Paris's rubber-tyred Métro trains. You can see in this photograph the main wheels and horizontal guide wheels, together with the broad tracks on which they run.

Bottom right: The suspension monorail which has been operating in Tokyo's Ueno Park zoological gardens since 1957. The passenger carriages are suspended from wheeled bogies that run along the top of a concrete beam, which constitutes the monorail track. The bogies are kept on the track by means of guide wheels that run along the sides of the track.

Left: The other main monorail set-up – the straddle design, exemplified by the Disney monorail that runs through the Disneyland pleasure park on the outskirts of Los Angeles.

Bottom left: A monorail of sorts, an aerial ropeway, or téléphérique. In this simple arrangement the passenger car hangs from a cable which is slung between steel towers and driven by a stationary engine. In the great Alpine téléphériques, however, the passenger car hangs from a trolley that runs along one cable and is hauled by another cable.

Going Underground

The future in urban mass transportation rather seems to lie underground. Underground-railway, or subway, systems have already proved their worth in cities throughout the world and are capable of handling prodigious volumes of rush-hour traffic. The world's busiest subway system, in New York City, carries over 2000 million passengers a year between its 462 stations.

The oldest (since 1863), and still the largest subway system is that in London, which has over 400 km (250 miles) of track (though not all of it stays underground). Like many established subway systems the London Underground is gradually expanding its routes and in the 1970s added the Victoria and Jubilee lines. Among modern subway networks San Francisco's Bay Area Rapid Transit (BART) system (1975) is noteworthy, though it has been plagued with troubles and has not even begun to attract the traffic envisaged.

Electric traction is now universal on subway systems, with the notable exception of the one in Haifa in Israel, which is cable-worked. In the majority of cases the electricity is picked up from a third rail, though in some an overhead conductor is used. Subway trains do not have separate locomotives; the power units are incorporated in some of the passenger cars. The cars usually have flanged wheels like ordinary trains which, in the confines of the tunnels, creates an awful noise. The Paris Métro, however, runs softly on rubber-tyred wheels. Only at crossing points do the trains descend onto orthodox flanged wheels.

The latest subway lines make extensive use of automation. Automatic ticket-dispensing and checking machines, lifts, escalators and even travelling platforms may be used to speed the transfer of passengers into and out of the stations. The sliding doors of the driverless, lightweight trains open and close automatically. The trains follow one another at intervals and at speeds dictated by a flexibly programmed computer, which responds to data continuously fed to it from sensors and controllers along the route.

The sailing boat uses most primitive, elemental technology compared with the space-age Concorde (opposite) and 'hovering' air-cushion craft (below). But it will outlive these technological upstarts because it is powered by a renewable energy source – the wind.

It is difficult to imagine yourself travelling faster than a rifle bullet and more than twice as high as Everest, but you do when you ride in Concorde. This silvery dart of a supersonic airliner can whisk you across the Atlantic in under four hours. If you fly west, you travel so fast that you beat the Sun, and according to local time you arrive before you set out!

Concorde represents the pinnacle of aeronautical engineering, but it is too costly to manufacture and to operate on a large-scale. The skies today are dominated by the wide-bodied 'jumbo' jets. They are much slower than Concorde, if you can call 960 km/h (600 mph) slow, but they are propelled by more efficient fan-jet engines and can carry over 400 passengers at a time.

Compared with aircraft, and indeed most other forms of transport, conventional ships are incredibly slow. No significant increase in their speed occurred for centuries, until recently. Today novel kinds of craft are taking to the seas which are capable of speeds up to 150 km/h (90 mph). They can variously be described as low-flying boats or surface-skimming aircraft. They are the hydrofoil, which 'flies' on underwater wings, and the hovercraft, which rides on a cushion of air.

But it is as long-distance freighters that ships excel. Nothing can transport cargo in bulk as economically as ships. Fortunately the ship, one of man's early great inventions, is here to stay. It will be with us long after the car, lorry, train and plane have quit the transportation scene, due to lack of fuel. It will simply revert once again to sail for propulsion, tapping the energy blowing in the wind.

On the Move——Sea and Air

Shipshape

Considering it was one of Man's earliest great inventions, the ship took a long time to evolve into an efficient form of transportation. Broadly speaking there were few significant differences between the design of an ancient Egyptian vessel of, say, 2000 BC and a Viking longship of AD 1000. They were open, fairly flat-bottomed vessels propelled principally by oars. They had a single mast set with a square sail for use when running with the wind. They were steered by means of a large oar over one side.

Over the next 400 years, however, a succession of innovations transformed the ship into a truly ocean-going craft and prompted sailors to blaze new trails across the oceans, to discover 'new' worlds. These innovations included the adoption of the triangular, lateen sail, which made it possible for ships to sail close to the wind: the stern rudder; extra masts; and raised platforms, or 'castles' in the bows and stern.

Like the lateen sail and the castle, the multimast configuration was common in the Mediterranean regions long before it reached northern waters. But by the early fifteenth century a three-masted arrangement was near universal. In early designs there was a short foremast in the bows set with a small square sail. Amidships was the tall mainmast set with a large square sail. Aft of the mainmast was a mizzenmast, carrying a lateen sail. Alternatively, all three masts carried lateen sails.

Carracks, Caravels and Clippers

Typical of the ships of the period were the broad-beamed carracks of Italy and Spain and the slighter-built caravels of Portugal. The word 'caravel' is derived from the term carvel, the method of construction used for the ships. In this method, the hull was fashioned from planks butted edge to edge over a supporting skeleton of ribs. The joints were caulked with pitch to make them watertight. This method, developed in the Mediterranean, contrasted with the overlapping-plank, or 'clinker' construction of northern regions.

In the early 1500s much bigger carrack-type ships were built, such as the British *Great Harry*, properly named *Henri Grace de Dieu*, and Scotland's *Great Michael*, which measured 61 metres (200 ft) long. They were gradually superseded by the smaller galleons, whose high-stepped stern castle, or poop, made them look top heavy. The forecastle was set back behind a long projecting 'beak'. They usually had three decks and sometimes four masts. They also carried heavy armament of cannon, which were fired through gun ports in the sides of the ship.

In the 1500s and 1600s the Portuguese, Spaniards, Dutch, French and British fought fitfully for supremecy of the high seas. They opened up profitable trading routes particularly to the East. Heavily rigged cargo vessels called East Indiamen were built for the Far East routes. They resembled the fighting ships of the period and carried many cannon for defence. By now the high poop and forecastle of earlier vessels had given way to a more even deck line.

In the 1700s the rigging of ships was progressively refined and extended, culminating in the classical full-rigged ship of the early 1800s. In this rig the foremast, mainmast and mizzen carried up to six square sails each. The former lateen sail on the mizzenmast was converted to a gaff sail set on a boom. Triangular jib sails were set at the bows, and there were also triangular staysails set between the foremast and mainmast.

The Age of Sail reached its zenith in the mid-1800s in the shape of the swift tea clippers, which plied between the Far East and Europe and America. In favourable conditions they could reach speeds of over 20 knots.

Rise of the Steamers

But by then the days of sail were all but numbered. Ships powered by steam were already crossing the Atlantic. The steam revolution really got under way in 1801, when a simple paddle-wheel craft, the *Charlotte Dundas*, worked for a time in Scotland towing barges. Six years later in the United States Robert Fulton inaugurated with his *Clermont* the first passenger-carrying steamboat service, between Albany and New York.

Soon steam engines driving side paddle wheels were fitted to ocean-going ships. In 1819 the US ship *Savannah* made a transatlantic crossing partly under steam. In 1838 the British ships *Sirius* and *Great Western* made the crossing entirely under steam. But like many steamships to follow they still carried copious sails.

In 1845 the British ship *Great Britain* ushered in the age of the modern ship. It was built of iron and was propelled by screw propeller, the invention of Swedish inventor John Ericsson six years previously. It crossed the Atlantic at an average speed of 9 knots. Designed by Isambard Kingdom Brunel, it was the largest ship afloat until Brunel built *Great Eastern*, a remarkable vessel propelled by both paddle wheel and propeller. It measured no less than 211 metres (692 ft) long, had a beam of 25 metres (82 ft) and displaced over 18,000 tonnes.

The early steamships were powered by reciprocating piston-in-cylinder steam engines. In 1894, however, Sir Charles Parsons demonstrated with his speedy craft *Turbinia*, the superiority of his steam turbine for marine propulsion. By 1910 the majority of ocean liners were turbine-powered. A few years later oil began to replace coal as the preferred fuel for shipping. The oil engine, or diesel, also began to be used to power small vessels.

The 1950s saw the introduction of nuclear-powered vessels. First was the American submarine *Nautilus* (1954). The first non-military nuclear ship was the Russian icebreaker *Lenin* (1959), followed three years later by the US merchant ship *Savannah*. The high capital cost of nuclear-power vessels has prevented their large-scale use.

AD 800s–1000s
Viking longship
1492
Christopher Columbus's
'Santa Maria'
1620
Pilgrim Fathers'
'Mayflower'
1700s
East Indiaman
1807
Robert Fulton's 'Clermont'
1845
Brunel's
'Great Britain'
1894
Parson's speedy
'Turbinia'
Mid-1800s
British tea clipper
1912
The 'unsinkable'
'Titanic'
1959
Cockerell's SRN1 hovercraft
1952
The 'United States',
holder of the Blue Riband
1953
Piccard's
bathyscaphe
'Trieste'.
1962
Nuclear merchant ship 'Savannah'

Above: The German vessel 'Gorch Fock', one of the handful of 'tall ships' remaining on the high seas. She (ships are always feminine) is a barque, like most of her sister ships, characterized by the lateen sail on the rear, or mizzenmast.

Under Sail

Only a small number of large sailing ships exist on the high seas today. The majority are training ships used to introduce young people to the art of sailing, often as part of their naval training. Only a handful are full-rigged – have square sails set on all masts. Strictly speaking, in sailing parlance, these are the only ones that should be termed 'ships'. Many of the large sailing vessels are barques. They have a lateen-type, or fore-and-aft sail, on the mizzenmast. This makes them easier to handle than the full-rigged ships.

Smaller commercial sailing vessels still operate in many parts of the world, however. They include designs that have remained virtually unchanged for many centuries. The felucca, or Nile boat, still retains the characteristic long lateen sail introduced a thousand years ago. The Chinese junk remains one of the most efficient sailing vessels ever developed. It has a flat bottom, high stern and bamboo-stiffened sails.

The many small sailing vessels that once went fishing or carried cargo around the coasts have all but disappeared. But their diverse rigging is preserved in many privately owned pleasure craft, or yachts.

Below: In complete contrast is the Chinese junk, which predates the barque by several centuries and is still in widespread use, as here in Hong Kong harbour.

Sailing Rigs

There is often much confusion over the correct terminology for the rigging of sailing yachts – that is, the arrangement of masts and sails – but the following is broadly accepted. A sloop is a single-mast boat with a mainsail and a headsail, or jib. A cutter has in addition a second headsail. The schooner, ketch and yawl have two masts and carry three or more sails. A schooner has the foremast the same size as, or shorter than, the mizzenmast. A ketch has the mizzenmast smaller than the foremast and well forward of the stern, while the yawl has the mizzenmast right in the stern.

On each mast the sail may be gaff or Bermudan. In a gaff rig the sail is supported at the top by a spar called a gaff, which extends at an angle from the mast. In a Bermudan rig the sail is a tall and narrow triangle extending right up the mast. The Bermudan sail is easier to handle and enables a boat to sail closer to the wind. It has become the standard rig for racing yachts.

The main parts of a modern yacht are shown on the right. It is a Bermudan-rigged cutter. The ropes and wires that are attached to the mast and sails have various names. 'Sheets' are the ropes used to trim, or adjust, the sails. 'Halyards' are the ropes that raise and lower the sails. 'Stays' support the mast for and aft, while 'shrouds' support it at the sides.

The sails of most yachts are now made of man-made polyester fabric such as Dacron and Terylene, which is rot-proof and keeps its shape well. The sheets and halyards may be made of polyester or nylon. While some yachts are still built traditionally with wooden hulls, the majority have hulls made of fibreglass, or glass-reinforced plastic (GRP). They are fashioned by building up successive layers of synthetic resin and glass-fibre matting on a polished wooden mould. A catalyst is mixed with the resin to cure, or harden it. Fibreglass hulls are strong and light and free from rot and rust.

Some hulls are built in steel or aluminium; others in ferrocement. Ferrocement hulls are made by squeezing smooth concrete through a close-mesh frame of steel pipe, rods and chicken wire.

Large yachts have a fixed keel, which is weighted, or ballasted with lead or concrete. The keel of a yacht is designed to resist the sideways thrust on the hull caused by the action of the wind on the sails. It is ballasted so that it counteracts the overturning effect of the wind on the sails. Small yachts, such as single-sailed dinghies or catboats, have a movable keel, or centreboard, which can be lowered or raised as need be. They have no ballast, and when they heel over in the wind, their crew must lean out over the side to counterbalance the wind pressure.

Flexible Wings

The action of a sail propelling a yacht is quite complicated. In some respects it behaves like a turbine, in others like a wing. When the wind is

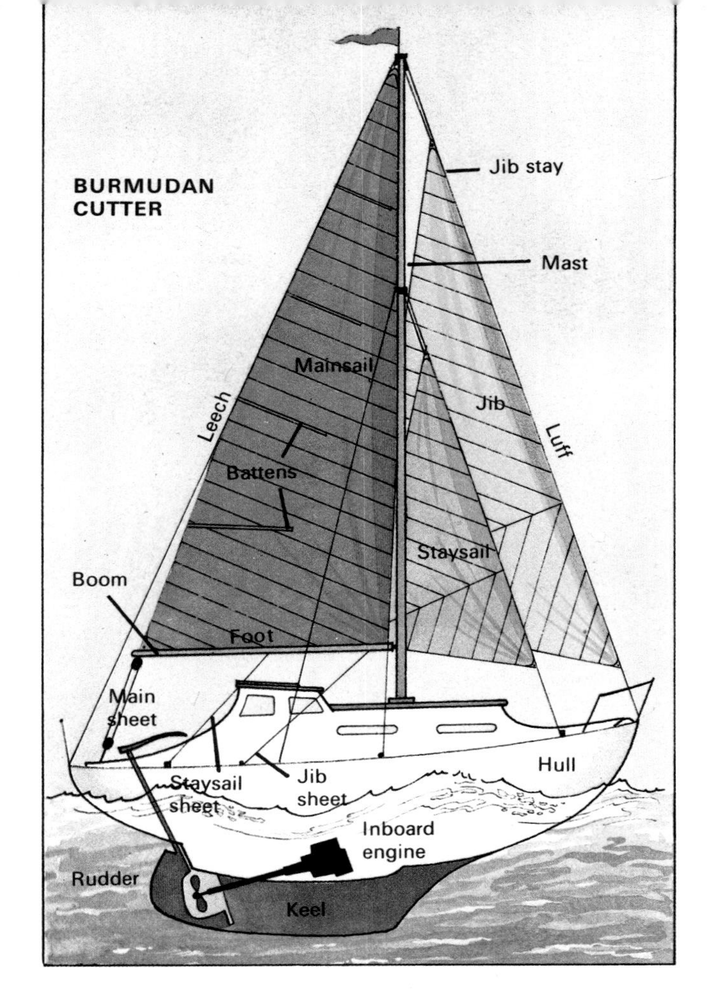

immediately behind the direction of travel and the yacht is said to be running, the sail is extended nearly at right-angles to the boat and the yacht is pushed along by the wind, rather like an impulse turbine is spun by a water jet (see page 48). When the yacht is sailing at an angle to the wind direction, the sail deflects the airstream and achieves some of its forward thrust by reaction, rather like a reaction turbine.

When sailing at an angle to the wind, however, the main forward propulsion comes from the aerodynamic properties of the sail. Under the pressure of the wind, it flexes to form a curved aerofoil shape like a plane's wing. And it develops a thrust, just as a plane's wing develops lift. The thrust on the sail can be resolved by the parallelogram of forces into two forces at right-angles to one another. One (the drive) acts in the direction the boat is travelling: the other (the leeward) at right-angles to it. The resistance of the water on the hull and the keel helps prevent the yacht moving to leeward, while the drive propels the boat forwards.

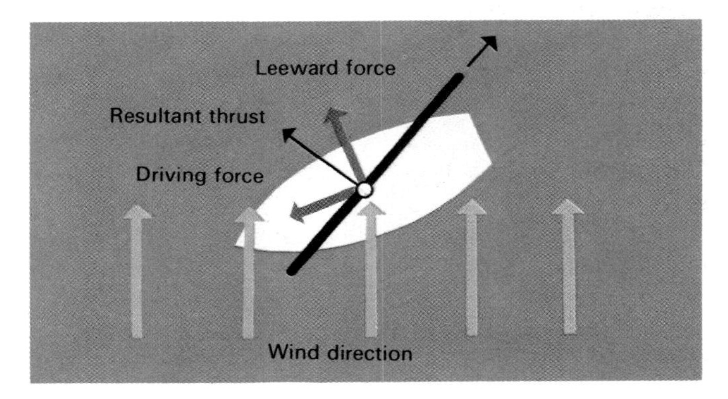

A yacht moves fastest when sailing across the wind – with the wind abeam. This motion is called reaching. It cannot sail directly into the wind because in this condition the sail flaps uselessly like a flag. But it can sail indirectly against the wind, following a zig-zag path, by a method known as tacking. This movement is called beating to windward or beating-up. The yacht zig-zags approximately at 45° to the direction of the wind. It travels first with the wind blowing from starboard (starboard tack), then 'comes about', or turns through a right-angle so that the wind now blows from the port side (port tack). The sail boom swings quickly across the yacht as it comes about and must be avoided at all costs!

Under Steam

Though many far-sighted mariners predict a gradual return to sail for ship propulsion in the fuel-hungry years to come, most ships for the meantime will continue to be driven by screw propeller. Steam-turbine or diesel engines will provide the motive power.

As a visit to a busy port will show, ships come in very many shapes and sizes. They are each designed for a particular role on the high seas – carrying passengers or general or specialist cargo; trawling for fish; towing larger vessels into port; dredging harbour channels; and so on. Each specialist use dictates the form of superstructure and deckline, the below-decks layout, and the kind of propulsion and auxiliary machinery.

The superstructure of a cruise liner, devoted entirely to carrying passengers, is extensive to accommodate the bulk of the cabins and public rooms above the waterline. Lifeboats line the sides. General cargo vessels, or freighters, on the other hand have comparatively little superstructure, except in perhaps one or two places, typically amidships above the engine room and aft near the stern. The former carries the navigation bridge and main accommodation, while there is extra accommodation aft. Most of the deck area is relatively uncluttered and carries numerous derricks, which are simple cranes used to swing cargo into and out of the holds beneath the deck. Each derrick is powered by a steam or electric winch.

Bulk-cargo carriers have even less superstructure, usually placed with the engines, aft. They carry such things as ore, coal, oil and liquefied gas. Of paramount importance and of the largest dimensions are the oil tankers which transport crude oil from the Middle East to Europe and North America. Tankers are by far the world's largest ships. They are built very long because the longer the ship, the more efficiently it can be propelled. The French tanker *Batilus* in 1976 became the first ship to exceed 500,000 tonnes deadweight (deadweight being the term used to denote cargo-carrying capacity). Some

Right: Fishing boats moored in sheltered waters. At sea it is often a very different story, with the fishermen battling against gale-force winds and mountainous seas to net a worthwhile catch.

Far right: The loading and unloading of cargo at ports continues day and night. Here cargo is being unloaded from a container ship by specialized handling gear.

Top right: The 'Queen Mary', one of the most famous of the great transatlantic liners, is now moored at Long Beach in California, where she serves as a hotel and conference centre. Of 81,237 gross tons and 310·6 metres (1019 ft) long, she was launched by Queen Mary herself in 1934 and made her maiden voyage two years later.

Below: Passenger liners these days are used mainly for holiday cruising.

BUOYANCY AND STABILITY

It is a mystery to some people how a ship, made of steel and carrying thousands of tonnes of cargo, can remain afloat. The answer lies with Archimedes' principle. Archimedes stated that when a body is immersed in water (or any other fluid for that matter), it experiences a force called upthrust. The force is equal to the weight of water displaced by the body.

If a body displaces a volume of water that weighs *less* than it does, that body will sink because the upthrust on it is not enough to support it. If a body displaces a volume of water that weighs *the same* as it does, then that body will float. The upthrust on it will equal its weight. So for a ship to float, it must have an underwater volume large enough to displace a weight of water equal to its own weight. The size of warships is always given in terms of their displacement – the tonnage of water they displace.

The weight of a ship acts vertically downwards through an imaginary point called the centre of gravity, while the upthrust on it acts vertically upwards through a centre of buoyancy. When the ship is at rest, the forces are in equilibrium, and the two centres lie on the same vertical line (Diagram 1). For a ship to be stable, it must always tend to return to an upright position after it has been forced to heel over by, say, wave motion. The heeled-over state is shown in Diagram 2, which shows the centre of buoyancy shifted to a new position. If we extend a vertical line upwards from the new centre of buoyancy to the original vertical line through the centre of gravity and original centre of buoyancy, it meets at a point called the metacentre. As long as the metacentre remains above the centre of gravity, then the weight of the ship acting downwards and the upthrust on it acting upwards at the new centre of buoyancy form a couple that tends to right the ship. If the metacentre should fall below the centre of gravity, the resulting couple will tend to make the ship heel over more and capsize.

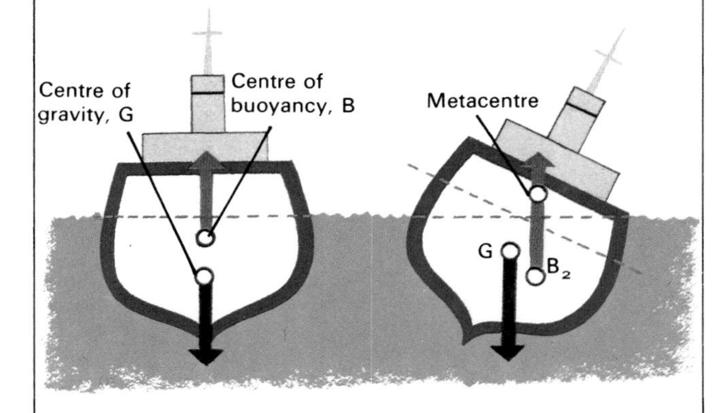

400 metres (1312 ft) long, she has a beam of 63 metres (207 ft) and a draught of 29 metres (95 ft).

Such is the bulk of supertankers that their inertia at cruising speed (about 16 knots) is enormous, and they require several kilometres in which to stop. It makes them impossible to manoeuvre suddenly and at close-quarters and increases the risk of collision at sea.

Much cargo is now transported worldwide in boxed, standard-sized containers, and special ships designed to carry containers have evolved. Containerization is a very efficient way of handling mixed cargo, the containers being readily transferred between ship and lorry or rail car by purpose-built container-handling equipment. An extension of the same principle is provided by the development of the more recent LASH (lighter-aboard-ship) vessels. These are designed to carry standard-sized lighters or barges, containing any cargo. They are equipped with powerful travelling cranes, which can lift the lighters bodily out of the water and stow them on deck.

Basic Design and Construction

Practically all ships these days have hulls made of steel. They are constructed by welding together thousands of steel plates on a skeletal steel framework. Whereas in the past a ship was built from the keel upwards, these days it is usually built in separate sub-assemblies which are then welded together to form the whole. The sub-assemblies will include the outer hull, horizontal deck plates, and vertical bulkheads.

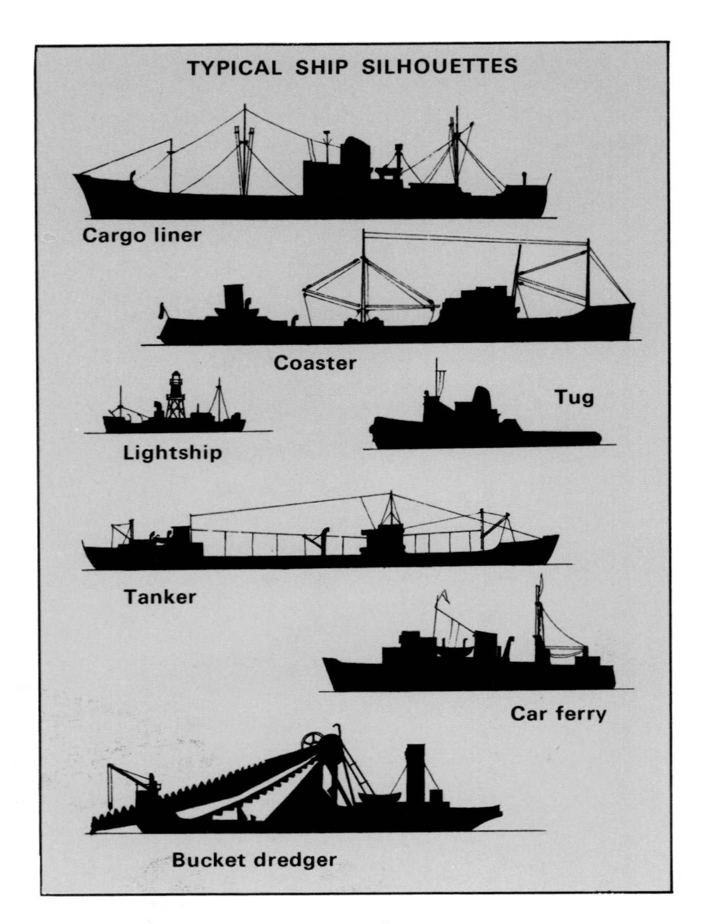

Above: From the silhouette of a ship we can often tell what its function is. For example, the cargo liner has accommodation in the superstructure amidships, with a clear deck forward, with derricks, for handling cargo. The tug has a flat stern, to facilitate towing.

The bulkheads not only embue the hull with strength but also provide watertight seals between the various compartments. If the hull of one compartment is holed, then the compartment can be sealed, preventing the rest of the ship being flooded. For a similar reason the hull is given a double bottom. This confers great strength at the base and also provides storage space for fuel oil, fresh water, or water ballast.

The deck and bulkhead method of construction give the completed hull great structural strength. This is abundantly needed in a ship at the mercy of the elemental forces of wind, wave and current. The hull is subjected to stresses caused by its own weight, the upthrust of the water, wave impact, vibration and the ship's continuous motion of rolling (side to side), pitching (rocking back and forth) and yawing (slewing left to right).

The undulations of the waves also create stresses for they cause some parts of the ship to be lower in the water than others. The parts more submerged will experience more upthrust than the others, setting up stresses. With a wave crest amidships, the ship will experience greater buoyancy there than at bow and stern, which will tend to drop, the tendency being called sagging. Conversely when a wave trough is amidships, the ship will tend to droop there while bow and stern are buoyant, this tendency being called hogging.

The shape of the hull, particularly at and below the water-line, has a marked effect on the performance of a ship and dictates the power requirements for propulsion. When a ship is in motion, skin friction between the hull and the water causes resistance to that motion. The movement of the ship also creates pressure in the water, which manifests itself as the characteristic bow wave. This is a fine spectacle but wasteful of energy. Energy is also dissipated by eddy currents formed in the ship's wake. Optimum designs for a hull shape are usually obtained after extensive hydrodynamic testing of scale models in large water tanks. In these tanks the models are towed by a travelling bridge, while wave-making devices simulate a variety of surface conditions.

Propulsion Units

The commonest forms of marine power units are the steam turbine and diesel engine. Their mode of operation is explained in the chapter 'Power to the People'. A large liner would normally be powered by twin steam turbines, developing between them tens of thousands of horsepower. The turbines of Cunard passenger liner *Queen Elizabeth 2*, for example, develop a total of 110,000 horsepower.

The steam turbines are supplied with steam by marine boilers, of which the water-tube type is most efficient. In this type the water is heated as it passes through tubes located in a furnace. It contrasts with the older type of fire-tube, or Scotch boiler, in which hot furnace gases are drawn through tubes surrounded by water in a boiler, as in a steam locomotive (see page 94).

In a typical water-tube steam generator, oil is atomized and burned in a supply of air in a furnace surrounded by arrays of water-filled tubes. The

Right: Construction scene in a Japanese shipyard, where oil tankers are being built. As in all shipyards these days the hull is constructed by prefabrication. A tanker is divided into many separate tanks by longitudinal and cross bulkheads. Within each tank are further 'wash' bulkheads to prevent undue wave formation, which on a large scale could make the ship unstable.

Left: This is a purpose-built drilling vessel used to drill exploratory wells in offshore oil fields, as here in the North Sea. It has additional manoeuvring engines fore and aft to help it remain on station while drilling.

water absorbs heat and turns to steam in a steam drum. From there the steam is piped through the superheater located in the path of the outgoing furnace gases and thence into the turbines. In a typical installation some 80,000 kg (175,000 lb) of water are evaporated per hour, and the steam is delivered to the turbines at a temperature of up to 500°C and at a pressure of up to 70 atmospheres.

Because steam turbines need to rotate very rapidly for greatest efficiency but propellers need to turn relatively slowly, reduction gearing is required in the drive. Usually double reduction gearing is needed to bring about the required reduction, which may be as much as 50:1 (for a turbine operating at 4000 rpm driving a propeller at 80 rpm). Reduction gearing is also needed with diesel-engine power units, though a single reduction gear usually suffices since the speed of the diesel is much lower.

Experiments have also taken place with diesel-electric drive. Diesel-driven generators produce electricity, which then powers electric motors to spin the propellers. But the method has found little application. Neither has gas-turbine propulsion, except in a few fast naval patrol boats.

With a few notable exceptions nuclear propulsion, too, has been restricted to naval vessels because of high initial capital costs. Notable exceptions have included the US ship *Savannah*, the German ore carrier *Otto Hahn*, and the Russian icebreakers *Lenin, Artika* and *Sibir*. Nuclear propulsion has found most application in submarines (see below), though it is also used in a few surface vessels such as the giant American aircraft carrier *Nimitz*.

Beneath the Waves

With oceans covering over two-thirds of the globe, it was inevitable that Man should seek ways to explore the underwater world. The earliest means was by diving bell, a primitive form of which was mentioned by Aristotle in the 200s BC. But not until the 1700s did it become a practical tool. The diving suit soon

SHIPS' PROPELLERS

The design of ships' propellers is very critical. Though they can be of enormous size – measuring 10 metres (33 ft) or more across and weighing in excess of 30 tonnes – they are the sole means of propelling ships several hundred metres long and weighing tens of thousands of tonnes. The blades of a propeller usually have an aerofoil section rather like an aeroplane propeller, but they are very much broader. As they rotate they accelerate a column of water towards the rear. Reaction to this accelerating column (by Newton's third law) drives the propeller and thus the ship forwards.

The marine propeller in effect screws itself through the water much as a metal screw bites through wood. The tip of the propeller describes a helical curve as it advances through the water. The distance it would advance per revolution if it were passing through a solid medium is called its pitch. In practice, since water is a highly mobile medium, the screw 'slips' as it rotates. The amount of slip is about 20–40%, depending on the efficiency of the propeller.

Theoretically a given propeller performs at optimum efficiency at one ship speed only, but of course ships need to travel at different speeds in different conditions. The answer to the problem may be to fit a propeller whose blade angles can be varied. The pitch can then be made finer – for manoeuvring in harbour, for example – or coarser – for cruising – as required. This type of propeller, known as a variable-pitch propeller, is also widely used in aeroplanes.

Most marine propellers are made of corrosion-resistant manganese bronze and have between three and six blades. The speed at which they turn varies from ship to ship. Speeds of about 80–120 revolutions per minute are most common. Greater propulsive efficiency is achieved with larger, slower-speed propellers.

One phenomenon lower-speed propellers are less prone to is cavitation. This is a condition in which at high speeds of rotation bubbles of water vapour tend to form on the low-pressure, or suction, side of the propeller blades. The bubbles disrupt the smooth water-flow pattern around the blades, and the propeller becomes less efficient. Also, as the bubbles collapse, they tend to set up vibration and cause erosion of the blade surface.

Right: Pilot's eye view of the nuclear-powered aircraft carrier 'Nimitz', which has the colossal displacement of 92,800 tonnes. Note the angled landing deck now universally used on carriers.

Top right: The propeller is cast in manganese bronze and is corrosion resistant, unlike the ship's steel hull, which must therefore be painted.

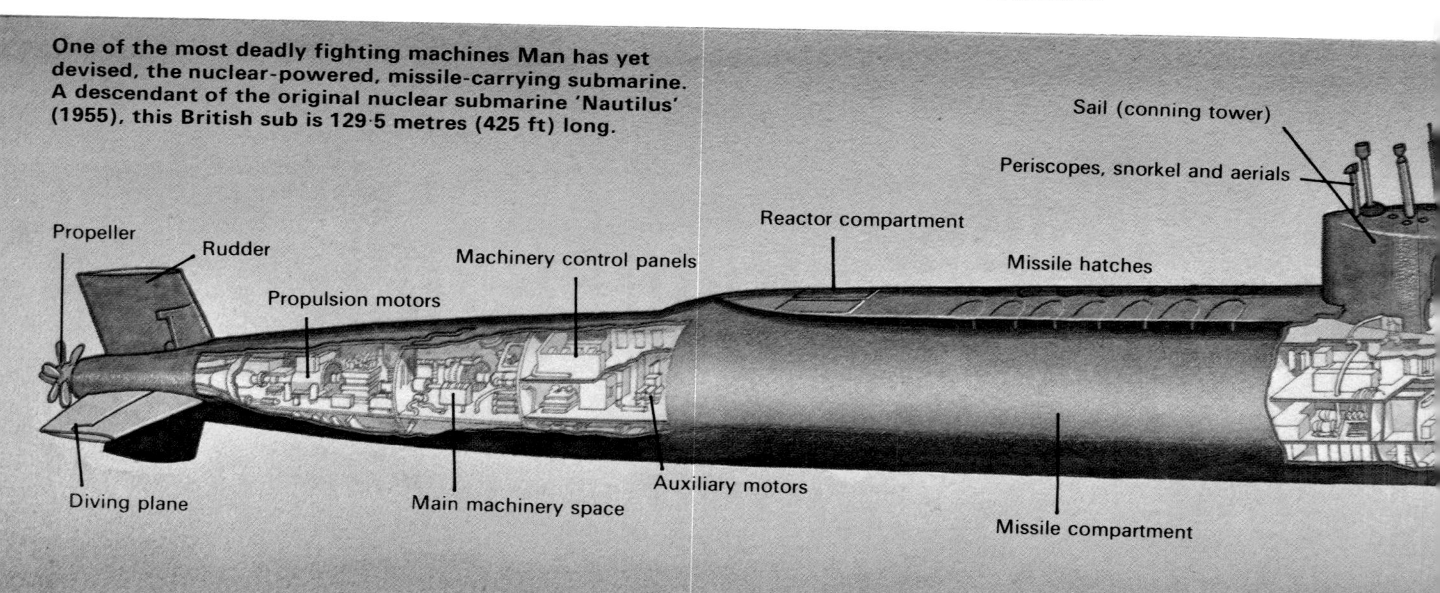

One of the most deadly fighting machines Man has yet devised, the nuclear-powered, missile-carrying submarine. A descendant of the original nuclear submarine 'Nautilus' (1955), this British sub is 129·5 metres (425 ft) long.

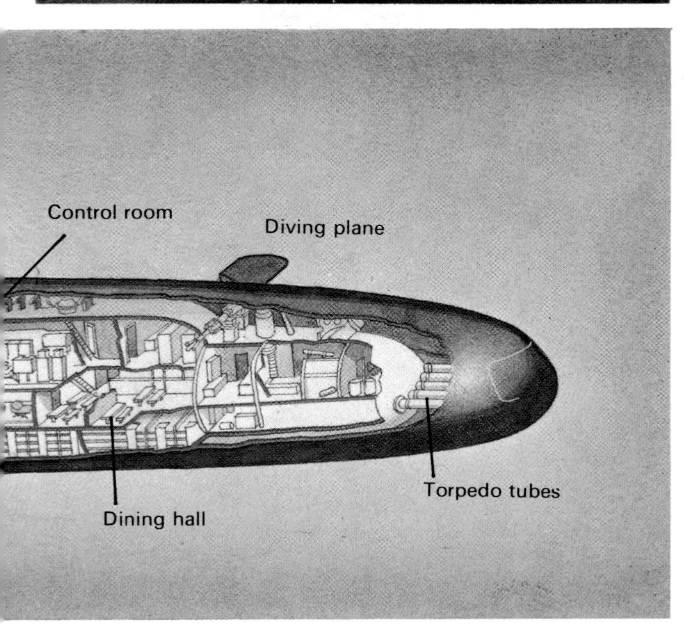

followed, but those who wanted greater underwater mobility began experimenting with submarines. Until comparatively recently submarines were operated almost entirely by the world's navies. But now new types of small submarine craft known as submersibles have evolved to further undersea exploration generally and to act as support vessels for the offshore oil industry in particular.

Submarines have a history dating back over 200 years. David Bushnell's *The Turtle* (1776) has the distinction of making the first underwater attack, during the American War of Independence. In World War 2 submarines created havoc among surface shipping and today they have developed into the most deadly weapon the world has ever known.

The most advanced vessel afloat in 1980 was the 170-metre (560-ft) long American submarine *Ohio*. Like all the latest naval submarines it is nuclear powered. It is armed with 24 nuclear missiles, which it can fire while still submerged. These missiles, known as Trident, carry a dozen nuclear warheads, each of which can be deployed to a different target and is capable of devastating a large city. The missiles are propelled to the surface by high-pressure gas, their rocket motors igniting when they enter the air.

Both the American and the Russian navies have built up large fleets of nuclear-powered submarines, and the Russians also possess a vast fleet of conventional, or non-nuclear vessels. But, nuclear or conventional, all submarines have many basic design features in common and function in essentially the same way.

In a conventional design the submarine consists of a welded steel pressure hull of circular cross-section for maximum strength. It is typically cigar-shaped. Outside the pressure hull is an outer hull, the space between providing room for water-ballast and fuel tanks. Alternatively the water-ballast tanks are slung like panniers in bulging tanks to the sides of the pressure hull.

Some nuclear submarines are of a rather different design. Many have their water-ballast tanks within the pressure hull. The American 'Thresher' and 'Sturgeon' class of nuclear submarines have a teardrop shaped hull, the bulbous part being at the bows. This design conforms more closely to an ideal streamlined profile than the cigar-shape, and allows the submarine a maximum speed well in excess of 30 knots. The latest Russian 'Alpha' class submarines are known to have an underwater speed of 40 knots. They are believed to be constructed in the strong, lightweight metal titanium and to have a greater operating depth than any Western submarines – down to 1200 metres (4000 ft).

Projecting above a submarine's hull is the structure known as the conning tower, or sail. It houses retractable periscopes, antennae and 'snorkel' (breathing) tubes, and provides a bridge for navigation on the surface. At the stern is a single propeller for propulsion. There also is the cruciform arrange-

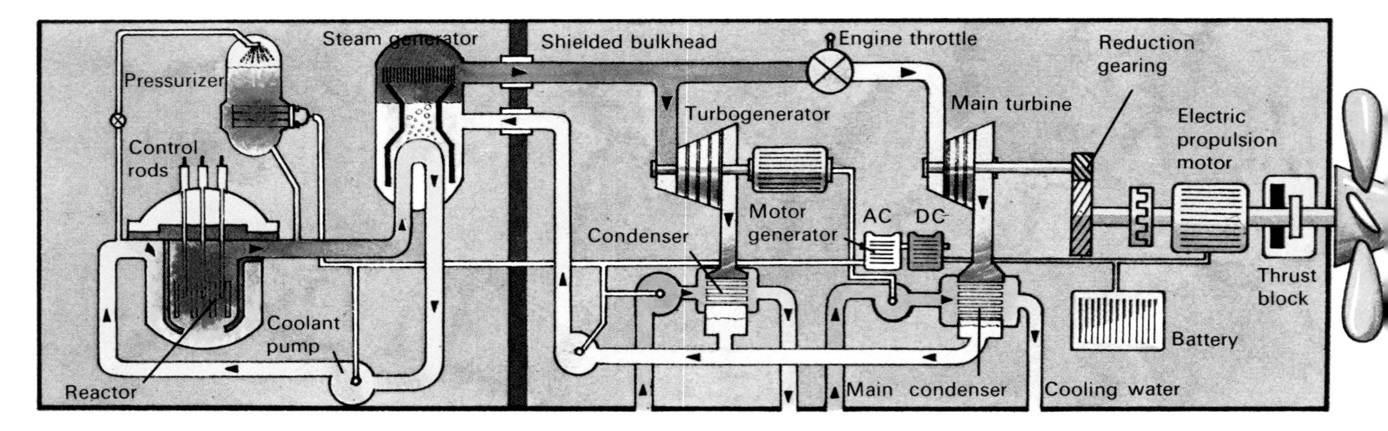

Diagram showing the principal components of the power-generating machinery of a nuclear submarine. Splitting uranium atoms provides the heat to raise steam, which is then fed to propulsion turbines and a turbogenerator.

ment of vertical rudders and horizontal diving planes, also called hydroplanes. Another pair of fin-like diving planes is located near the bows or on the sail. For diving and surfacing submarines use the flooding or venting of the main water-ballast tanks in combination with forward motion and inclination of the diving planes.

When a submarine is on the surface, its ballast tanks are filled with air, giving it positive buoyancy. To dive, the air is released from vents in the top of the tanks, and seawater floods in through 'free-flood' holes at the bottom. With the main tanks full of water the submarine achieves neutral buoyancy in which its weight is just balanced by its displacement. In this state forward thrust by the propellers and downward inclination of the forward diving planes cause it to dive. To maintain neutral buoyancy underwater as the submarine's weight changes – for example, after using fuel or firing torpedoes – water is displaced from smaller water-ballast tanks called trim tanks in various parts of the vessel. Water is also transferred between the tanks to keep the vessel on an even keel.

To surface the diving procedure is reversed. The main ballast-tank vents are closed, and compressed air (at some 25 atmospheres pressure) is forced in to expel the water through the free-flood holes. The diving planes are set for surfacing, and forward thrust by the propeller brings the vessel to the surface.

In conventional submarines motive power for propulsion on the surface is provided by a diesel engine. The diesel may drive the propeller directly or turn a generator to produce electricity. The electricity is then fed to electric motors to turn the propeller. Electricity is also fed to charge batteries which provide motive power when the submarine is submerged. The diesel engine, requiring air for combustion of its fuel, cannot of course be used then. But it can be used at periscope depth, 'breathing' through snorkel tubes.

Nuclear submarines are powered by a nuclear reactor of the pressurized-water type. The diagram above shows the schematic layout of the nuclear

Above: A North Sea diver checks his SCUBA (self-contained breathing apparatus) equipment before a dive. He will descend to working depth in the diving bell behind him. The diving bell is no longer open as it used to be. It is now sealed at the surface, retaining surface pressure until the diver is ready to start work. Then he opens a hatch and the bell compresses to ambient pressure.

Right: A Vickers 'lockout' submersible about to be launched from the gantry of its support vessel. A wet-suited diver stands near the crew compartment access hatch. In front is the large transparent dome of the viewing port.

propulsion system in American-designed submarines. Heat is produced in the reactor core by the fission of uranium atoms. Pressurized water coolant circulates through the core and extracts the heat. It passes through a heat-exchanger unit, the steam generator, where water is boiled into steam. The steam is piped to conventional steam-turbine machinery, which spins the propeller, and to a turbo-generator, which produces electricity. The steam is then condensed and recycled to the steam generator.

Since the heat-production process does not involve combustion, the system is independent of the outside air. Nuclear submarines can thus remain submerged indefinitely. Oxygen for the crew to breathe is readily available from the electrolysis of distilled sea water. The build-up of carbon dioxide and other potentially dangerous gases is prevented by circulating the atmosphere through various scrubbing agents and filters in a sophisticated air-conditioning system.

The Mini-Sub

The commercial mini-submarine, or submersible, now plays a prominent role in offshore oil operations, in the North Sea for example. It is used to ferry divers down to their site of work on the seabed and provides them with life-support facilities. It is equipped with manipulating arms and attached tools which enable it to perform a variety of tasks in support of the divers, or independently. It is equipped with closed-circuit television, videotape recorders, still cameras and floodlights. Many submersibles can operate down to depths of about 450 metres (1500 ft).

In addition, the submersible is widely used for underwater surveys which are required before laying submarine pipelines or choosing sites for drilling or production rigs. The survey is usually carried out in conjunction with a surface support ship. Both submersible and ship are fitted with sensitive sonar equipment which is employed to pinpoint the position of the submersible at all times. It makes use of devices called transponders – electronic units that are triggered by one signal into emitting another signal. Reference transponders are laid on the seabed along the submersible's route, and the submersible triggers them into emitting sound waves which are picked up by the ship. The signals from the three vary according to the submersible's position.

The latest type of submersible, such as Vickers Oceanics L-class vessels and Intersub's PC 1202, incorporates what is called diver lockout. It has two separate compartments – one houses the crew, who work in one-atmosphere 'shirt-sleeve' conditions. The other is a diving compartment, which can be pressurized to match the external pressure of the sea. Divers enter and leave it through a hatch.

INTERSUB'S PC1202 SUBMERSIBLE

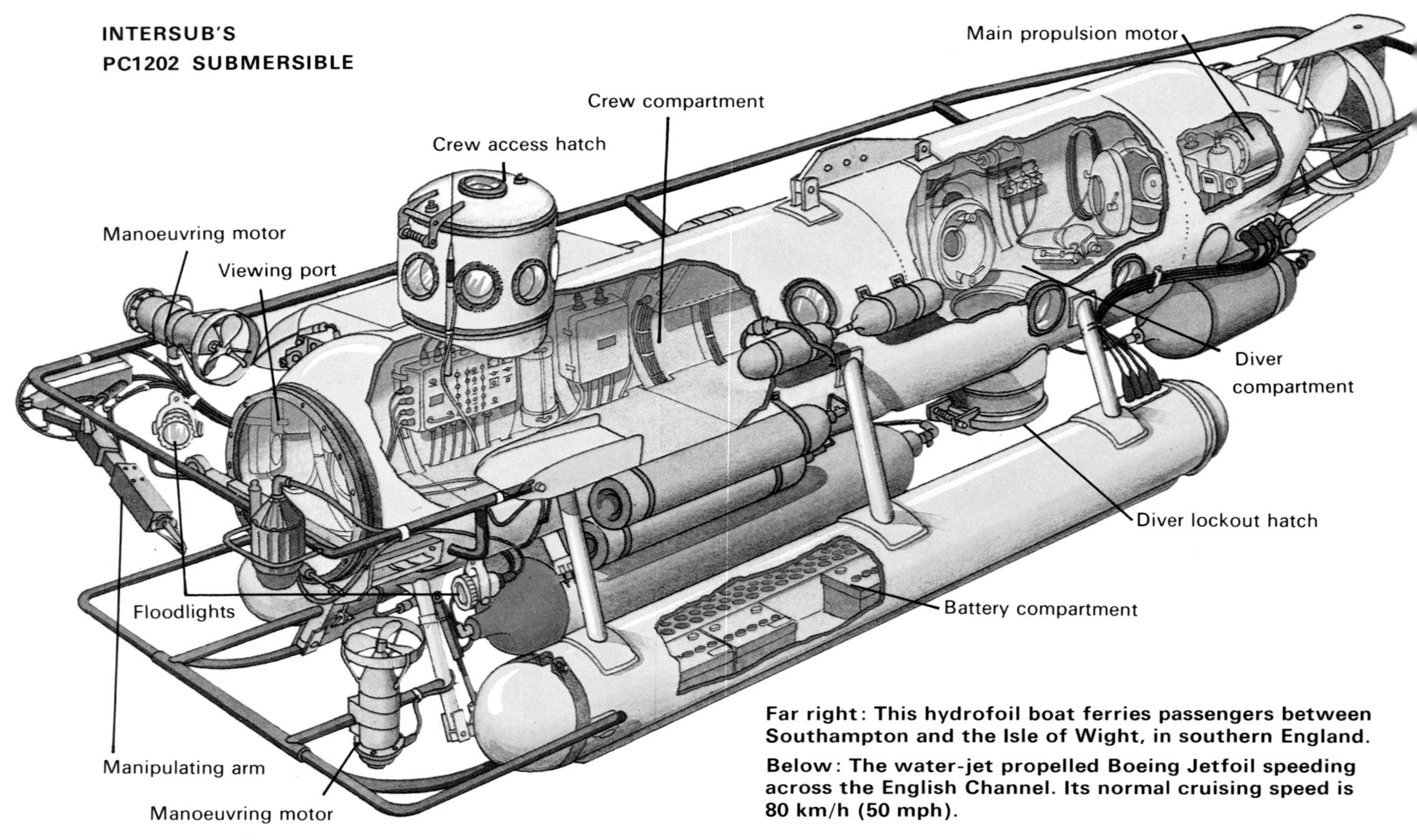

Far right: This hydrofoil boat ferries passengers between Southampton and the Isle of Wight, in southern England.

Below: The water-jet propelled Boeing Jetfoil speeding across the English Channel. Its normal cruising speed is 80 km/h (50 mph).

This kind of submersible has made possible the new breed of 'saturation' diver, who lives and works for long periods in a pressurized environment. 'Saturation' means that his blood is saturated with the gas he is breathing – which for deep dives is a mixture of helium and oxygen. When not diving, he lives in a compression chamber on the surface support ship. This has hatches through which he can transfer to the pressurized diving compartment of the submersible for transport to his underwater work site. By eliminating the lengthy decompression period which deep divers once had to endure, saturation divers can spend more of their time actually working.

Like conventional submarines, submersibles are propelled by a stern propeller, the motive power coming from batteries. In addition, for precise manoeuvring, they have small electric thruster units mounted horizontally and vertically at the bows and often at the stern as well. The majority of submersibles are made of metal, but in 1977 Vickers introduced a glass-fibre craft which has a much longer design life.

Surface Skimmers

Compared with other forms of transport, ships travel very slowly indeed. Even crack liners can manage little more than 30 knots (about 55 km/h, 35 mph), while freighters are often much, more slower. The drawback to travel at sea is that a ship expends most of its power in overcoming the surface friction, or drag, between the ship's hull and the water. Minimize that and the ship will go faster.

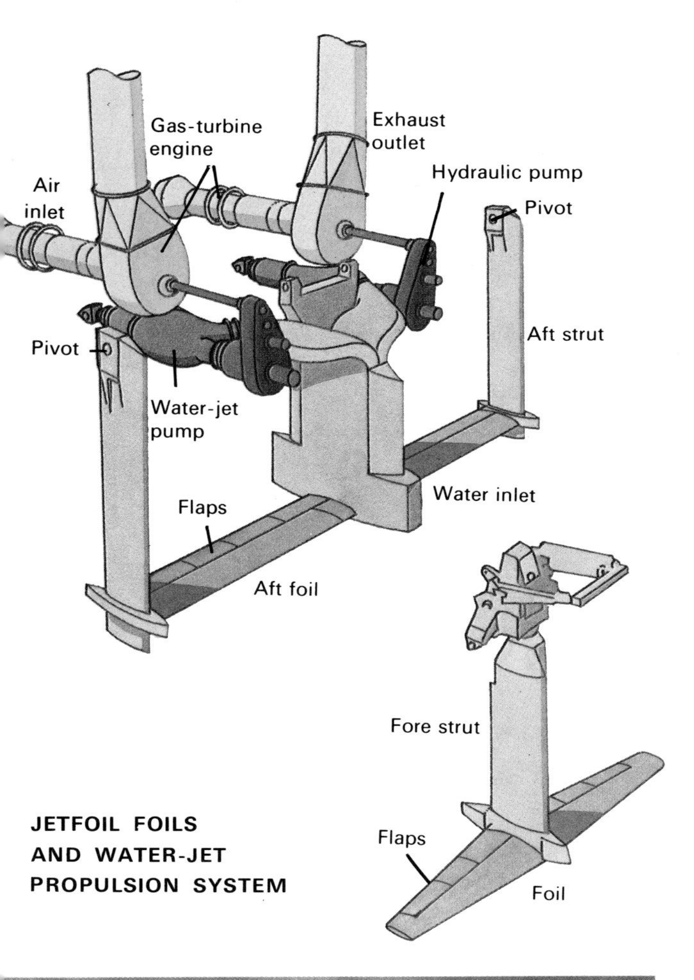

JETFOIL FOILS AND WATER-JET PROPULSION SYSTEM

Speedboats do this by hydroplaning, going so fast that most of the hull rides up out of the water. But this is hardly practical for larger craft. One answer is to fit the hull with underwater wings, and propel it through the water at such a speed that the wings develop enough 'lift' to raise the hull above the surface. However fanciful it sounds, this method is workable and has enabled surface vessels to reach speeds of up to 80 knots (150 km/h, 93 mph). The underwater 'wings' are known as hydrofoils, as are the boats they are fitted to. It is interesting to note that the earliest successful hydrofoil was built by inventor of the telephone, Alexander Graham Bell, whose craft HD4 set the water speed record of 60 knots (110 km/h, 70 mph) in 1918.

The foils are of similar cross-section to an aeroplane wing and develop lift in the same way (see page 124). Because water is a much denser medium than air, the hydrofoils are much smaller than aerofoils. Two main types of foils are used – surface-piercing and fully-submerged. These are illustrated in the two photographs shown on this page.

The hydrofoil in the small picture has surface-piercing foils of V-section, while that in the large picture has its foils fully submerged. The surface-piercing design has greater inherent stability than the submerged type but is affected by wave disturbance. In rough water the foils tend to follow the contours of the waves, pitching the boat up and down. Submerged foils are much less affected, but require an automatic control system to keep them stable. Both pictures vividly demonstrate the great advantage of hydrofoils that they produce scarcely any wash, which is particularly important for inland waterways.

The Russians have had the most operating experience with hydrofoils and currently have upwards of a thousand craft in regular use. Their 'Kometa' class vessels have been particularly successful. In recent years, however, the Americans have come to the fore in hydrofoil design with the revolutionary Jetfoil built by the aircraft manufacturers Boeing. Passengers of the Jetfoil may indeed be forgiven if they think they are flying because they sit in airline type

seats, are served by stewardesses, and are subjected to subdued jet whine. And the ride when foil-borne is uncannily smooth.

Water Jetting

The Jetfoil has its foils arranged in a so-called canard arrangement with a single foil and strut forward and a double foil-strut aft. Both foils are retractable for operation in shallow water. They carry control surfaces, or flaps, at the trailing edge. The forward and aft flaps operate differentially (one up, one down) to provide height control. The aft flaps themselves operate differentially to provide roll control for changes in direction. The forward strut is also steerable.

The control surfaces are operated by a sophisticated computerized automatic control system, which acts upon information supplied to it by sensitive gyroscopes, accelerometers and height sensors.

Propulsion is by water jet. Water is taken up by a scoop in the centre of the aft foil and pumped out through twin nozzles in the form of jets (see page 117). Two pumps are used, driven by gas-turbine engines. Hull-borne control is provided by deflectors at the pump outlet nozzles which deflect the jet stream.

Riding on Air

Another surface skimmer of much more recent origin than the hydrofoil rides on a cushion of air. It is generally known as an air-cushion vehicle (ACV) or in Britain, where it was conceived in 1959, a hovercraft. ACVs of many types are in operation throughout the world, from single-seat 'do-it-yourself' models built by enthusiasts to massive car ferries capable of carrying hundreds of passengers. Many ACVs have found application in military roles, since they are fully amphibious and speedy over terrains that would be impassable to other craft.

ACVs can operate at speeds comparable with those of hydrofoils. The ride they give is not so smooth as that of hydrofoils and they generate a lot of spray. But they can operate more economically.

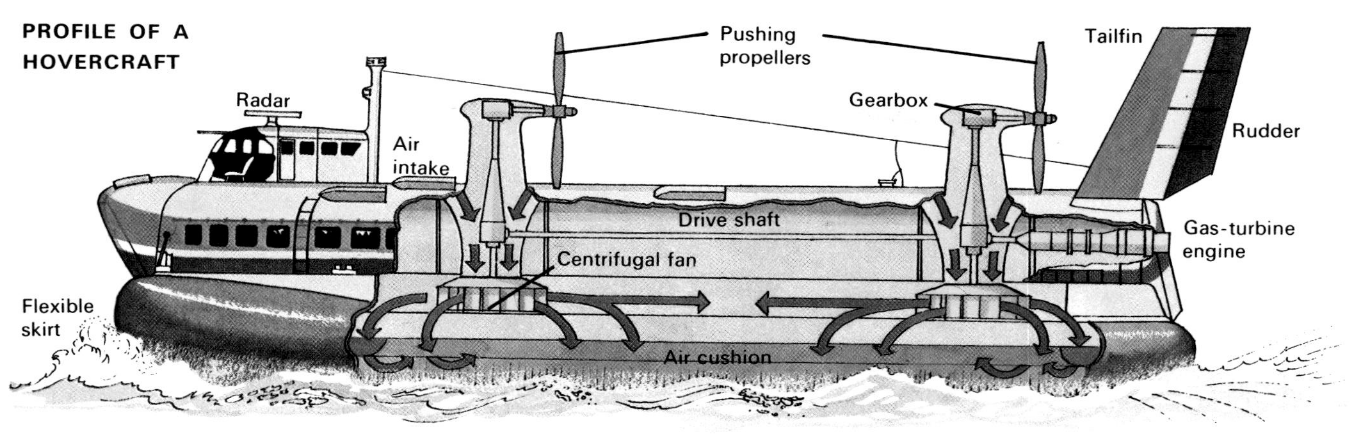

The most successful ACV to date has been the SRN4 car ferry operating across the English Channel. By 1978, after 10 years service, SRN4s had carried 12 million passengers and 2 million vehicles. In that year a 'stretched' SRN4, the Super 4, entered service, becoming the world's largest ACV. With a length of 56 metres (185 ft) and beam of 28 metres (92 ft) it can carry over 400 passengers and 55 cars at speeds up to 65 knots (120 km/h, 75 mph).

Hover and Thrust

Many of us have unwittingly observed the principle of the ACV while washing up. Upturn a hot cup on to a wet smooth surface, and the slightest touch can send the cup skimming away. Heat from the cup expands the air inside to form a 'cushion' which starts to escape around the edges. Friction between cup and surfaces is dramatically reduced, allowing the cup to move easily.

In a simple 'plenum-chamber' ACV air is blown by a powerful fan into a cavity underneath to form an air cushion. Air must be continually supplied to make up for leakage under the side walls. Greater 'hover' height is obtained by fitting a flexible skirt around the walls. The skirt has developed in the latest ACVs into a double-walled bag which is inflated by the air. Air escapes from the bag inwards to maintain the air cushion. Flexible extensions beneath the bag, called fingers, make contact with the surface over which the craft is travelling, forming an air seal for the cushion.

In the SRN4 four powerful centrifugal fans 3·5 metres (11·5 ft) in diameter provide the lift. They are incorporated in an integral system with the four propellers which provide propulsion for the craft. The propellers, which are of the variable-pitch type and operate in a 'pushing' mode, are 6·4 metres (21 ft) in diameter. Directional control is provided by means of the rudders on the trail fins. In some ACVs it is achieved by swivelling the propellers.

The ACV is an example of one type of what are often called ground-effect machines (GEM). They require the proximity of the ground in order to work, unlike true aircraft. Another interesting type of GEM currently achieving attention is the ram-wing craft, also called an aerofoil boat. It achieves an appreciable part of its lift from a build-up of pressure between its wings and the ground. Craft designed to utilize the effect can cruise much more economically than conventional planes.

Left: With skirts deflated, the largest air-cushion vehicle in the world, the Super-4, is seen here being loaded. An elongated version of the SRN4, it has proved ideal for its role as a cross-Channel car ferry.

Below: Germany's amphibious aerofoil boat X114, which flies by ground effect. Propelled by a 'pushing' propeller in the rear, it cruises at about 145 km/h (90 mph) only a few metres above the waves.

Flights of Fancy

Men seem always to have envied the birds their ability to fly and speculated how they themselves could do so. The consensus of opinion favoured strapping wings onto the arms and flapping them like a bird does. But men's pectoral and arm muscles are not, pro rata, as strong as those of a bird. No amount of flapping can provide enough lift to support a man's weight, as many erstwhile aviators discovered to their cost.

In the event it was not flapping wings but hot air that provided man with a means to fly. Two French brothers, Jacques Étienne and Joseph Michel Montgolfier, in June 1783 launched their 'globe aerostatique' in the square of their home town of Annonay. It was a large spherical bag of fabric and paper, the open neck of which was held over a smoky fire. The bag was inflated by the hot fumes and, becoming lighter than the surrounding air, proceeded to rise and drift away. The hot-air balloon was born. In October 1783 the French physicist J. A. C. Charles demonstrated the superiority of a different kind of balloon, inflated by hydrogen, the lightest of all gases.

Balloons are subject to the caprices of the winds and as a mode of transport thus leave much to be desired. Attempts were soon made, therefore, to design a self-propelled, steerable balloon – a dirigible or airship. The first honours went to Henri Giffard in 1852. Drawing upon the designs and experience of others, he built a successful steam-driven craft.

Attempts with electrical propulsion also had fitful success, but from 1897 on the petrol engine was the favoured power source in airships. The stage was set for the airship era. In 1900 Count von Zeppelin launched his first metal-framed, rigid craft, 128 m (420 ft) long and 11·5 m (38 ft) in diameter. Ten years later Zeppelins began carrying passengers and mail. In 1919 the British airship 'R34' flew across the Atlantic and back. In 1920 the giant 'Graf Zeppelin' circumnavigated the globe, in a total flying time of 12½ days. In the 1930s, however, a spate of accidents all over the world brought airship development to a rapid and tragic halt. From thenceforth the skies belonged to the aeroplane.

Wrongs and Wrights

The aeroplane is a heavier-than-air machine, which remains airborne because of aerodynamics. It relies on forces set up when its wings move through the air.

The first person to appreciate this point, and well-named the 'father of aeronautics', was Sir George Cayley. In 1799 he sketched a diagram of the forces involved in flight on a silver disc which is now in London's Science Museum. In 1852 Cayley built the first heavier-than-air machine to fly – a curious contraption that looked like a cross between a dinghy and a kite. It was the first glider.

Other glider pioneers followed in Cayley's footsteps, of whom the greatest was Otto Lilienthal in Germany. At first he experimented with flapping-wing devices, or ornithopters, but soon discarded them for fixed-wing gliders. He inspired Octave Chanute and the brothers Wilbur and Orville Wright in the United States to take up gliding and eventually to attempt powered flight.

The Wright brothers, who owned a cycle shop in Dayton, Ohio, suspected that existing data on aerodynamics was unreliable. So they systematically began to accumulate their own data by conducting experiments in a crude wind tunnel. By 1902 they had developed a practical and controllable man-carrying glider. In the following year they built a light-weight petrol engine to power it. On 17 December 1903, at Kitty Hawk, North Carolina, Orville flew under power for the first time for a fleeting 12 seconds.

The Wright's plane, 'Flyer 1', was a biplane with a wingspan of about 12 m (40 ft). It had twin 'pusher' propellers, front elevators and rear rudder. Flying became the latest craze, especially in France, and a good many curious craft joined the Wright biplane in the skies. Among the early pioneers in Europe were Alberto Santos-Dumont, a colourful character already famous as an airship designer; Henri Farman; the Voisin brothers; and Louis Blériot.

Distant Horizons

Blériot designed a monoplane of a much more modern appearance, and in 1909 flew one of his machines across the English Channel. In general, though, aircraft development was slow until World War I, when the combatants soon realised the potential of aerial warfare. The Great War brought about improvements in the reliability and power of aircraft, which after the war prompted optimistic aviators to attempt longer and more hazardous flights. In 1919 the Atlantic was conquered. During the 1920s and early '30s aviation expanded at an ever-increasing rate. Scheduled flights began, first to carry airmail, and then passengers as well in more comfortable airliners. Metal began to replace wood and fabric as the main material of construction. Biplanes gave way to monoplanes, which proved themselves superior in the air races that were then in vogue, such as the Schneider Trophy. The streamlined Boeing 247 of 1933 set the pattern for future airliner design, boasting retractable undercarriage, trim tabs, automatic pilot and variable-pitch propellers. Two years later the most successful aircraft of all time, the Douglas DC-3, took to the skies.

But in Britain and Germany research was accelerating on a new method of propulsion that was to transform the aviation scene. This was the jet engine, the construction of which was outlined by British airman Frank Whittle in a 1930 patent. In the event the Germans tested a jet-propelled plane first, the Heinkel He-178, in 1939. But jet propulsion did not come into its own until after World War 2. The rocket-powered research craft, Bell X-1, pioneered supersonic flight in 1947, proving that a sound 'barrier' does not exist if a plane is suitably designed. First fighters and then airliners – the Russian Tu-144 and the Anglo–French Concorde – went supersonic.

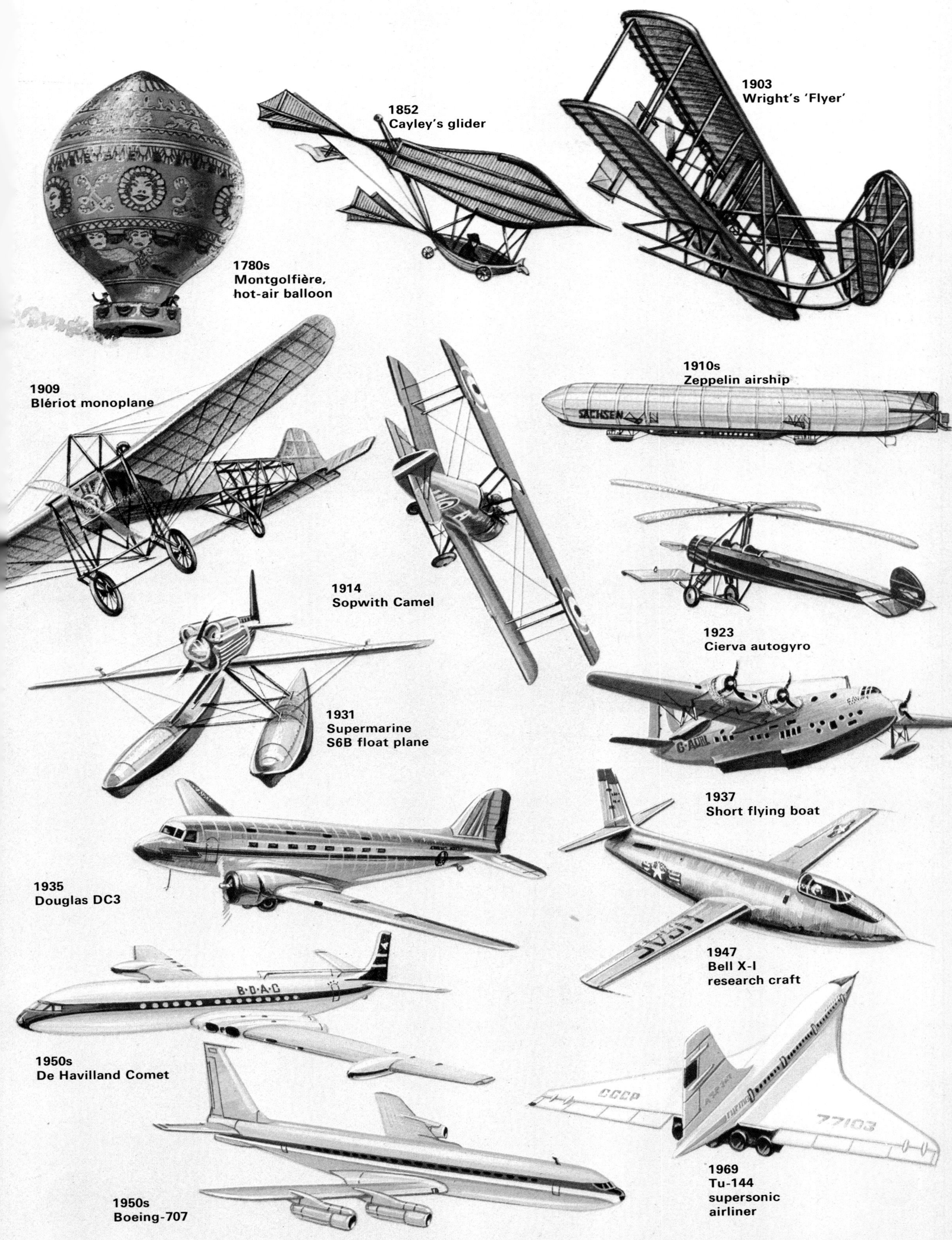
1780s
Montgolfière,
hot-air balloon
1852
Cayley's glider
1903
Wright's 'Flyer'
1909
Blériot monoplane
1914
Sopwith Camel
1910s
Zeppelin airship
SACHSEN
1923
Cierva autogyro
1931
Supermarine
S6B float plane
1937
Short flying boat
1935
Douglas DC3
1947
Bell X-I
research craft
B·O·A·C
1950s
De Havilland Comet
CCCP
77103
1969
Tu-144
supersonic
airliner
1950s
Boeing-707

Left: A hot-air balloon soars above the treetops after take-off. The crew are still heating the air inside the envelope with the propane burner to generate more lift.

'DOUBLE EAGLE' TRIUMPHS

On 17 August, 1978, three American balloonists completed the first transatlantic balloon crossing in the 11-storey high helium-filled balloon, 'Double Eagle II'. 'Double Eagle II' lifted off from Presque Isle, Maine, in the United States, on the evening of August 11. Riding on a west-to-east moving ridge of high pressure, it floated for most of the 5000-km (3100-mile) passage at an altitude of 3000 metres (10,000 ft) or more, though it plummeted at one time as low as 1310 metres (4000 ft) and shortly afterwards soared to a gasping 8200 metres (25,000 ft). 'Double Eagle II' made its landfall near Miserey, west of Paris, after a little over 137 hours in the air.

Right: Details of the 'Europa' airship, one of several operated by Goodyear for publicity purposes. It is a regular visitor to international air displays.

Lighter-than-Air Craft

Although the heyday of the balloon and airship has long since past, they are still occasionally to be seen in the skies. Ballooning has become a popular minority sport, and international balloon meetings are held at various centres each year. Hot-air, hydrogen and helium balloons all have their devotees.

The overall design of balloons has changed little over the years. They have a spherical or pear-shaped bag, or envelope, made of gas-tight material. This is now generally a coated polyester plastic like Mylar or Dacron. From the envelope is suspended a basket which carries the crew.

The principle upon which balloon flight depends was first stated by the Greek scientist Archimedes. Any object will float in a fluid (liquid or gas) if it displaces a weight of fluid equal to its own weight. If it displaces a weight of fluid greater than its own weight, it will experience an upthrust which will carry it upwards. So you need to fill a balloon with the lightest possible gas for maximum uplift.

First contender is hydrogen, the simplest element and the lightest gas of all, whose density is only one-fourteenth that of air. Hydrogen is readily available. It can be prepared in numerous ways – by the electrolysis of water; by the reduction of steam by red-hot coke; and as a by-product of numerous industrial processes, for example, the cracking of hydrocarbons in petroleum refining. The one great drawback with hydrogen is that it forms an explosive mixture with air. A naked flame, static electricity, or lightning can ignite it. The spectacular and tragic conflagration of the German airship 'Hindenburg' in New Jersey in 1937 bears witness to this.

The next lightest gas is helium. This is one of the so-called rare noble gases, which are characterized by their inertness. This makes helium an ideal balloon gas. The drawbacks are that it is twice as heavy as hydrogen and relatively scarce. It occurs in the air in minute amounts and can be obtained from liquid air by fractional distillation. But most helium is extracted from the natural gas emanating from oil wells in North America. Helium is now widely used in balloons. Three Americans made the first crossing of the Atlantic, in 1978, in a helium balloon.

The pilot of a hydrogen or helium balloon has scant control over his craft. Once airborne he can go upwards only by throwing out ballast, usually sand. He can go down by opening a valve to let some of the inflating gas escape from the balloon. When landing he pulls a cord to open a long seam, or ripping panel, which allows the gas to escape quickly. This reduces the risk of the balloon being blown away.

The modern hot-air balloon is rather more versatile. The pear-shaped envelope is open at the lower end, and beneath it is slung a blow-torch burner. This is fuelled by propane from a bottle of liquid gas. The burner is fired initially to inflate the envelope with hot air, and then intermittently during flight to maintain the air within the envelope at a suitable temperature. In this way a controlled rate of descent can be achieved without too much difficulty.

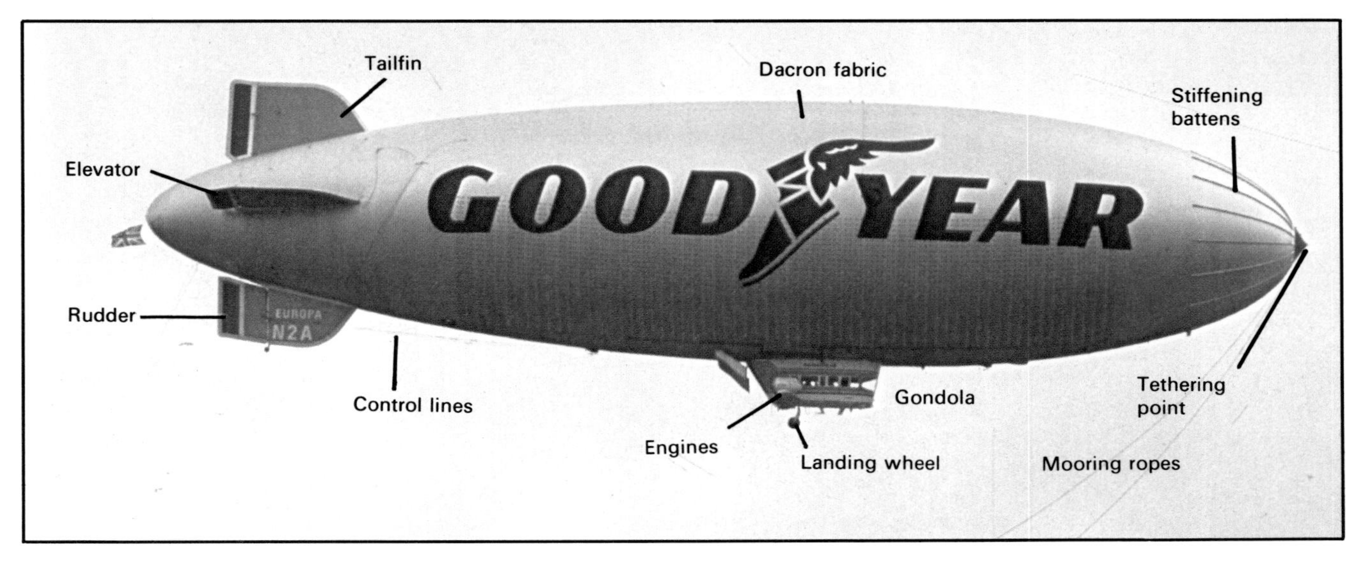

Airship Revival

In recent years renewed interest has been shown in airships, and people in the know are forecasting that they could once more become commercially successful, operating as both passenger and freight carriers. With the costs of conventional airline travel ever-rising, airships should be able to offer a competitive service. And, although they would fly much slower than aeroplanes, they would save time by operating directly from sites in city centres.

In the 1970s the only successful airships operating were those made by Goodyear. By 1980 a fleet of six were in service throughout the world. Though their function is to assist in the public relations work of the company, they nevertheless are accumulating valuable flight experience in what is now a novel form of transport.

Typical of the fleet is 'Europa', pictured above. It is of non-rigid design. This means that its shape is maintained almost totally by the gas it contains, which is helium. The envelope, which is made of double-ply neoprene-coated Dacron, measures 58 m (190 ft) long and has a maximum diameter of 14 m (46 ft). The total volume of the envelope is 5750 cu m (203,000 cu ft), of which nearly a third is taken up with ballonets. These are air-filled bags inside the envelope in the front and rear of the craft. Air can be forced into the ballonets or released from them to correct the 'trim' of the airship and alter its buoyancy.

The small cabin, or gondola, beneath the envelope can carry a pilot and up to six passengers. Mounted behind it are two six-cylinder, horizontally-opposed and air-cooled petrol engines, which drive 'pusher' propellers. These give it a maximum speed of 80 km/h (50 mph). Europa normally cruises at a height of between about 300 and 600 m (about 1000–2000 ft), with a ceiling of some 2500 m (about 8000 ft). Directional control is provided by rudders and elevators on the fins at the tail, which are operated by cable from inside the gondola.

Less conventional designs have been suggested for the airships of the future. One has a pair of stubby wings with engines mounted on them. Another is shaped like a saucer, a design which appears to be more stable in flight than the traditional cigar-shape.

The Aerofoil

Whether 'flying-saucer' airships take to the skies or no, aviation will be dominated by the aeroplane for the foreseeable future. Whereas an airship obtains its lift statically by means of lighter-than-air gases, a plane obtains its lift aerodynamically, that is, by moving through the air. When any object moves through the air, it interacts with the air and experiences forces that cause it to move this way and that. This will be familiar to anyone who has put his hand out of the window of a moving car.

By inclining your hand upwards slightly into the airstream, you notice that the aerodynamic forces cause it to rise. Herein lies the key to aeroplane flight. To make a plane fly you must attach to it surfaces which are inclined slightly into the airstream so that they experience a net upward force, or

Left: Soaring in a sailplane, or glider, is an exhilarating experience. There is no noisy engine to shred the nerves; you glide like a bird on currents of air. The sailplane is a severely stripped-down version of an aeroplane, with minimal controls and instrumentation. Like the plane, it has ailerons, elevators and rudder for manoeuvring in the air. It is constructed in lightweight materials such as plywood and glass-fibre reinforced plastic (GRP). The striking thing about sailplane design is the relatively large wingspan, often more than double the body length.

Right: Pedalling hard, Bryan Allen propels his diaphanous craft 'Gossamer Albatross' at a steady 19 km/h (12 mph) on his way across the English Channel in June 1979. His 2 hour 49 minute flight won for Gossamer's designer, Paul MacCready a £100,000 prize. MacCready also designed 'Gossamer Condor', again piloted by Allen, which became the first successful man-powered craft to fly two years before.

lift. The surfaces that generate the lift in a plane are the wings.

Experience, and experiments in wind tunnels, have shown that to generate maximum lift, the wings must have a special shape. They must be relatively flat underneath but curved above. The front, or leading edge should be rounded, while the rear, or trailing edge should end in a narrow taper. Surfaces of such shape are known as aerofoils (American, 'airfoils').

When air flows around an aerofoil (or an aerofoil moves through the air, which comes to the same thing), it tends to travel faster over the curved upper surface and slower underneath. But a basic law of fluid dynamics, first advanced by Daniel Bernoulli in the 1700s, states that when the velocity of a fluid increases, the pressure of that fluid decreases and vice versa. Applying this law to the aerofoil, the pressure above the aerofoil decreases (the air speeds up), while the pressure below the aerofoil increases (the air slows down). With low pressure above and high pressure below, the aerofoil thus tends to lift.

The faster the air flows around the aerofoil, the greater will be the lift it experiences. On take-off a plane accelerates through the air at ever-increasing speed until the lift generated by its aerofoil wings exceeds its weight. Then it becomes airborne.

The amount of lift a wing experiences increases with the speed of travel. It also increases with wing area and with wing camber, or curvature. To generate enough lift, low-speed planes need to have large, thick wings with pronounced camber. On the other hand, high-speed craft can achieve enough lift with smaller wings of much flatter and thinner section.

In a given plane lift can be increased by increasing the angle the wings present to the airstream – the so-called angle of attack or incidence. In level flight the wings are arranged to give an angle of attack of about 3°. If a pilot wishes to slow down but maintain his height, he can increase the angle of attack by putting his nose up. And as the angle increases, so does the lift. But he can continue increasing the angle of attack, and slowing down, only to a certain point. This occurs at an angle of attack of about 15°. If this angle is exceeded, the airflow around the wing, which has hitherto been smooth, suddenly breaks up into eddies. Lift drops abruptly, and the wing is said to stall. In a stalling condition a plane drops rapidly out of control, often in a spin.

The four main forces acting on a plane.

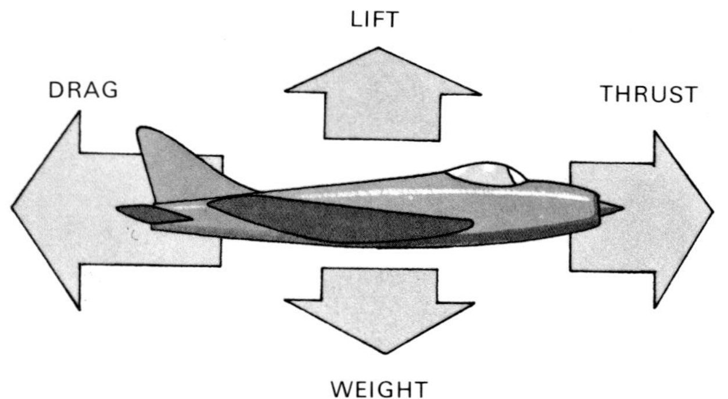

Flow around an aerofoil at low angle of attack.

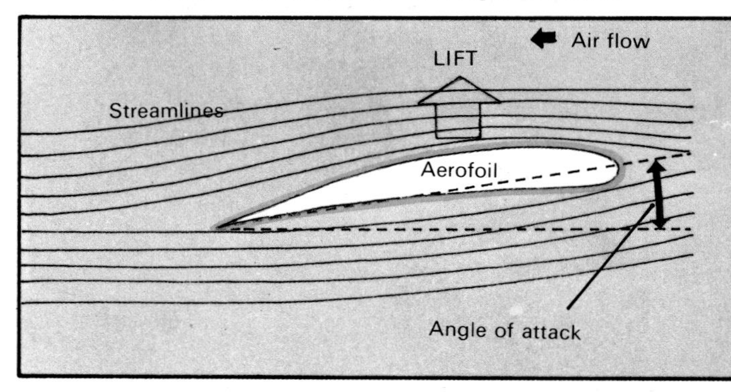

Quit Stalling

To delay the onset of stalling, the wing can be fitted with various devices, including flaps and slats. Flaps are hinged panels at the trailing edge of the wings which are angled downwards to increase lift. They work by effectively increasing the wing camber. They may be slotted for greater effect. Many aeroplanes have small flaps on the leading edge too. Flaps are lowered at take-off and prior to landing when maximum lift is required at low speeds.

The turbulence that accompanies stalling can be delayed by means of another device, the leading-edge slat. This is a curved panel at the front of the wing which can move downwards. In this position it allows a layer of high-pressure air from the underside of the wing to flow up and over the upper side. By so doing it prevents the boundary layer of air breaking away from the wing and thus becoming turbulent.

In contrast it is often necessary for a plane to lose height rapidly under control, and devices to destroy lift are also usually fitted to the wings. They are known as spoilers. They are hinged panels located in front of the flaps. When activated they move upwards and cause turbulence in the air flow over the wings, thus reducing lift.

The very successful Boeing 727 medium-range airliner, for example, has on each wing two triple-slotted trailing-edge flaps, in front of which are seven spoilers. At the leading edge it has four slats located on the outer two-thirds of the wings and three flaps inboard of them.

Thrust and Drag

So far we have considered the opposing vertical forces acting on a plane – the downward force due to its weight, which in level flight is balanced by the upward force of lift generated by its wings. Opposing horizontal forces also act on a plane in flight. The forward force that keeps the plane in motion is the thrust provided by the engine. This is opposed by the resistance of the air, or the drag.

The engines used in most planes today are jet engines. They develop thrust by burning kerosene fuel and expelling the hot gases produced at high speed from a rear nozzle. Several kinds of jet engines are employed, including pure turbojets for high-speed combat aircraft; economical turbofans for

Flow becomes turbulent at the stalling angle.

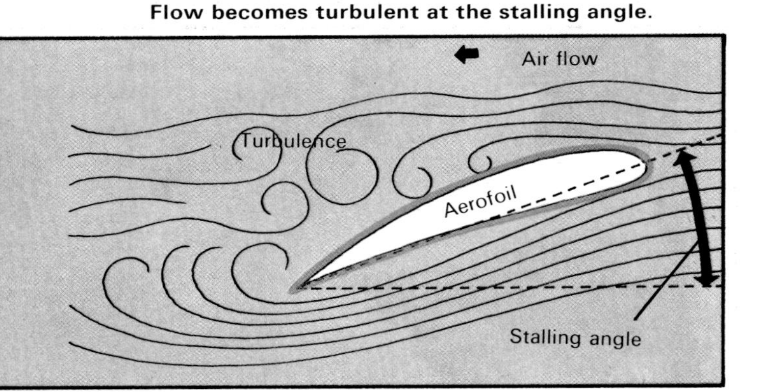

Devices used to increase or decrease wing lift.

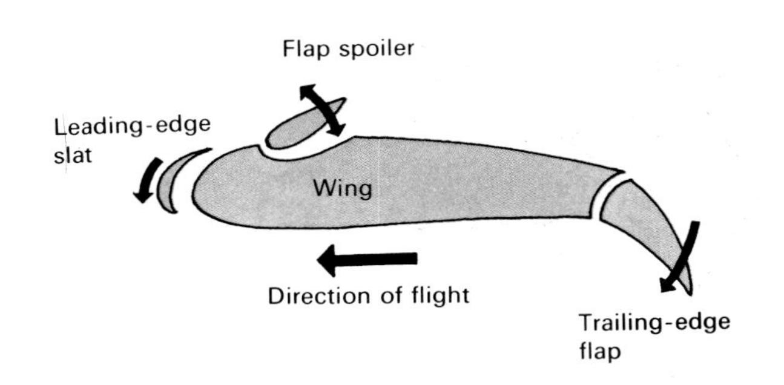

THE WAY IT WORKS

For any object to move easily through a fluid – gas or liquid – it must be suitably shaped. It must be streamlined so that the fluid flows past it smoothly, without turbulence. If turbulence, or eddying, does set in, the object experiences greatly increased resistance to motion, or drag, which consumes extra power. In aeroplane flight turbulence around the wings can destroy the lift which supports the plane in the air. To get an idea of the aerodynamic (airflow) problems they are faced with, designers conduct extensive tests on scale models of their aircraft before finalizing their designs. They carry out these tests both in water tunnels and wind tunnels.

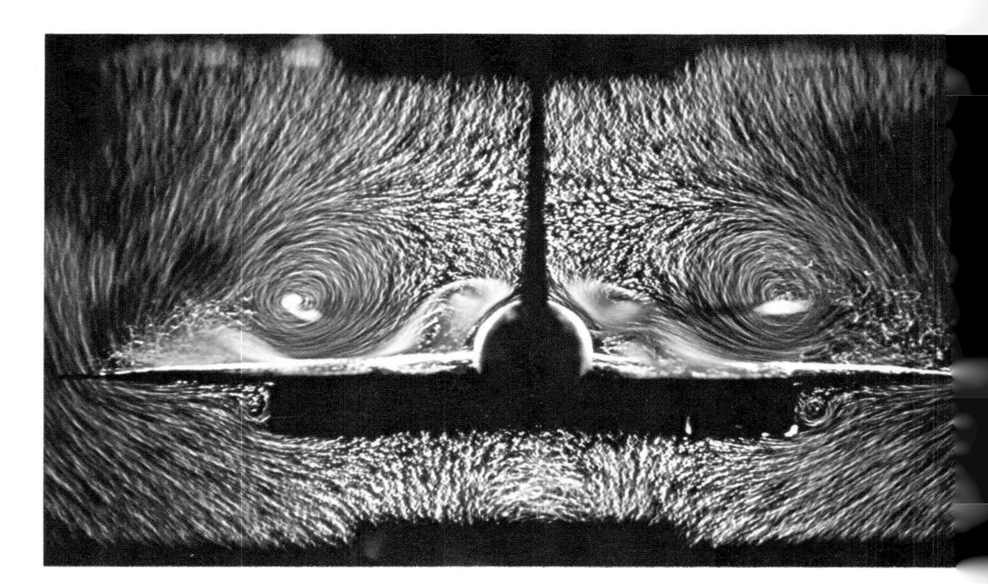

Above: A photograph showing flow patterns around a model aircraft tested in a water tunnel. Prominent are the vortices (whirlpools) generated over both wings.

Left: Some of the wind tunnels used for testing are vast, like this one at the Royal Aircraft Establishment, Farnborough. The huge fan in the background sucks air past suspended models. Instruments connected to the suspension wires register aerodynamic forces acting upon the model. They are analogous to the actual forces that will act on the full-scale plane when it flies.

Below: By means of what is called Schlieren photography, the shock waves around a model can be viewed. It relies on the change in refractive index caused by the waves.

airliners; and turboprops for low-speed transports. (For a detailed discussion of these engines see information on page 45.)

Turboprops derive most of their thrust from the propeller spun by the jet engine. Many light aircraft are still propelled by propeller driven by piston engines. These are generally air-cooled petrol engines which operate in a similar way to car engines (see page 36), though they are of different configuration – often horizontally opposed or radial.

One major source of drag is the friction between the surface of an aeroplane and the air, often called skin friction. This is at a minimum when the surface is smooth. Another source of drag is related to the lifting force created by the air flow around the wings; This lifting force does not act vertically, but at an angle. By the parallelogram of forces, this force can be resolved into a vertical component and a horizontal component. It is the vertical component we call lift. The horizontal component, which opposes forward motion, is called induced drag.

The induced drag on a wing depends both on its cross-sectional shape and its shape in plan. Minimal drag is developed by a wing which is of aerofoil section and is long in relation to its width. The length-to-width, or span-to-chord, ratio needs to be high in low-speed aircraft. That is why racing sailplanes, or gliders, have enormously long wings. The shape of the aircraft as a whole, not only the wings, naturally affects the total drag, and must be as streamlined as possible.

The aerofoil section and span-to-chord ratio of the wings and the overall aerodynamic shape of a plane vary according to the designed operating speed. The wing shape also is critical. Low-speed craft can have wings projecting from the fuselage at right-angles. But this arrangement is not suitable for craft which travel at speeds of more than half the speed of sound.

Right: Twin-propeller piston engines power this aircraft. The two propellers are contra-rotating – they rotate in opposite directions. This arrangement makes for a more balanced engine. Propellers, or 'airscrews', are of aerofoil cross-section and develop forward thrust when they spin, in much the same way that aeroplane wings develop lift when they move through the air. On most propellers the pitch, or angle, of the blades can be adjusted in flight. This is necessary to keep the propeller functioning at optimum efficiency as flight conditions change. A variable-pitch propeller would be set to fine pitch at take-off, for example, and to coarse pitch at cruising speed. It would be set to reverse pitch after landing so it could act as a brake.

GOING SUPERSONIC

When a plane travels through the air, it generates pressure waves in front of it. The waves travel away from the plane at the speed of sound – sonic speed, which is about 1220 km/h (760 mph) at sea level. When, however, the plane flies at speeds close to sonic speeds, the pressure waves cannot 'escape', but bunch up to produce shock waves. The air flow becomes very turbulent and unless a plane is specially designed, it is subject to severe buffeting and greatly increased drag. This effect, experienced by diving fighter aircraft in World War 2, for example, was once thought to present a barrier to flight at speeds beyond sonic speed, which led to the popular concept of the 'sound barrier'.

However, a suitably designed plane – with needle nose, swept-back wings and a wasp-waist fuselage – can slip through the 'barrier' with scarcely a tremor. In fact people on the ground beneath a supersonic plane are more affected. They hear the shock waves triggered off by the plane as an explosion like a heavy gun firing, which is called the sonic boom. The speed of sound decreases with increasing altitude as the air pressure decreases. And aircraft speeds are now usually expressed in terms of Mach number – this is the aircraft's speed relative to the local speed of sound. So Mach 1 is sonic speed, Mach 2, twice sonic speed, and so on.

Sweeping the Wings

When a plane with straight wings flies at more than half sonic speed, the air flowing over its wings will actually start moving at supersonic speeds in places. When this happens, shock waves form. The streamlined boundary layer starts to break away, and a considerable increase in drag occurs. A solution to the problem is to sweep back the wings at an angle from the fuselage. When this is done, the onset of supersonic shock waves is delayed to a much higher flying speed. The wings of modern airliners are swept back at an angle of about 35°. This is the sweep angle of the DC-10 and the Trident; the Boeing 747 has a sweep angle of $37\frac{1}{2}$°. With such a sweep back, they can attain speeds up to 960 km/h (600 mph) without difficulty.

Sweep-back, however, causes design problems, and research has come up with a better solution – the supercritical wing. First developed by NASA, the supercritical wing is flatter on top than a conventional aerofoil and has a downward camber at the trailing edge. It should allow planes to fly very close to the speed of sound without creating drag-inducing turbulence.

Planes flying at supersonic speeds need wings even more swept back. The wing span, however, can be narrower. This has led to the development of the triangular, or delta-wing, to be seen in such combat aircraft as the Saab Draken and the Mirage 4000, and the supersonic airliners Concorde (Anglo-French) and Tu-144 (Russian). Delta-winged craft, however, have high take-off and landing speeds. The need to reduce these speeds led to the development of the variable-geometry wing. Craft with these wings include the Grumman Tomcat, the Panavia Tornado and the General Dynamics F-111. The wings of these planes are designed to swing horizontally while the planes are moving. For take-off and landing the wings are extended outwards to give maximum lift at the low speeds. While accelerating to cruising speed, the wings are swung back towards the fuselage where they form with the tailplane a delta-wing shape (see above).

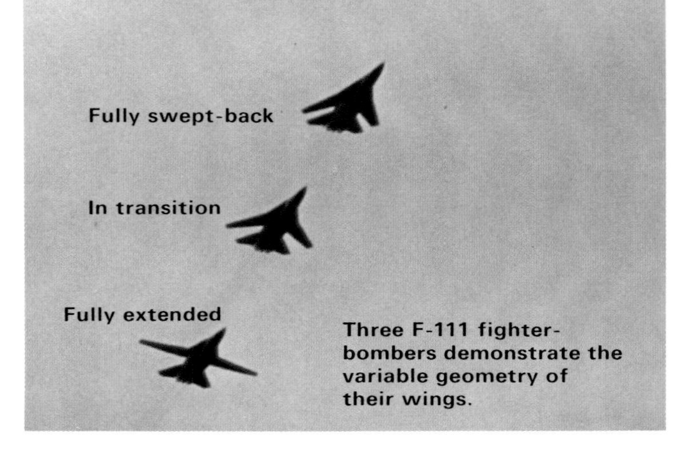

Three F-111 fighter-bombers demonstrate the variable geometry of their wings.

Stabilizing Flight

The wings of most planes do not project from the fuselage horizontally. They are usually angled upwards slightly so that the wings are higher off the ground at the tip than at the root. The angle of inclination is called the dihedral angle. The purpose of this kind of design is to give the plane an inherent resistance to rolling, or rocking from side to side.

In most planes the dihedral angle is only a few degrees. For example, the twin-jet Dassault Mercure airliner has 5° dihedral; the Boeing-707 6°. Other planes either have no dihedral or have wings that dip the other way, that is, downwards. The slight angle of dip is called the anhedral angle. The gigantic Lockheed Galaxy transport, for example, has $5\frac{1}{2}$° anhedral wings.

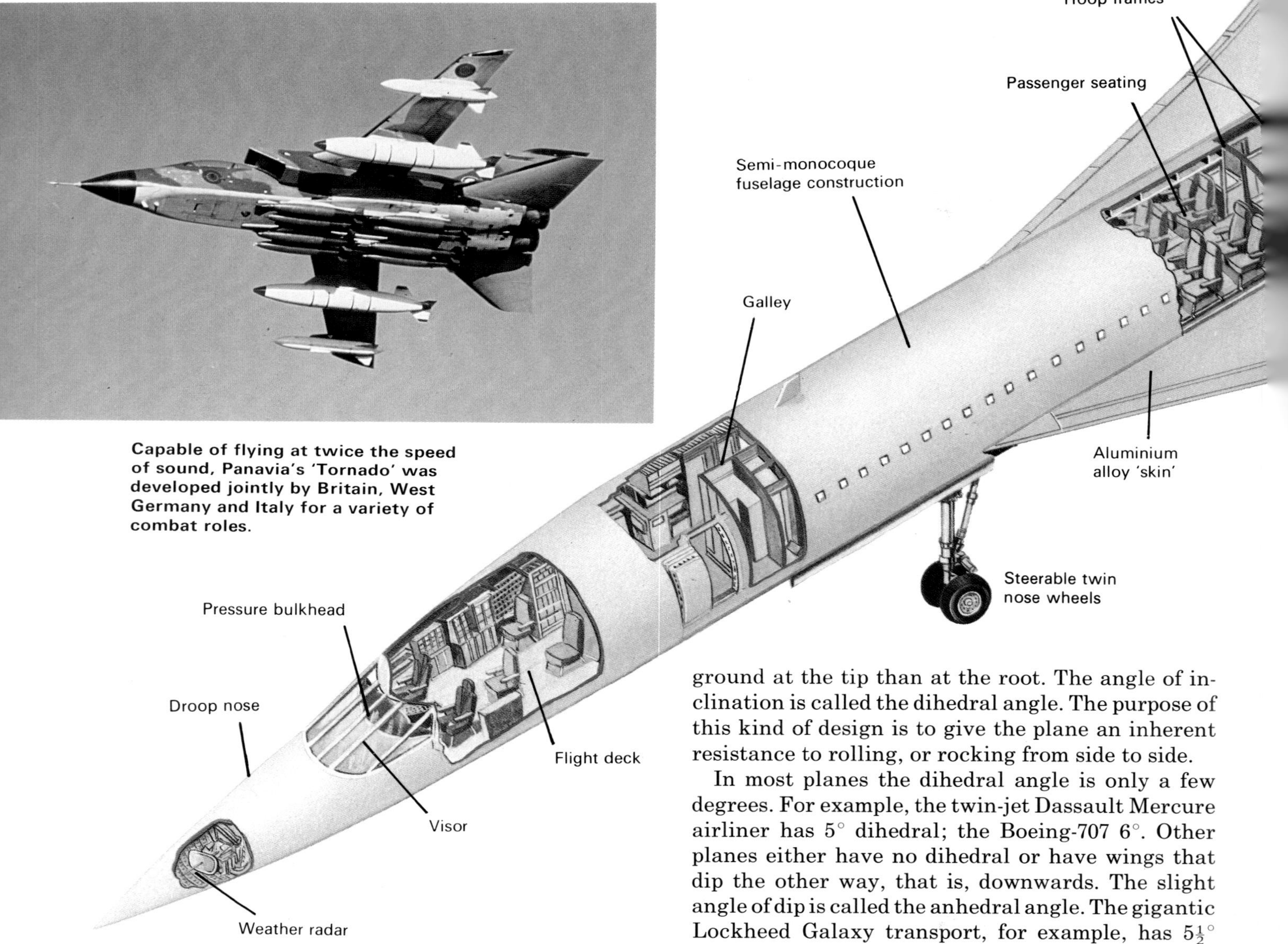

Capable of flying at twice the speed of sound, Panavia's 'Tornado' was developed jointly by Britain, West Germany and Italy for a variety of combat roles.

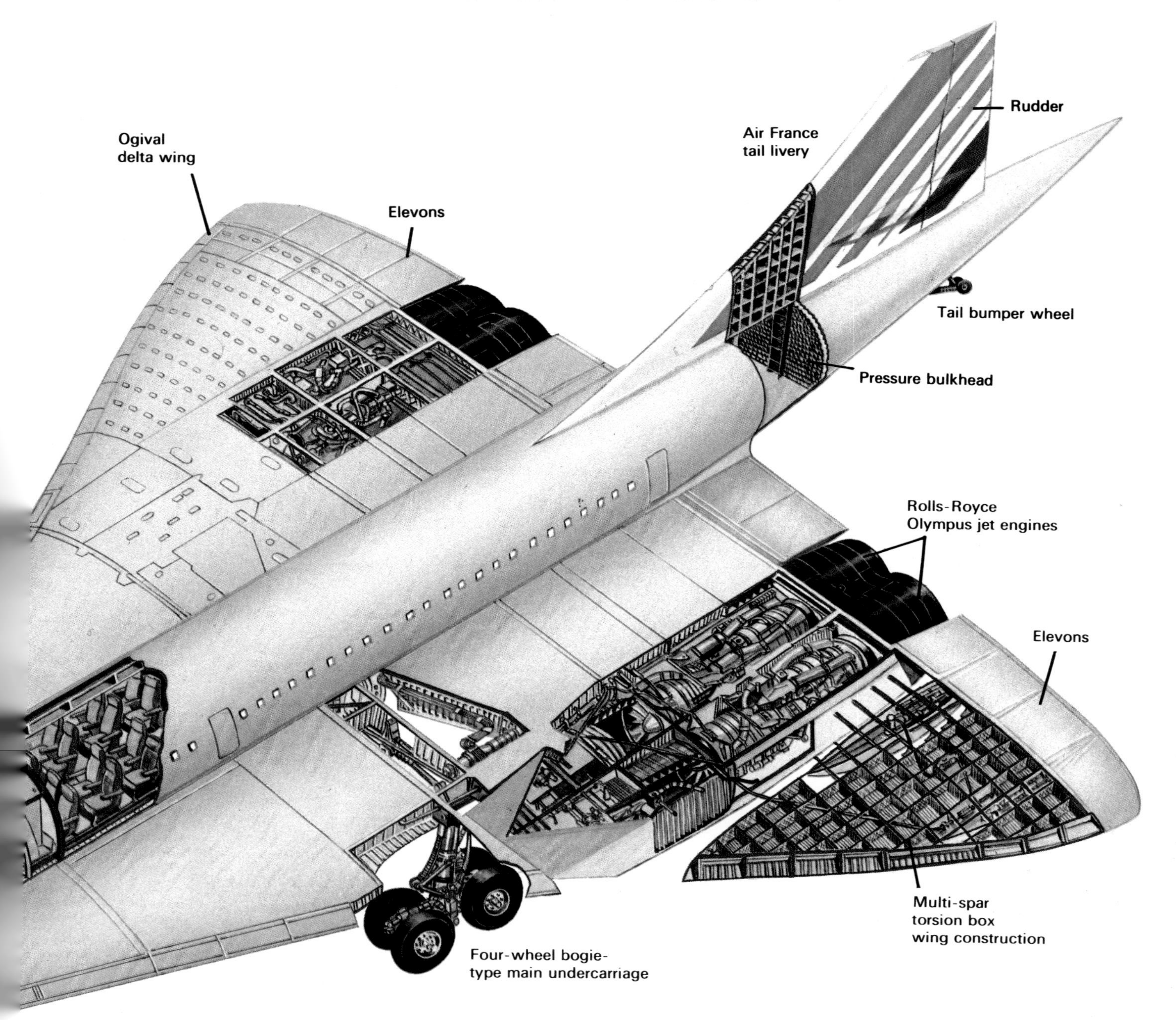

The Anglo-French 'Concorde', the most advanced passenger aircraft ever built. It cruises at twice the speed of sound – about 2170 km/h (1350 mph) at its cruising altitude of some 16,700 metres (55,000 ft). It is 61·7 metres (202·3 ft) long and its ogival delta wing has a span of 25·6 metres (83·8 ft).

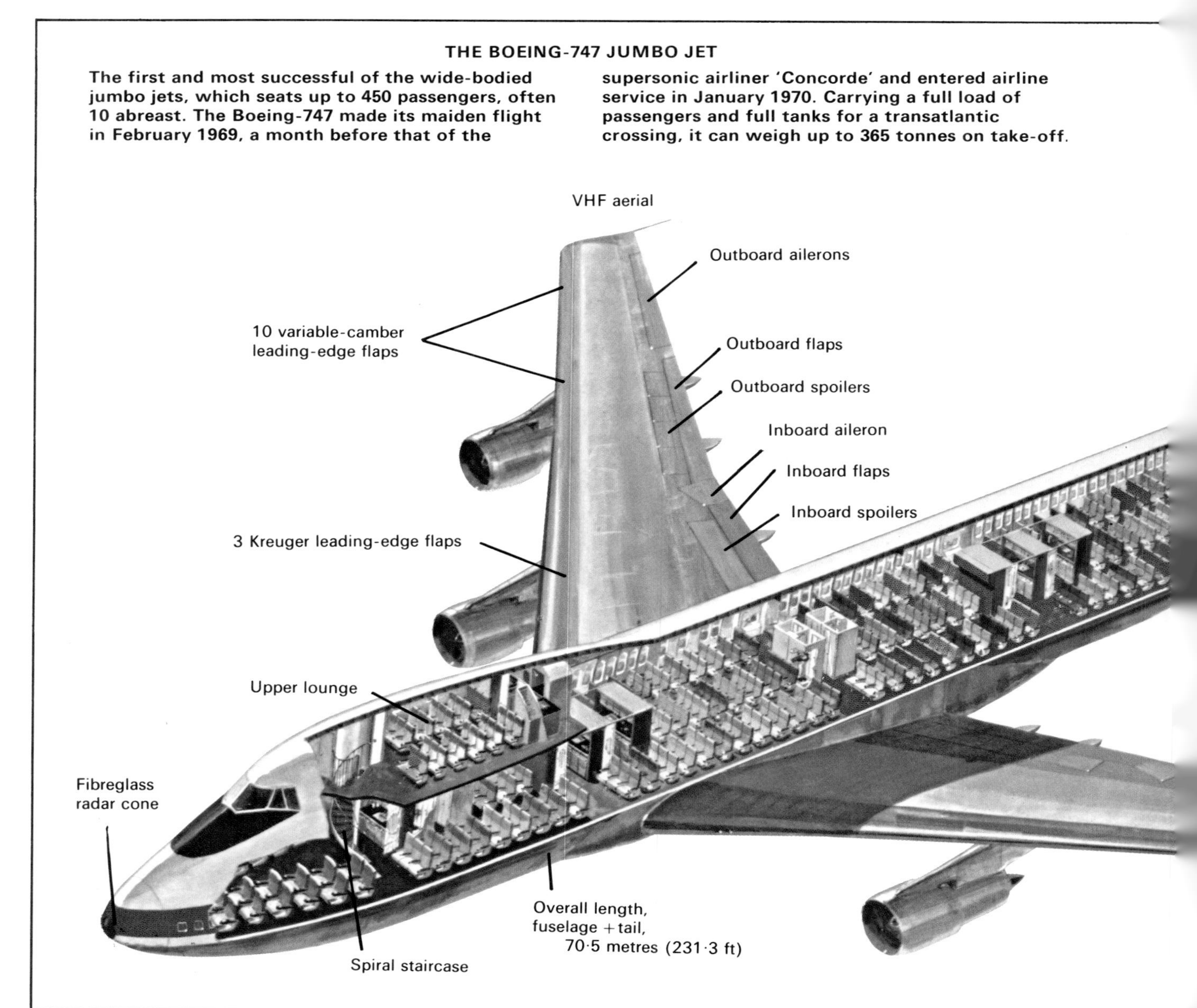

THE BOEING-747 JUMBO JET
The first and most successful of the wide-bodied jumbo jets, which seats up to 450 passengers, often 10 abreast. The Boeing-747 made its maiden flight in February 1969, a month before that of the supersonic airliner 'Concorde' and entered airline service in January 1970. Carrying a full load of passengers and full tanks for a transatlantic crossing, it can weigh up to 365 tonnes on take-off.

A plane also has built-in design characteristics to make it inherently stable about the other two axes. Two other motions that tend to afflict a body moving through the air are pitching and yawing. Pitching is a tendency for the nose of a body to oscillate up and down. It is prevented in a plane by having a horizontal surface at the tail – the horizontal stabilizer, or tailplane. The tailplane is usually mounted horizontally or inclined upwards at a slight dihedral angle. A few planes, however, have a tailplane with a pronounced anhedral. The McDonnell-Douglas Phantom, for example, has a tail with 23° of anhedral.

Yawing is the tendency of a body to slew from side to side. It is corrected in a plane by a vertical surface at the tail – the vertical stabilizer, or tailfin. In most planes the tailfin and tailplane are of cruciform design (cross-shaped). But in some jet planes the tailplane is raised to the top of the tailfin. It is positioned there so as to be clear of the jet exhausts, which would otherwise hinder its action. This arrangement is known as a T-tail. Rear-engined jets like the BAC-111, the Fokker Fellowship and the Boeing 727 have T-tails.

There are many other kinds of tail configurations. The Grumman Tomcat and the MiG-23 Foxbat have twin tailfins, with the tailplane split on either side of the engine cowlings, or nacelles. Delta-wing designs have no separate tailplane. The rear part of the delta doubles as a tailplane.

Controlling Flight

To control, or steer, a plane in flight the pilot moves a control column or wheel which moves control surfaces on the wings, the tailplane and the tailfin. These surfaces are hinged panels along the trailing edge, which can be moved upwards and downwards or from side to side as the case may be. The control surfaces deflect the airstream flowing past them and cause the plane to change its attitude in the air.

The control surfaces on the wings are the ailerons. The ailerons are located near the wing tips, beyond the flaps. They are operated by moving the control

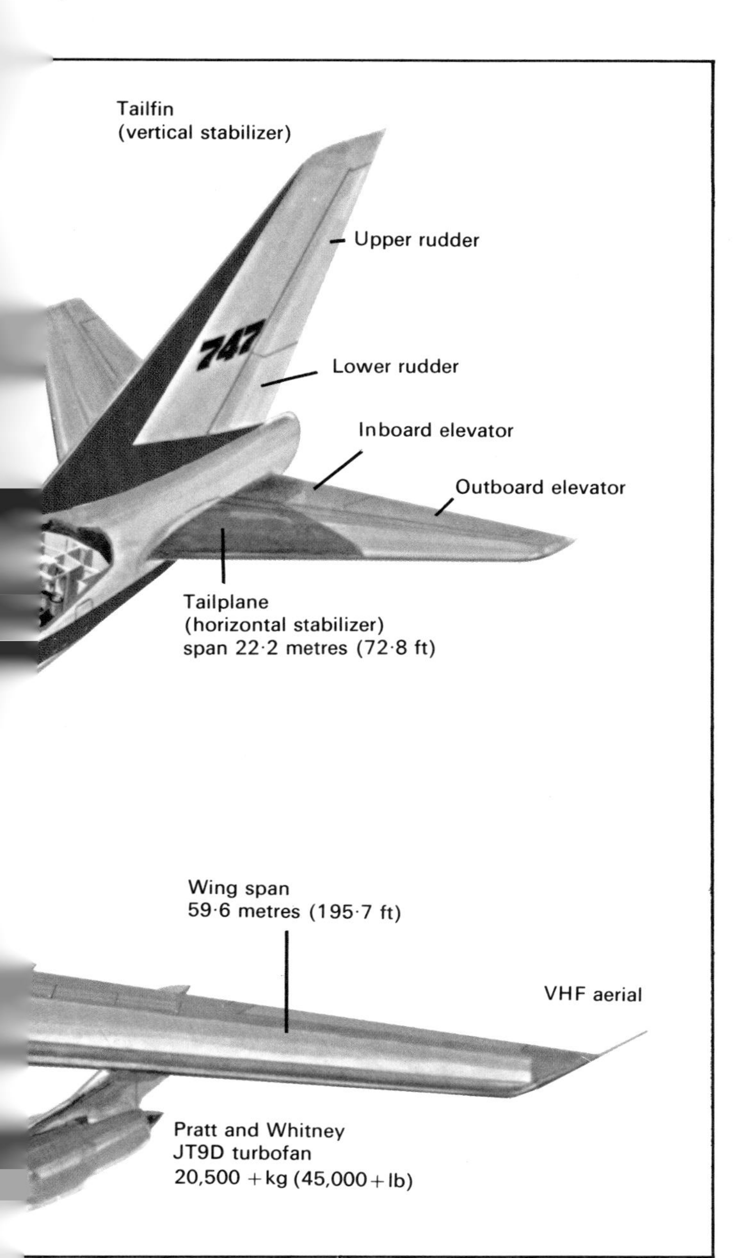

column from side to side. Moving it to the left raises the aileron on the left wing and simultaneously lowers the aileron on the right wing. This makes the left wing drop and the right one rise. This kind of movement is called banking. Moving the column to the right has the reverse effect.

The control column also operates the control surfaces on the tailplane, called elevators. Pushing the column forwards lowers the elevators and the nose of the plane tends to go down and the tail up. Pulling at the column has the reverse effect. In many planes, particularly the T-tailed type, the whole tailplane can be moved to aid control.

The control surface on the tailfin is the rudder, which is operated by foot pedals. Pressing the left pedal kicks the rudder to the left and turns the plane's nose to the left. Pressing the right pedal has the reverse effect.

Because air is a very tenuous fluid and the plane has considerable inertia, a simple turn cannot be executed merely by moving the rudder. This causes the plane to point in a different direction but still tend to continue in the same direction, effectively performing an aerial skid. To turn smoothly without side-slipping the pilot uses the ailerons as well as the rudder. He banks the plane at the same time as he turns, dipping the right wing when turning right and the left when turning left.

The pilot can vary the plane's altitude simply by opening or closing the engine throttle. If the plane is in level flight, increasing its speed increases the lift on the wings, and the plane gains altitude. Conversely, decreasing its speed reduces lift and the plane loses altitude. By controlling the throttle and the elevators in combination a smooth climb or descent can be executed.

The control surfaces on the trailing edge of a delta-wing plane are rather different because the wing and tailplane are combined. They perform the

Right: With inboard and outboard flaps lowered, a Boeing-747 comes in to land over the runway approach lights. The influx of passengers from 'jumbos' landing in quick succession can cause headaches for airport staff.

Below: This view of a propeller-driven plane shows the prominent dihedral uplift of its wings, and its 'T-tail'.

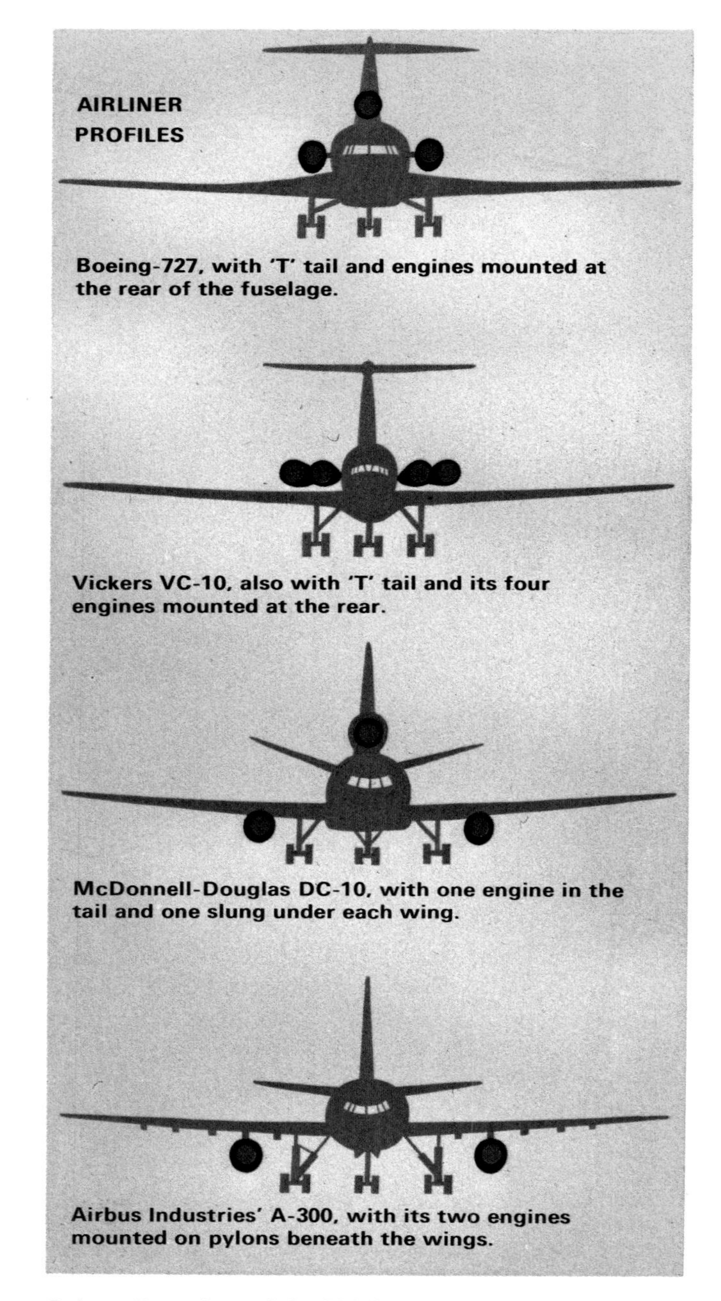

AIRLINER PROFILES

Boeing-727, with 'T' tail and engines mounted at the rear of the fuselage.

Vickers VC-10, also with 'T' tail and its four engines mounted at the rear.

McDonnell-Douglas DC-10, with one engine in the tail and one slung under each wing.

Airbus Industries' A-300, with its two engines mounted on pylons beneath the wings.

Below: Rear view of the F14 Grumman 'Tomcat' shows its twin tailfins and split tailplane. This 'swing-wing' fighter is designed for carrier-based duty.

function of both elevators and ailerons, and are appropriately called elevons. As elevators they act in unison (both sides up, both down) and as ailerons differentially (one side up, the other down).

The conditions of the air through which a plane flies is continually changing, requiring slight alterations to the control surfaces if a steady flight path is to be maintained. To avoid constant adjustment to the main control surfaces, subsidiary surfaces are provided in most planes in the rudder and elevators. They are narrow panels within the main surfaces which can be moved independently to finely tune, or trim, the plane to the prevailing flight conditions. They are known as trim tabs.

In sailplanes and some light planes the control surfaces are linked to the control column and rudder pedals mechanically by cables, rods and levers. But in other planes they are operated hydraulically, or by means of liquid-pressure. So are the flaps, slats and spoilers on the wings, and the landing gear, or undercarriage, which is retracted into the fuselage in flight. Hydraulic systems are used because they can transmit and apply large forces while taking up little space. In outline a hydraulic system works like this. The pilot pulls the control lever, say, for lowering the wing flaps. This causes an engine-driven pump to build up pressure in the system. This pressure is transmitted through pipes to hydraulic cylinders, or rams, whose pistons are connected to the flaps. The pressure forces the pistons down the cylinders, thus actuating the flaps. Pressure is switched to the other side of the pistons to raise the flaps.

In practice a hydraulic system is quite complex, containing numerous selector valves, non-return valves, relief valves and locking devices. The fluid is usually mineral oil, and pressures in the system are of the order of 210 kg/sq cm (3000 lb/sq in). Most planes are fitted with two, three and even more independent hydraulic systems, so that if one springs a leak or otherwise fails, there are others to take its place.

Structural Strength

In the air weight is at a premium, and a plane must be made as light as possible and at the same time strong enough to withstand the stresses set up in flight. The favoured materials for airframe construction are aluminium alloys, which combine lightness with strength. Typical is duralumin, which contains copper, manganese, magnesium and often silicon as alloying elements. It has much the same strength and hardness as mild steel but is substantially lighter.

The airframe is constructed of sheets of aluminium alloy over a supporting framework. In the wings the aluminium 'skin' is supported by two or more girder-shaped spars which cantilever out from the fuselage. Usually they are formed of individual plates and sections joined together. The skin of the fuselage itself is supported by cross-wise circular frames, or ribs, and lengthwise stiffeners, or stringers. Since the skin of the plane forms a near-continuous shell over

Left: The oldest, yet newest Lockheed 'Tristar'. It made the first test flight in 1970, but is now equipped with the latest 23,000-kg (50,000-lb) thrust RB-211 turbofan engines and sophisticated instrumentation. It incorporates auto-landing system, moving-map display, digital auto-pilot, automatic brakes and many other advanced features.

Below: The instrumentation in a modern airliner is extensive and frankly bewildering to the layman. A modern pilot can no longer fly 'by the seat of his pants' but must rely on instrumental data to guide him.

Below: The McDonnell-Douglas 'Phantom', which first flew in 1958, is still in widespread service worldwide and is capable of Mach 2·5.

Bottom: The most successful vertical take-off plane, the Hawker-Siddeley 'Harrier' jump jet.

the supporting skeleton, this kind of structure is called semi-monocoque, 'monocoque' meaning 'single-shell'.

In places where high strength is vital, such as the wings in highly manoeuvrable combat aircraft, the skin and its reinforcing stiffeners are machined from solid sheet instead of being made up of a number of sheets joined together. The spars may also be forged or machined in one piece.

The individual parts of the airframe may be joined by riveting, welding, or bonding with high-strength epoxy adhesives. Extensive use is now made of sandwich, or honeycomb construction, particularly for the control surfaces and for localized reinforcement elsewhere. In this type of construction skin plates of aluminium are bonded to an aluminium honeycomb core.

Over a long period of service cracks may appear in the skin due to localized stresses, such as at rivet holes, or because of metal fatigue. It could be disastrous if these cracks spread from plate to plate across the skin. So in an airframe joints are deliberately introduced in various areas to prevent such cracks from spreading. This is known as fail-safe construction.

There are regions in a plane where very high temperatures are developed, such as near the jet exhausts. In these places titanium-based alloys or stainless steel are used, which have better high-temperature strength than aluminium.

Vertical Take-Off

Modern high-speed planes in general require a long runway to operate from, for they need to accelerate to a fairly high speed before they can take off. Some combat planes may use rocket assistance at take-off. Other planes are specially designed for a short take-off and landing (STOL). They incorporate devices such as 'blown' flaps on the wings, which permit smooth airflow past the wings at low speeds.

A few planes, however, are designed for a vertical take-off and landing (VTOL). Some swivel their engines bodily, directing them downwards at take-off and then swivelling them into usual fore-and-aft position for normal flight. They have never successfully overcome the problems of transition from vertical to level flight.

One design has achieved spectacular success, however – the Hawker-Siddeley Harrier. It can take-off and land more or less in its own length, yet can exceed the speed of sound in level flight. The Harrier has its single jet engine mounted in the normal position but directs the jet exhaust in different directions by means of swivelling nozzles. It has four nozzles which can rotate through 98° from their fully aft position.

Since there is no effective air flow over the wings in vertical mode, it has a jet reaction control system to keep it stable. This is fed by air bled from the high-

The Boeing 'Chinook', one of the most successful twin-rotor helicopters. Powered by two gas-turbine engines, it can carry more than 30 passengers.

pressure compressor in the engine. The bleed air also powers the motor that actuates the swivelling nozzles.

Perfect Flight

More versatile than any VTOL plane is the helicopter, which can hover or fly in any direction with consummate ease. Its versatility is due to the kind of wing it has, which is a rotary wing. The 'wing' in fact consists of a number of rotor blades, which have an aerofoil cross-section. They are attached to a central hub, which is rotated by the helicopter's engine. As the rotor blades spin round they develop aerodynamic lift to support the helicopter in the air.

Things are not quite that simple, however. When a blade is advancing upwind, it develops more lift than when it is returning downwind. Uncorrected, this would create a serious imbalance which would make for unstable flight. The answer is to make the rotor blades flexible or hinge them to the hub. Then they automatically accommodate themselves to the instantaneous aerodynamic state they find themselves in.

The rotation of the rotor in one direction by reaction tends to make the body of the helicopter turn in the opposite direction. In single-rotor helicopters this tendency is countered by siting a small sideways-facing propeller at the tail. In twin-rotor helicopters the rotors are designed to rotate in opposite directions, so that they counterbalance one another.

Historically the helicopter was preceded by a different kind of rotary-wing craft – the autogyro. This differs from the helicopter in that its wing blades are not powered. They are autorotating – they begin rotating when the autogyro travels through the air. So the machine also requires a conventional means of forward propulsion, for example, a propeller. A few privately owned autogyros exist, but they have long-since passed out of commercial use.

STAR TURNS

The flight controls of a helicopter work by altering the pitch of the rotor blades. The diagram below shows the way in which the pitch is altered. The blades are flexibly attached to the transmission shaft and spun round by it. Each blade is linked by a rod to a so-called rotating star. Beneath this is a stationary star, which can be moved bodily up and down or tilted in any direction by means of pivoting control linkages. The linkages are activated by two main flight controls, via a so-called mixing unit.

One is the collective-pitch control, which increases or decreases vertical lift by altering the pitch of all the blades in unison. To increase lift, for example, the control stick is pulled up. This causes the control linkages to lift the stationary star, and thus the rotating star, thereby increasing the pitch of the blades.

For horizontal flight the rotor blades must provide horizontal thrust as well as lift. This is done by making the pitch of the blades vary during their cycle of rotation. Control over this operation is provided by the cyclic-pitch control, which tilts the stationary star in the appropriate direction. For forward motion, for example, the stars tilt so that the blades have their maximum pitch when moving downwind and their least when moving upwind.

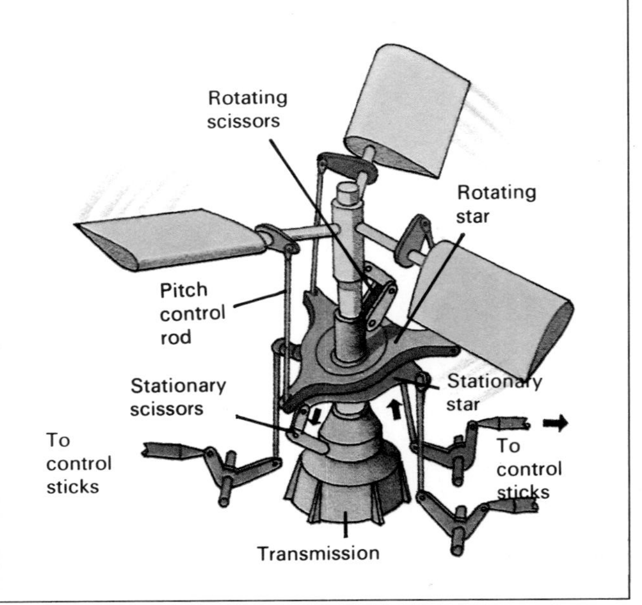

Communications

Left: The TV and radio mast on Twin Peaks beams broadcasts to the San Francisco Bay area.

Right: A 'page' from the TV magazine presented by BBC's Ceefax teletext service.

Top right: One of the most sophisticated of the present generation of 'compact' 35 mm single-lens reflex cameras, the Nikon FE, seen here coupled with its motor-drive unit.

Bottom right: The more efficient push-button telephones are gradually replacing dial phones, even in call boxes.

Never have people en masse been better informed than they are today. Thanks to modern technology, we are all bombarded with news and views about people and places anywhere in the world. These come through the mass media – the press, radio and television. By means of television we can see events as they actually happen, maybe thousands of kilometres away, maybe even on another world. The sheer immediacy of television and its presence in the living room gives it more impact and influence than any other communications medium. Its true value, as an educational tool, has yet to be fully realized, its main function in the average home at present being as a purveyor of entertainment.

It was said that the television would kill the oldest form of mass communication – the newspaper – but this has not happened. However, the latest developments in teletext could eventually have this effect. Teletext effectively offers viewers an electronic news and information magazine, moreover a magazine that can change minute by minute as material is updated. When fully exploited, teletext will be able to provide viewers with all kinds of instant information they cannot readily get elsewhere – up-to-date weather 'now-casts' for any district, menus from popular restaurants, railway timetables, and so on.

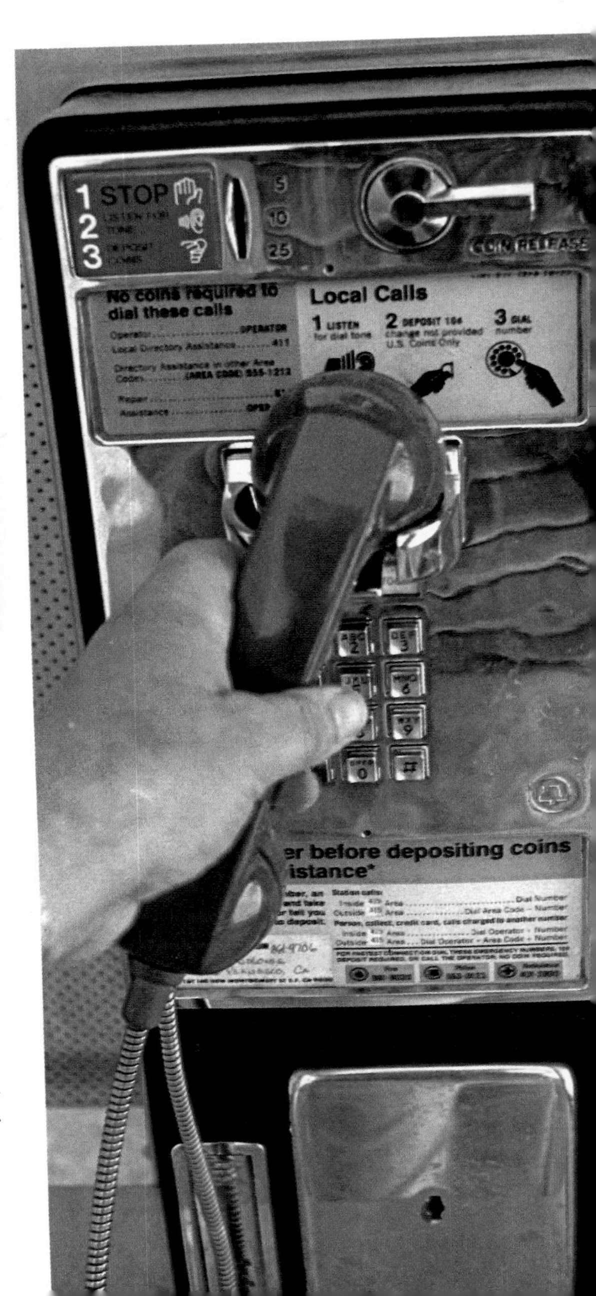

The teletext television of today, made more versatile with a coupled video recorder, is a miracle of electronic wizardry. It is the end product of a stream of nineteenth century innovation and development that encompassed the telegraph, telephone, phonograph (gramophone), tape recorder and wireless telegraph (radio). The modern versions of these latter-day wonders use the same sophisticated technology as the television. They all utilize compact components made from deliberately impure crystals (semiconductors), and many incorporate the miracle chip (microprocessor), which apparently seems to offer all things to all men. Composite 'videocasseivers' are already available that receive radio programmes on all wavebands and in stereo, and television programmes on a microvision screen; record onto and playback from an integral cassette; and incorporate a stopwatch and alarm clock. It is all a mite mind-boggling.

Power of the Press

The introduction of mechanical printing following the re-invention of movable type in the 1450s by Johannes Gutenberg (see page 29) could not have been more timely. It coincided with the reawakening of interest in philosophy, the arts and the sciences during the period we call the Renaissance.

Few significant improvements to the original screw printing press occurred until the end of the eighteenth century. In about 1795 the Earl of Stanhope in England constructed the first all-metal press, improving the quality and quantity of production.

In the early 1800s Friedrich König in Germany experimented with using steam power to drive printing presses, and by 1814 had developed an effective machine that was able to print for London's *Times* 1100 sheets an hour. It used a cylinder to press the paper against the inked type.

The first means of mechanically illustrating books was by xylography, or printing from woodcuts, designs being cut in relief in wooden blocks. This complemented the method of printing (letterpress or relief). By the 1600s books were being illustrated by means of engraved or etched metal plates, printing being from the inked recesses of the plate (intaglio). In the 1790s the first steps were made in the art of lithography. This means the printing from a flat surface (originally stone = Latin 'lithos') on which a design is drawn by greasy ink. When wetted and inked, only the design retains the ink, which can then be transferred to paper.

Writing with Light

It was while experimenting with means of inscribing an image on lithographic stone that Frenchman Joseph Nicéphore Niépce, in about 1826, pioneered the art of photography. To develop photographic techniques, he joined in partnership with Louis Daguerre, who accidentally discovered a method of developing a photograph image on copper plates coated with silver iodide, using mercury vapour. Daguerre announced his 'daguerreotype' process in 1839, making photography practical. At the same time, in England, William Henry Fox Talbot was experimenting with the negative-positive process of photography which was eventually to become dominant. George Eastman in the United States introduced photography for the masses by marketing his snapshot Kodak camera in 1888. It was the prototype box camera with a 1/25th second shutter. It contained a roll film long enough to make 100 exposures. It had to be returned to Eastman's factory for processing. 'You press the button, we do the rest', was Eastman's slogan.

While some photographers were perfecting photography as an art form, others were experimenting with devices to show moving pictures. 'Movies' came from the experiments of such pioneers as William Friese-Green in Britain and the Lumiere brothers, Auguste and Louis, in France. The Lumières shot their first film in 1895, using a portable hand-cranked camera of their own invention, which also served as projector. It used the principle of intermittent projection of still pictures, which is still current.

Calling Long Distance

The 1800s also witnessed the beginnings of long-distance communications and the birth of our modern electronic age. William Cooke and Charles Wheatstone in England and Samuel F. B. Morse in the United States developed different forms of electric telegraph, both in 1837, using the recently discovered phenomenon of electromagnetism.

In 1866 the telegraph began to go global with the laying of the first transatlantic cable between Ireland and the United States by Brunel's giant *Great Eastern*, the largest ship then afloat. Ten years later painstaking experiments by a teacher of the deaf, Alexander Graham Bell, resulted in the first telephone.

In 1887 German physicist Heinrich Hertz demonstrated that the electromagnetic waves which came to be called radio waves could be deliberately transmitted and received through the air. This prompted Italian engineer Guglielmo Marconi, among others, to experiment with sending coded signals from point to point by means of radio waves rather than wires. In 1895 he succeeded in doing so over a distance of 1·6 km (1 mile), and pioneered wireless telegraphy.

In 1901 Marconi's equipment transmitted the first wireless message across the Atlantic. A few months previously Reginald A. Fessenden demonstrated voice transmission by wireless, though exploitation of this 'wireless telephony' had to wait for improvements in transmitting equipment. These included particularly the invention of the thermionic valve (electron tube) by John Ambrose Fleming in Britain in 1904 and of the amplifying triode valve by American Lee De Forest three years later.

Meanwhile, inventors were experimenting with a new means of communication, involving the transmission of pictures by radio. In the forefront of this new 'television' technology was English inventor John Logie Baird, who succeeded in 1925 in transmitting a recognizable picture through the air using a hotchpotch of make-shift apparatus.

The 1930s saw radio waves being exploited in a different way – to locate objects in the sky. A research team in Britain under Robert Watson-Watt could by 1938 detect the presence of aircraft in the sky over 160 km (100 miles) away by bouncing pulses of short radio waves from them. This technique, which became known as radar, proved a decisive factor in Britain's air defence in World War 2.

In 1948 came the invention of the transistor. Because of its tiny size and low power consumption, the transistor revolutionized the design of communications equipment. By the 1960s further scaling-down of transistor-type components on silicon chips led to the birth of the communications satellite.

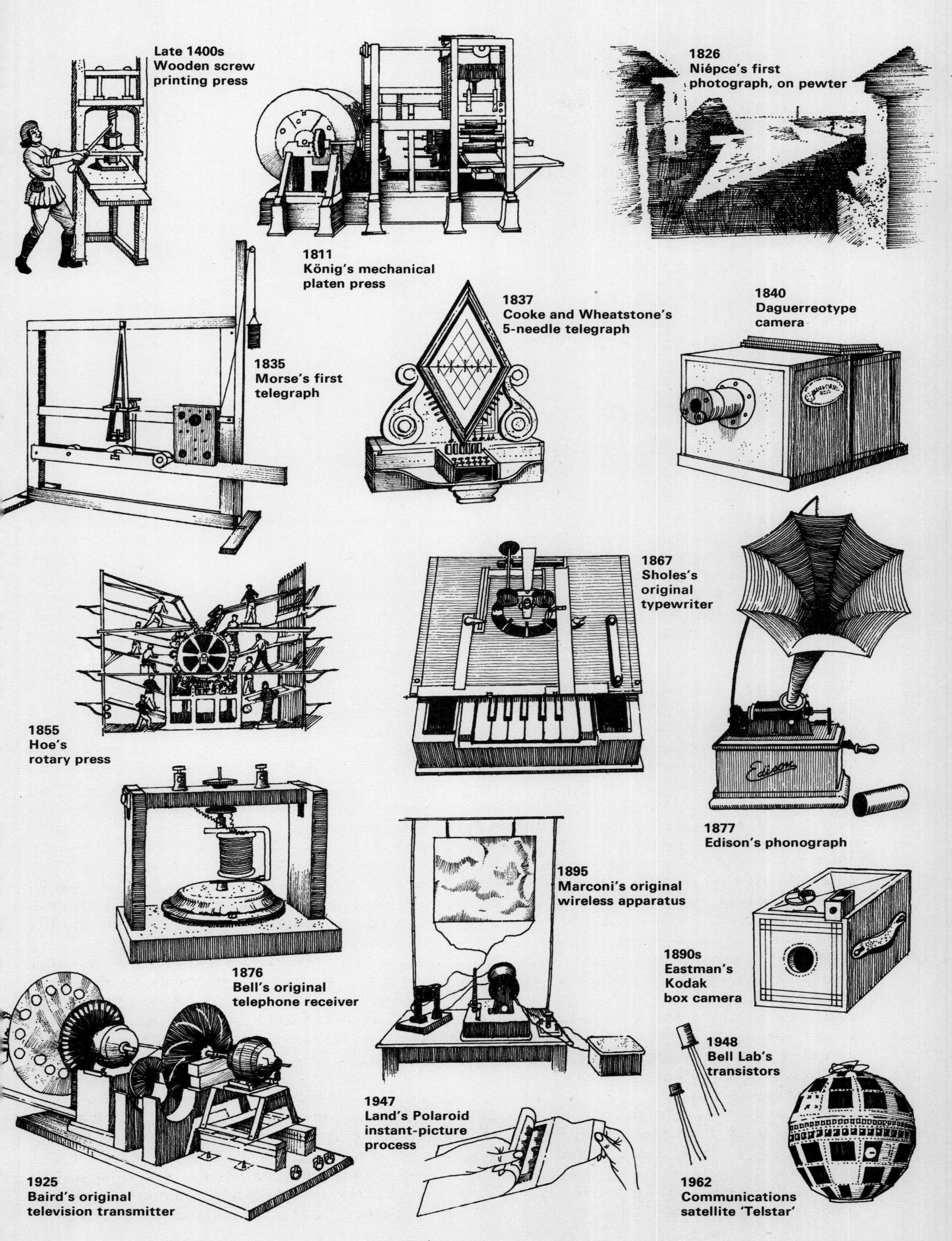
Late 1400s
Wooden screw
printing press
1811
König's mechanical
platen press
1826
Niépce's first
photograph, on pewter
1835
Morse's first
telegraph
1837
Cooke and Wheatstone's
5-needle telegraph
1840
Daguerreotype
camera
1855
Hoe's
rotary press
1867
Sholes's
original
typewriter
Edison
1877
Edison's phonograph
1876
Bell's original
telephone receiver
1895
Marconi's original
wireless apparatus
1890s
Eastman's
Kodak
box camera
1948
Bell Lab's
transistors
1925
Baird's original
television transmitter
1947
Land's Polaroid
instant-picture
process
1962
Communications
satellite 'Telstar'

The Printed Word

The traditional method of printing pioneered by Gutenberg, known as letterpress, is very widely used even today. It is still dominant in printing newspapers, but it is beginning to be superseded, particularly for book printing, by offset lithography, or litho. A third method is also used – rotogravure, which is particularly suited to the production of good quality illustrations.

Letterpress is a relief printing method, in which the printing surface is raised. Rotogravure is an intaglio method, in which the printing surface is recessed. Litho is a planographic method, in which the printing surface is flat. In modern practice, printing by either method is done on high-speed rotary cylinder machines, which may print on separate sheets, or on continuous rolls of paper, which are then cut to size. The former is called a sheet-fed press, the latter a web-fed. Modern web-fed rotary presses can operate at a fantastic speed, with an output of over 50,000 copies of a 32-page newspaper per hour.

In letterpress printing type may be cast in hot typemetal – a tin-lead alloy – either in full-line 'slugs' by a Linotype machine, or in separate characters on a Monotype machine which are then assembled into lines. The type is then made up into pages, and from these the printing plates are produced. A sheet of pasteboard is pressed heavily against the type to produce an impression. Then molten typemetal is poured into it inside a curved casting box. After trimming, this casting is electroplated with nickel to make it harder wearing, and is ready to be fitted on the press cylinders. This curved printing plate is known as a stereotype. Better quality plates can be made by a slightly different process involving electroplating with copper as well as nickel, the plate being known as an electrotype.

In rotogravure a separate printing plate may be made, but often the surface of the printing cylinder itself does the actual printing. In this process pits are etched in the cylinder by acid solution, preliminary masking treatments having been carried out. The cylinder itself is made of steel on which has been deposited a thick layer of copper, and it is the copper layer that is etched. The acid bites into the copper more or less depending on how it has been masked, to produce pits of various depths.

In printing, the whole cylinder is inked, then a doctor knife scrapes off the excess ink from the upper surface leaving it only in the pits. When the cylinder is pressed against the printing paper, the ink is transferred. The deeper pits retain more ink and print darker than the shallower ones, thus rendering different shades.

Printing in the Flat

Offset-litho printing uses a flat printing surface treated so that only the printing areas attract the printing ink. It works really on the principle that oil and water do not mix. It is known as offset because the printing cylinder does not itself come into contact with the paper, but transfers, or offsets, the printed image to an intermediate cylinder, the blanket cylinder, which in turn transfers the image to the paper. This method produces a sharper image than direct transfer would.

Litho lends itself particularly to typesetting by photocomposition, or film-setting, since a negative image of the text is needed to prepare the plate. In film-setting the typesetter first prepares a perforated tape carrying details of the text setting required. And this is fed to the composing machine. The perforations control the positioning of a matrix case containing transparent type characters. A beam of light shines through the matrix to form a latent image of the type, on a piece of film, which can then be developed.

In the plate-making process, the type negative is laid on a photosensitized plate made out of coated zinc or aluminium. It is then exposed to strong light and under the type areas that are exposed, the photosensitive coating is hardened. The rest of the coating is unaffected and is washed away. The exposed metal has an affinity for water and is easily wetted, while the hardened coating has an affinity for grease and repels water. Since printing inks are greasy, they adhere to the hardened coating which carries an impression of the type.

Telecommunications

Long-distance (tele-) communications have undergone revolution after revolution in the century and a half since Samuel Morse first devised the electric telegraph. Today intercontinental communications are as likely to be routed through space as through wires via the Intelsat communications satellites in geostationary orbit (see page 206). This global satellite system is run by Intelsat, the International Telecommunications Satellite Organization, made up of 80 countries. Many countries and organizations (such as NATO) have their own communications systems.

Far left: A printed sheet emerging from the final cylinder of a litho press.

Left: Pictures showing the separate colours printed in succession to reproduce a photograph.

Below: Diagrams illustrating the three basic printing processes of letterpress, offset litho and gravure. The printing surface is, in order, raised, flat and recessed.

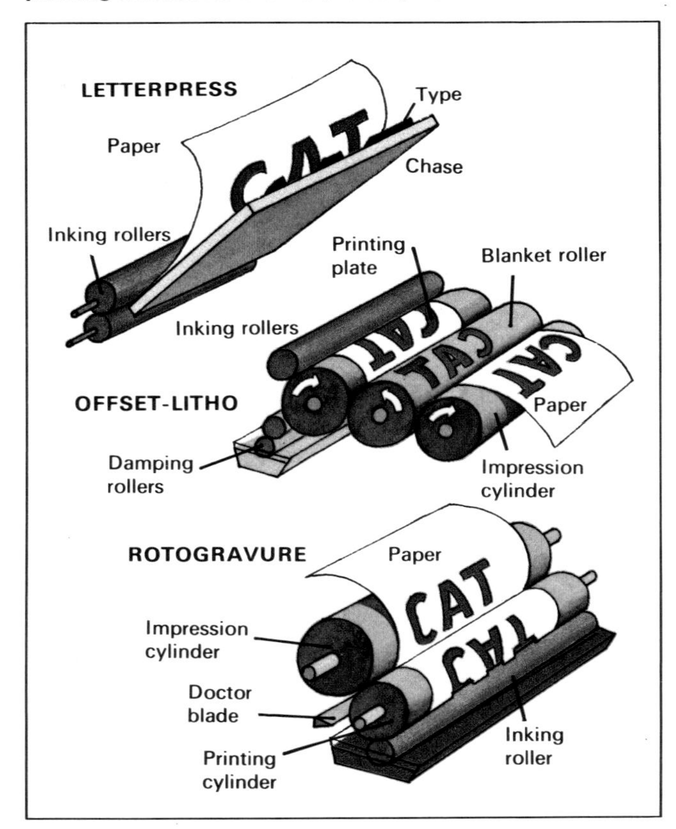

In a modern telecommunications set-up the message to be sent – telegram, computer data, voice, picture – goes through essentially the same stages between transmitter and receiver. First they have to be converted into fluctuating electric signals. In telegraphy this is done by a teleprinter, the modern derivative of the Morse key; in telephony and radio by a microphone; and in television by a TV camera. The signals may then be encoded into digital form. The converted, and if necessary coded signals are then superimposed on a high-frequency carrier wave, which can transmit them most efficiently. For radio – wireless – transmission the carrier wave is of radio or microwave frequencies; for transmission through cables, which are of the concentric, co-axial type, the carrier wave is of much lower frequencies.

The process of superimposing signals on the carrier wave is known as modulation. Three distinct methods of modulation are used in telecommunications. One is amplitude modulation (AM), in which the height, or amplitude of the carrier wave is modulated. Another is frequency modulation (FM), in which the frequency, or rate of alternation of the wave is modulated. The third is pulsed-code modulation (PCM) in which signals are transmitted in descrete pulses, not continuously.

Different carrier wave frequencies can be sent simultaneously through cables or through the air, making it possible to transmit many messages along the same path. This technique, called multiplexing, makes it possible to transmit several thousand two-way telephone conversations, for example, along a conventional co-axial cable.

At the receiving end of a telecommunications system a demodulator and decoder sort out the carried signals from the carrier wave and feed them to the appropriate receiver. This would be another teleprinter, a telephone handset, a radio or TV set.

Messages by Wire

For transmitting telegraphic code the teleprinter or teletypewriter is used. This is used to send messages via the telephone cable network by post offices and, via Telex, by private subscribers. To operate a Telex machine, a subscriber's number is dialled and a connection made between his and the sender's teleprinters. The sender then transmits a coded message along the line, which is simultaneously typed out by the receiving machine.

More familiar to the ordinary telecommunications user is the telephone. This houses in the handset both a transmitter (mouthpiece) and receiver (earpiece). The transmitter consists of a carbon microphone. It has a diaphragm which is in contact with lightly packed grains of carbon. The carbon forms part of a circuit through which a low-voltage (10

Finger plate

Gear

Return spring

Cam

Contacts

Above: The dialling apparatus of a telephone exploded. On the right is a regulating governor, which controls the speed at which the dial returns.

Above: A Telex machine. Operation of the keyboard encodes the message on punched paper tape. This is fed through the machine to transmit, and the message is simultaneously typed out at the receiving terminal.

Electromagnets

Carbon granules

Diaphragm

Diaphragm

Receiver (earpiece)

Microphone (mouthpiece)

A telephone handset, modern in appearance but relying on nineteenth-century technology.

volts) current is passed. When you speak into the microphone, your voice vibrates the diaphragm, and packs the carbon grains more tightly or less tightly together. This has the effect of altering the electrical resistance of the grains, and it thereby causes the current passing through the grains to fluctuate. These fluctuations are the voice signals.

After transmission, decoding, demodulating, and so on, similar signals enter the receiver of the person you are talking to. There they pass through the coils of a pair of electromagnets and make their magnetism fluctuate in sympathy. This causes a metal diaphragm in the receiver to vibrate, which in turn generates sound waves. These are a replica of the sounds you uttered into your microphone.

To contact a particular number, you first pick up the handset from the cradle. The cradle switch rises, applying the line voltage and connecting the telephone to the exchange. When you dial a number, the returning dial plate rotates a train of gears that open and close a pair of contacts and so generate electrical pulses. Each figure dialled generates that number of pulses. In the push-button phone an electronic circuit memorizes the sequence of pulses until the end and then transmits them together. These pulses travel along the telephone wire to the exchange, where they activate the switching mechanism, which will eventually connect your line with that of the person you are calling. The former electromechanical switching mechanisms are now being rapidly superseded by electronic ones, making for faster, more silent and more reliable connections.

Tuning In

As outlined earlier, in radio transmission voice signals are converted by a microphone into varying electrical (audio) signals. These are then superimposed on a radio-frequency carrier wave by modulation and transmitted. The radio set has to receive and then sort out the signals. It picks up the radio signals, which are electromagnetic vibrations, with an aerial. The aerial is bombarded as it were, with radio waves of many frequencies, each one carrying a different programme. And to select particular frequencies the radio has to be tuned.

Tuning usually involves altering the capacitance of a circuit linked to the aerial by means of a multi-leaved variable condenser (capacitor). When correctly tuned, the circuit allows in the desired radio carrier wave to the exclusion of others. The incoming signal is very feeble and so is next amplified by means, these days, of transistor circuits. Transistors perform the same task as electron tubes, or valves once did. Following amplification comes detection, or demodulation, which results in a weak replica of the original audio signals generated by the transmitting microphone. These signals are again amplified and fed to a loudspeaker, which converts them into audible sounds.

The standard loudspeaker is not like that in a telephone handset receiver but is of the moving-coil type. It consists of a large paper cone, to the apex of which is attached a small paper cylinder. Around this cylinder, which is located within the field of a magnet, is wound a fine wire coil. When the audio

signals pass through the coil, they make it faintly magnetic, causing it to interact with the field of the magnet. And it is attracted or repelled in sympathy with the incoming signal. The paper cone vibrates and creates sound waves replicating those originally fed to the transmitting microphone.

View From Afar

The process of television – the transmission of pictures by radio – is a relatively complicated one and can only be given in outline here, but the principles can be readily understood. The TV camera views a scene as a whole, as (for black-and-white TV) patterns of light and dark. It then proceeds to dissect that pattern into a series of closely spaced horizontal lines, and to generate variable electrical signals to represent the light and dark shades in each line. These line-by-line signals are generated rhythmically and fed to the transmitter. The TV receiver picks up the signals and rhythmically recreates the picture line by line on a fluorescent screen.

The process of line-by-line dissection and recreation is known as scanning. A picture is not scanned simply from top to bottom, but in two steps; first odd-numbered lines are scanned and then even ones. But the process occurs so rapidly that 25 complete pictures are dissected and recreated every second. This is analogous to motion pictures when a series of 'still' pictures is thrown on the screen. In each case our 'persistence of vision' fools our eyes into believing that movement is taking place. In most countries the standard TV system uses 625 lines to make the picture, but 405 lines is the standard in the United States.

In outline the TV camera (black and white) works like this. Light from the scene being televised is focused by a lens on to a vacuum tube called the camera tube. One of the latest types is called the Vidicon. In this tube the focused image passes through a so-called signal plate onto a photo-sensitive layer, which acquires electric charge according to the light falling on it. An electron beam from an 'electron gun' rhythmically scans the layer, which carries an image of the scene viewed in electric charges. As it scans, electric charge is transferred to the signal plate, from which picture (video) signals are taken.

On Screen

These video signals, representing shades of brightness in each line, are transmitted, together with a synchronizing scanning signal that details which line is being transmitted. In the TV receiver the brightness and scanning signals are eventually fed to a cathode-ray tube (CRT), whose fluorescent end

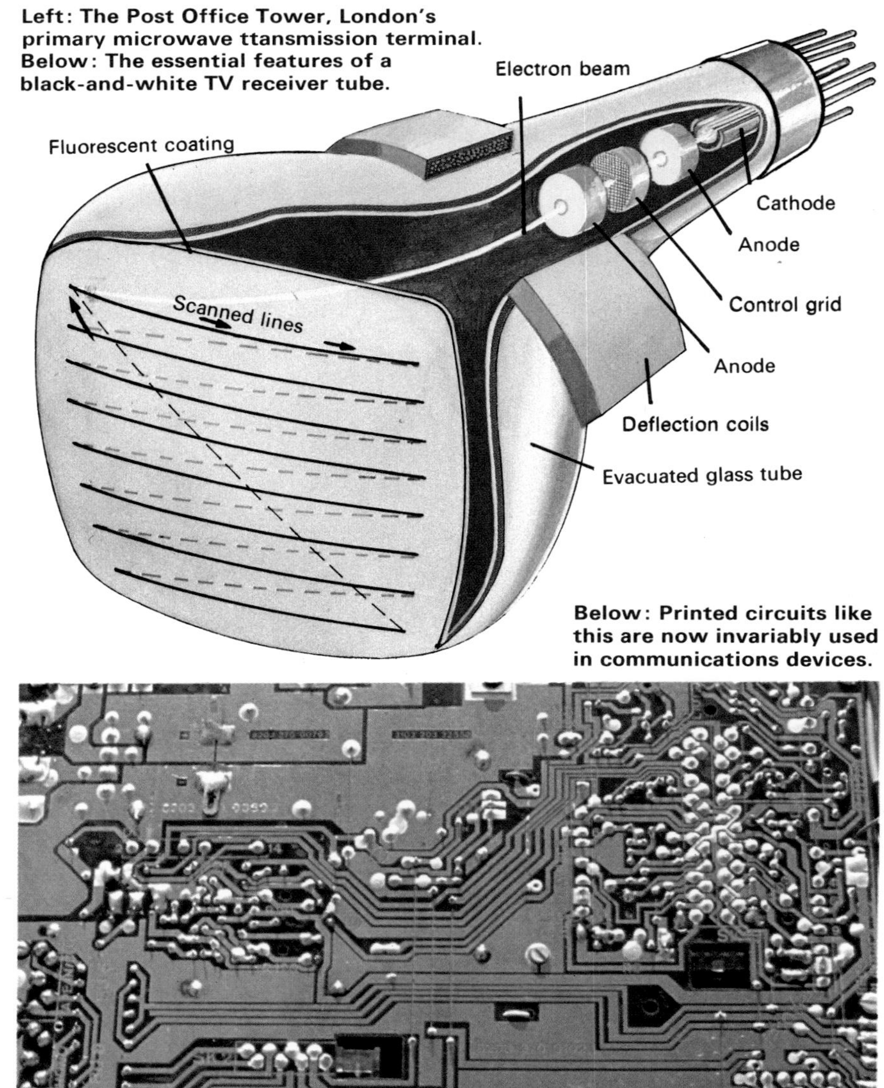

Left: The Post Office Tower, London's primary microwave ttansmission terminal. Below: The essential features of a black-and-white TV receiver tube.

Below: Printed circuits like this are now invariably used in communications devices.

forms the familiar TV screen. An electron gun in the CRT generates a beam of electrons. The scanning signals are fed to deflection coils around the tube which make the electron beam scan in the same manner as and in synchronization with the camera scanning beam.

The brightness signals are fed to an anode which causes the speed of the electron beam to vary. The faster it travels, the brighter spot it makes on the coated fluorescent screen. The coating consists of materials called phosphors, which glow when electrons hit them. Zinc and cadmium sulphides are common phosphors.

Glorious Colour

The principles of black-and-white TV transmission can be extended to give colour transmission. As in colour photography (page 149) the colour process depends on the fact that any colour can be made up by mixing together light of the three primary colours – blue, green and red – in different proportions. In the colour camera a system of prisms, mirrors and filters splits up the image formed by the camera lens into its primary colours. These coloured images are then scanned by three separate camera tubes, which convert the blue-ness, green-ness and red-ness in the scene into equivalent electrical signals.

These colour signals, and the appropriate scanning signals of course, are transmitted. They are eventually sorted out by the colour receiving set and fed to separate electron guns in the TV tube. These guns fire electrons representing the brightness of the various colours in the screen. By means of a perforated shadow mask it is arranged that the electrons carrying blue-ness fall on blue-emitting phosphors, those carrying green-ness on green phosphors, and those carrying red-ness on red phosphors. In each spot on the screen the colours effectively merge together to recreate the original colours.

Recording on Disc

Until recent years by far the most widely used means of playing pre-recorded music at home was by gramophone, or record player, using the familiar grooved, black plastic disc. This is now being challenged by the more portable and robust cassette player, using magnetic tape (see below). But the disc will be with us for some time to come in one form or another.

In outline the recording and playback process goes like this. A microphone in the recording studio picks up the sound vibrations in the air and converts them into analogous electrical vibrations. These are fed to an electromagnet which vibrates a cutting needle, or stylus, resting on a lacquer-coated disc which is rotating. The vibrating stylus cuts a wavy spiral groove in the disc. Copies are then made from this master disc in PVC plastic. To playback the recording, the stylus of the record player is placed in the spiral groove, and the disc is rotated at the same speed as the master was during recording. The wavy grooves on the disc make the stylus vibrate from side to side, and this vibration is converted into a variable electric current. This current exactly mirrors the current generated by the microphone originally and, when fed to a loudspeaker, causes the speaker cone to vibrate and recreate the original sounds.

In a practical record player the disc is played on a turntable driven by electric motor. It may be driven by rubber-tyred wheels from the motor spindle or by belt. Different sized wheels or belt pulleys come into play to give a choice of turntable speeds, usually $33\frac{1}{3}$ rpm (LP), 45 rpm (EP) and 78 rpm. Many modern players have only the first two speeds. The stylus that plays the disc is made of a tiny cone of synthetic sapphire or diamond with a spherical head. It is held in a cartridge, or pick-up head, on the end of a pick-up arm and rests lightly on the disc.

The electro-mechanical transducer (the device that changes the mechanical vibration of the stylus

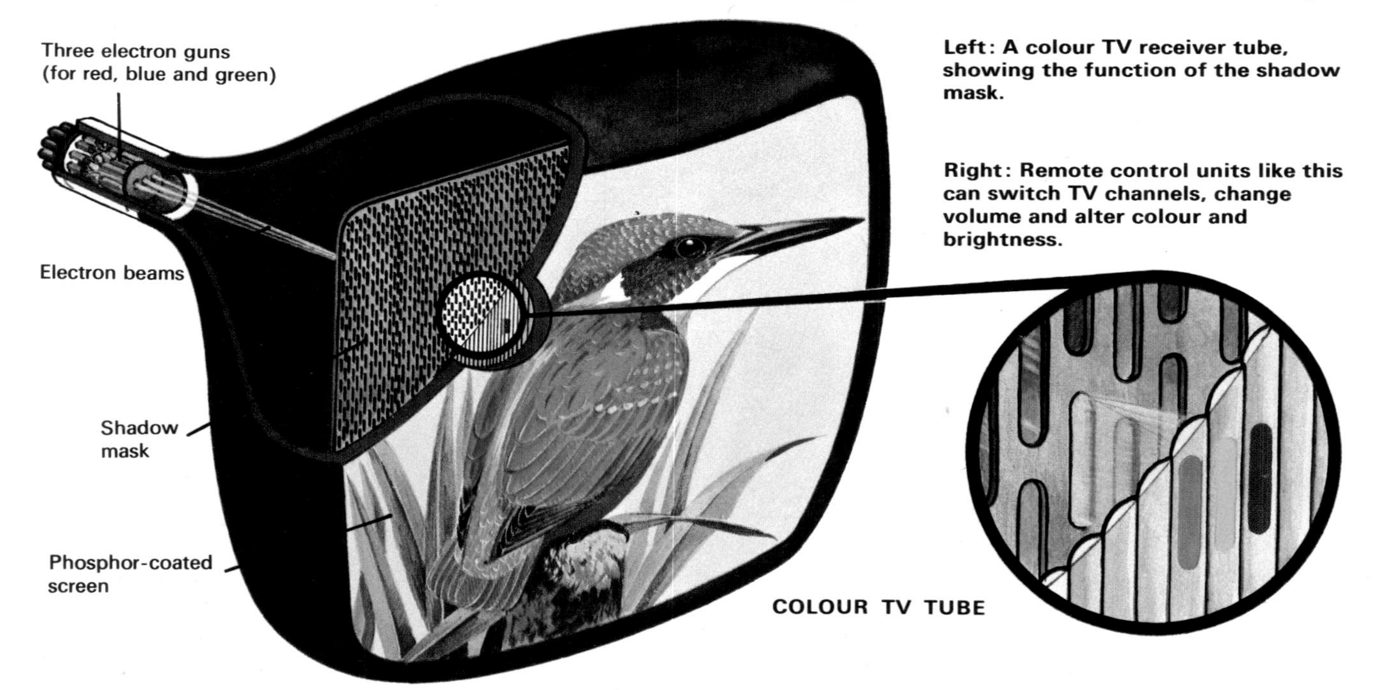

Left: A colour TV receiver tube, showing the function of the shadow mask.

Right: Remote control units like this can switch TV channels, change volume and alter colour and brightness.

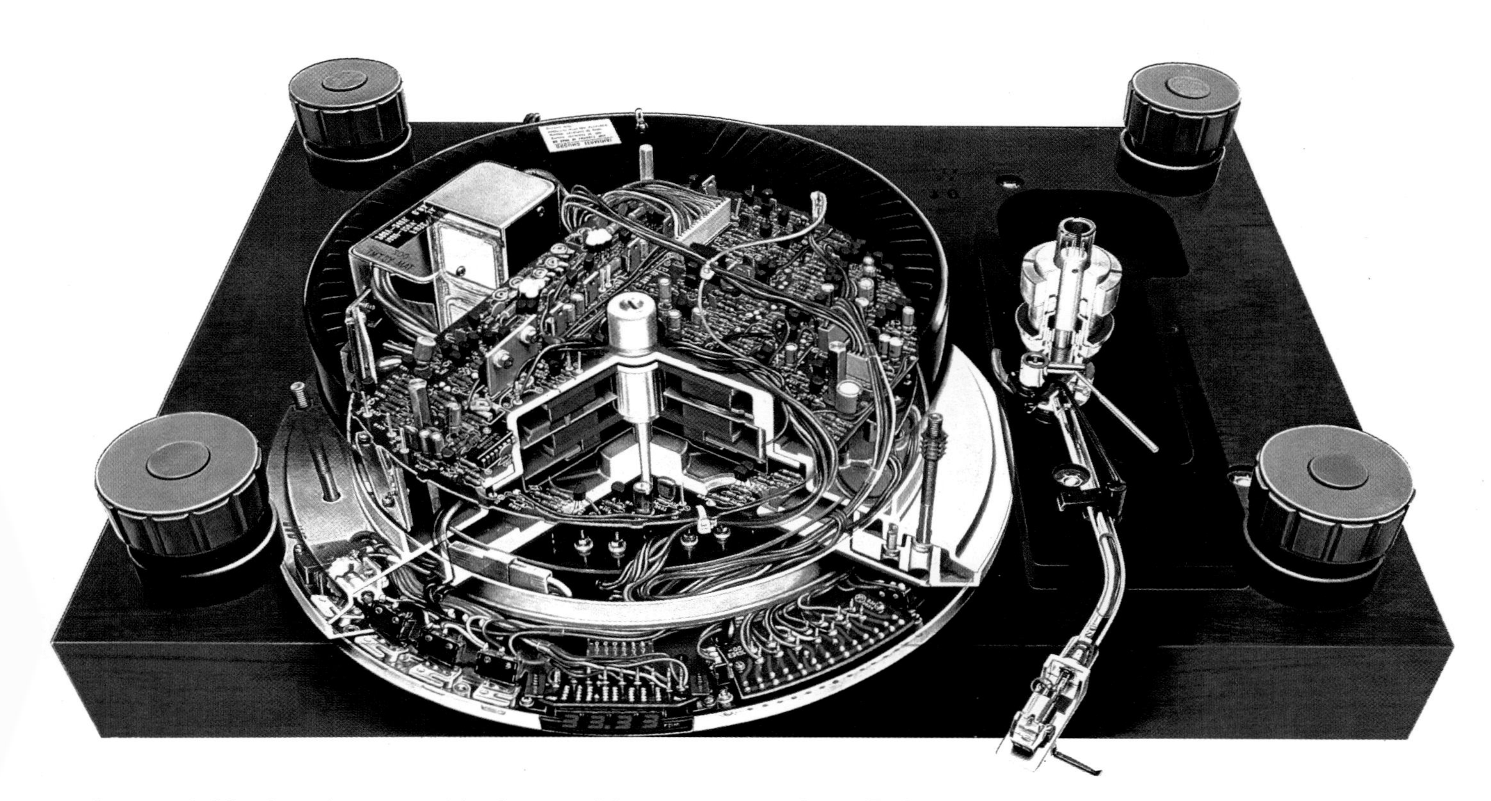

into variable electric current) in the cartridge can be one of several types – electromagnetic, ceramic or crystal. The widely used crystal type utilizes the phenomenon of piezoelectricity, in which certain crystals (such as Rochelle salt and barium titanate) produce tiny electric currents when they are deformed. The currents produced by either type of transducer are very feeble and need to be amplified before they can be fed to the loudspeaker.

Practically all modern records are in 'stereo' – stereophonic sound. Sounds are recorded in two microphones as 'right' (R) and 'left' (L) sounds and are played back through the record player as R and L sounds. In this way a more realistic depth of sound is achieved. In practice the R sounds are recorded on one side of the disc groove and the L sounds on the other. Twin sensors attached to the stylus feed separate signals to R and L speakers. In recent years four-speaker, quadrophonic reproduction has been pioneered to give a complete 'surround sound'.

Getting It Taped

Cassette tape-recorders are miniature versions of the magnetic tape-recorders that have been used for decades. In the full-scale machine the tape is about 6 mm ($\frac{1}{4}$-in) wide and is threaded externally through the recording head from one large spool to another. In the cassette player, however, the tape is much narrower (less than 4 mm, 0·15 in) and is permanently wound on reels within a handy plastic cassette, which slots into a snap-shut holder for playing.

In the holder it is automatically brought into

Below: Hi-fi enthusiasts prefer separate electronically matched units for sound reproduction rather than an all-in-one music centre, whose great virtue is compactness.

Above: The complex circuitry of a modern hi-fi record player is here revealed. Great care is taken to prevent motor-induced rumble, flutter and wow.

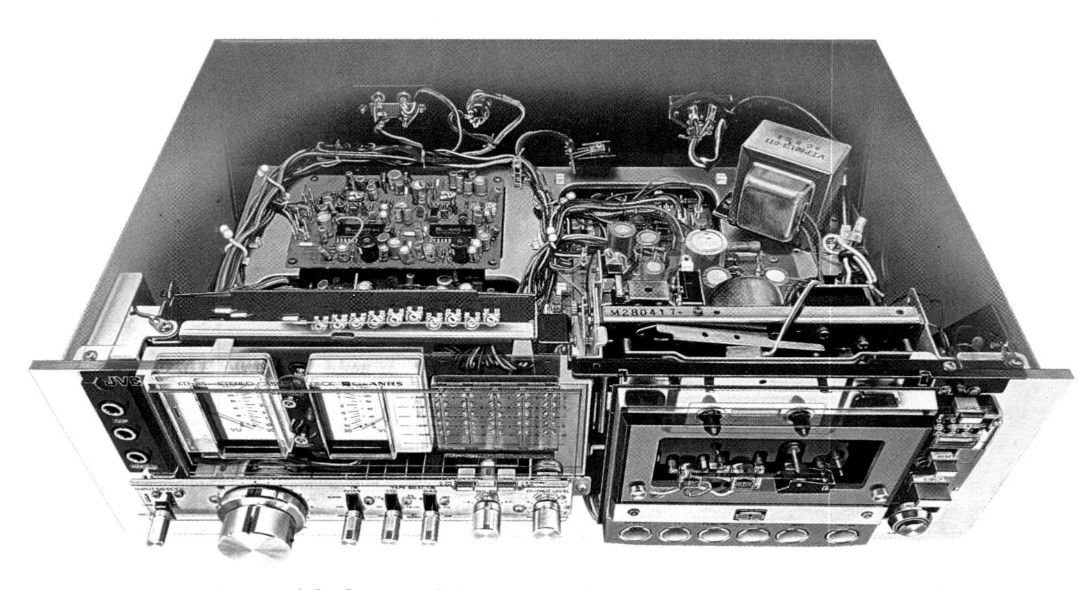

Left: A stereo cassette player with high-fidelity performance. This would form one of the modules in the kind of hi-fi set-up illustrated overleaf.

Below: Through the medium of photography we can capture the ephemeral image, the fleeting moment, and preserve it for posterity. We can record, with more or less artistry, our encounters with people and places, artefacts and architecture, and the world of nature.

contact with the spool drives and the recording head. In both large and small recorders, the tape is a plastic ribbon coated with either iron oxide (Fe_2O_3) or chromium oxide (CrO_2). The usual tape running speed for both is 47·6 mm ($1\frac{7}{8}$ in) per second, and they both work on exactly the same principles.

In recording the tape is passed across a recording head, which consists of an electromagnet – a core of iron with coils of wire surrounding it. The variable electrical signals from the recording microphone are passed through the head coils, creating a matching variable magnetic field. This field magnetizes the oxide particles on the tape moving past in a characteristic pattern. To play back the tape, it is rewound and passed across the head once more, but this time the head coils lead, via an amplifier, to a loudspeaker. The motion of the magnetic pattern on the tape past the head, induces in the coils a variable electric current identical to the one which created the pattern during recording. When this is fed to the loudspeaker, the original sounds are reproduced.

The latest thing in cassette recording relates not to audio but to video. It is the video tape recorder (VTR) known also as VCR (Video Cassette Recorder) or VHS (Video Home System) according to the type of recorder used. The systems are not compatible. The VTR uses a 13-mm ($\frac{1}{2}$-inch) wide magnetic tape on which are recorded magnetic patterns corresponding not only to sound signals, but vision signals as well – in colour. The VTR is plugged into an ordinary receiver to record and play back.

The VTR enables you to view one programme while recording another on a different channel; it has a built-in digital clock so that it can switch on and record programmes even when you are not at home. You can also buy pre-recorded tapes of films, for example, to play back at home. VTR manufacturers also offer a portable compatible video colour camera which enables you to make your own films for TV.

Permanent Pictures

Capturing permanent pictures rather than ephemeral television images is one of the most popular of all pastimes and hobbies. The practice of photography – 'writing with light' – has never been so foolproof, and excellent results can be obtained with even the cheapest cameras. And the general trend even with the more expensive ones has been towards automatic control. More sensitive 'electric-eye' exposure metering, coupled with 'microchip' technology, assures near-perfect pictures every time. And if you don't like waiting for your film to be processed, there are several 'instant-picture' cameras which will present you with a finished photograph in a minute or so. But first to basics.

Photography relies on the chemical action light has on various silver salts called halides (iodide and chloride). These salts are affected invisibly when light falls on them, but by the process of developing the invisible, or latent image they carry can be brought to light. The film is held in a light-tight box – the camera. So sensitive is it that it need only be

The essential elements of the simple box camera, which has its origins in the camera obscura ('dark room' in Latin) used by artists centuries before the birth of photography. The camera obscura was a darkened room to which light was admitted through a small hole. It threw an image of the outside scene, upside-down, on the wall opposite, which could then be sketched.

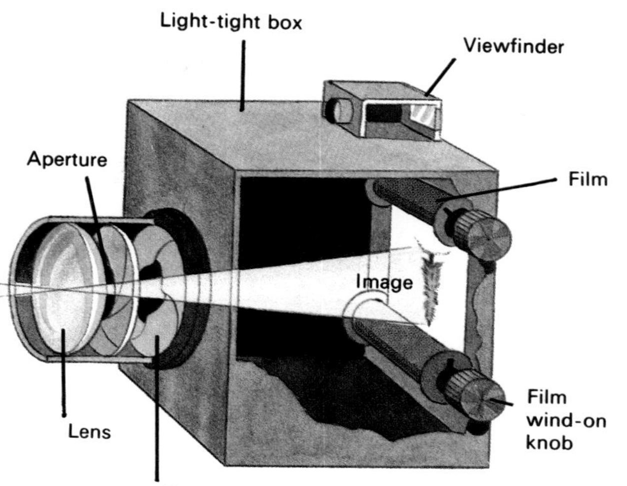

Below: Outline of a ciné projector, which is used to show movies – moving pictures. There is actually no such thing as a moving picture. It is an illusion created by our brain as a result of the persistence of vision of our eyes. What the movie projector does is throw onto a screen a succession of still photographs ('frames') taken in rapid succession. In the projector the film is fed in front of the lens, and light is shone through it from a lamp. During projection, the film is held firmly by the pressure plate. It is moved along one frame at a time by the claw feeding device. While the film is moving past the lens the rotary disc shutter blocks off the light from the lamp so that no blurring occurs. The standard projector projects at a film speed of 24 frames per second.

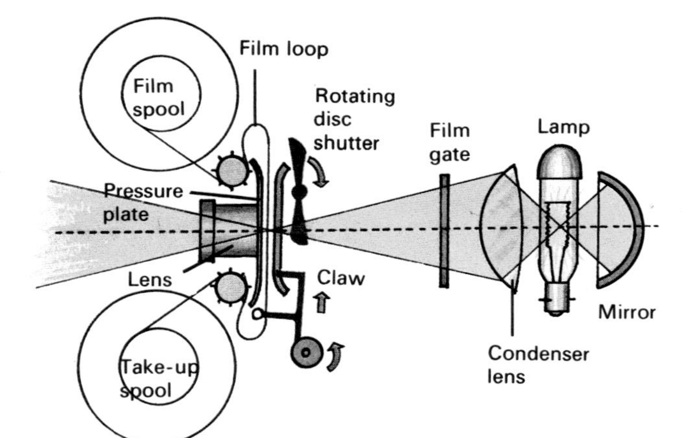

exposed for a fraction of a second to a minute amount of light. So the camera has only a narrow opening (aperture) to admit the light and a shutter over it which can open fleetingly to make the exposure. To ensure a sharp image, a lens is located in front of the aperture to focus light on the film. The final element of the camera is a viewfinder, which allows the photographer to view the scene he is taking.

Many practical cameras are almost as simple as this, except that they have provision for loading and winding on the piece of film, and a shutter-release button to trigger the shutter. They are modern versions of the original 'box camera', though they are seldom box-like any more. More advanced cameras have additional features which allow the photographer more control over the quality of the picture he is taking. He has provision to vary the exposure, either by adjusting the aperture of the lens or the speed of the shutter. He also has means of moving the lens in and out to focus sharply on objects at different distances from him, and usually a coupled range-finder to make focusing more precise. The lens itself is also more complex. It is a compound lens made up of a number of separate lenses of different curvature and of different glass. This design allows lens defects like chromatic (colour) aberration to be corrected.

f-Numbers and Shutter Speeds

The photographer is able to vary the aperture in distinct steps. Marked on a ring on the body or barrel

of the lens are numbers signifying the aperture, usually something like:

22 16 11 8 5·6 4 2·8 2 1·4

These are called *f*-numbers and are the ratio of the diameter of the aperture to the focal length of the lens. *f*22 signifies the smallest aperture on this scale, *f*1·4 the largest. The step from one *f*-number to the next, known as a stop, represents doubling or halving the exposure: *f*16 lets twice as much light in as *f*22, for example. Changing the aperture also affects the ability of the lens to focus at different distances, or in photographic parlance the 'depth of field'. A narrow aperture gives a deep depth of field, while a wide aperture gives a shallow depth of field.

The shutter speeds also vary in distinct steps, usually something like:

1 second, $\frac{1}{2}$, $\frac{1}{4}$, $\frac{1}{8}$, $\frac{1}{15}$, $\frac{1}{30}$, $\frac{1}{60}$, $\frac{1}{125}$, $\frac{1}{250}$, $\frac{1}{500}$, $\frac{1}{1000}$

Again, each one is twice or half the speed of the ones on either side, giving half or twice the exposure. The shutter speed is used in unison with the aperture to give optimum exposure, but the main reason for selecting a certain shutter speed is often to arrest any motion in the picture. While a speed of $\frac{1}{60}$ would be adequate for photographing a grazing horse, when the horse is galloping the same speed would result in the picture being blurred. In this instance $\frac{1}{500}$ or $\frac{1}{1000}$ would be needed to 'freeze' the action.

Reflex Cameras

In the simple box, or viewfinder camera, the viewing system presents to the photographer a slightly different view from that which falls on the film. This is because of parallax (different view from different viewpoint) between the viewfinder lens and the camera lens. This difficulty is overcome in the single-lens reflex (SLR) camera. The most popular types use cassettes of 35 mm (1·4 in) wide film. In the 35 mm SLR the same lens is used to view the scene in the eyepiece (viewfinder) as to take the picture, so there are no parallax errors. This is achieved by the use of a hinged mirror, a focusing screen and a reflecting pentaprism.

In the viewing mode the mirror intercepts the light entering the lens and reflects it onto a focusing screen at the same distance away as the film. This means that when the image is in focus on the screen, it will also be in focus on the film. The image on the focusing screen is reflected by two inner surfaces of the pentaprism into the eye at the eyepiece. When the shutter release is pressed to take the photograph, the mirror flips up, allowing light from the lens to expose the film when the shutter opens.

A more expensive type of SLR uses a larger format roll film, usually 55 mm ($2\frac{1}{4}$ in) square. They are more box-like and have an interchangeable back or magazine. They lack the pentaprism of the 35 mm SLR and are designed for top, or waist-level viewing. Large-format roll film is also used in the other type of reflex camera – the twin lens reflex (TLR).

Developing the Image

The film consists of an emulsion of fine grains of light-sensitive silver halides suspended in gelatine, coated on a base of transparent cellulose acetate plastic. During exposure of the film the silver halide grains are affected by the light falling on them – the more intense the light, the more grains are affected. In the process of developing the exposed film is treated with a chemical (developer) which converts the affected halide grains into black metallic silver.

This process must, of course, be carried out in the dark, otherwise all the unexposed grains will be affected and the film ruined. So must the next step, that of fixing the image. In fixing the film is treated with a solution which dissolves away the unexposed silver salts. This leaves the transparent film base with an image in black silver upon it. The image is a 'negative' one, since it is dark where there was light in the scene viewed, and light where there was dark.

To produce a replica, or positive picture of the original scene, the negative must be printed. This involves exposing a piece of photographic paper through the negative, so that where the negative is dark little light falls on the paper and where the negative is light much light falls on it. When the paper is developed, it then displays a light image where the negative was dark and a dark image where

The three main kinds of reflex cameras, so called because they incorporate a mirror to reflect incoming light onto the viewing screen. In the single-lens reflex cameras the mirror flips upwards during exposure.

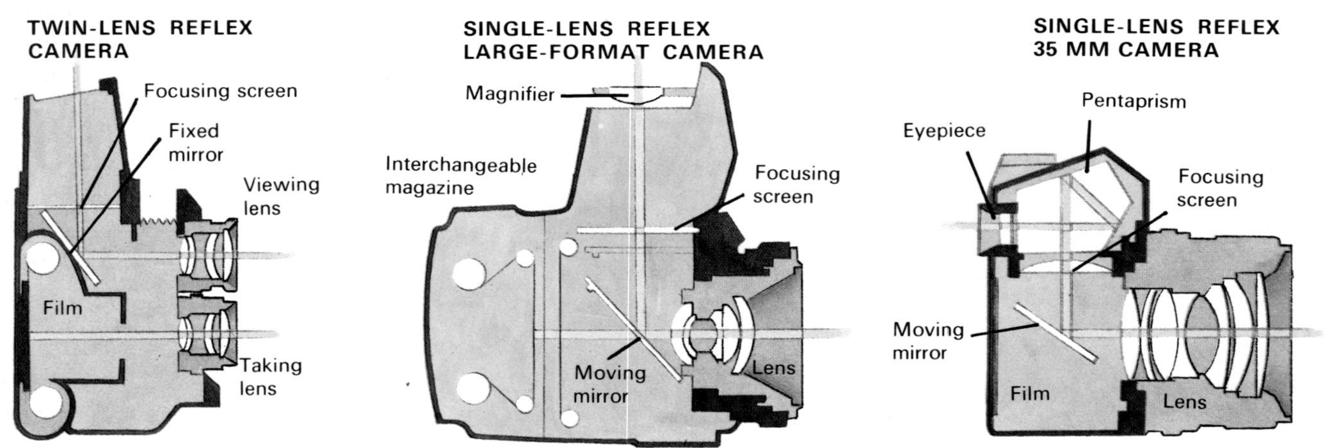

The Olympus OM compact single-lens reflex camera, available in two versions, the manual OM-1 and the manual/automatic OM-2. Like the Nikon FE illustrated on page 137 and similar models, they feature sensitive through-the-lens (TTL) exposure metering for 'spot-on' film exposure. This is provided by semiconductor photodiodes, chosen to match the spectral sensitivity of the human eye. A wide variety of multi-element lenses – close-up, wide-angle and telephoto – are available which are attached to the body with a bayonet mount.

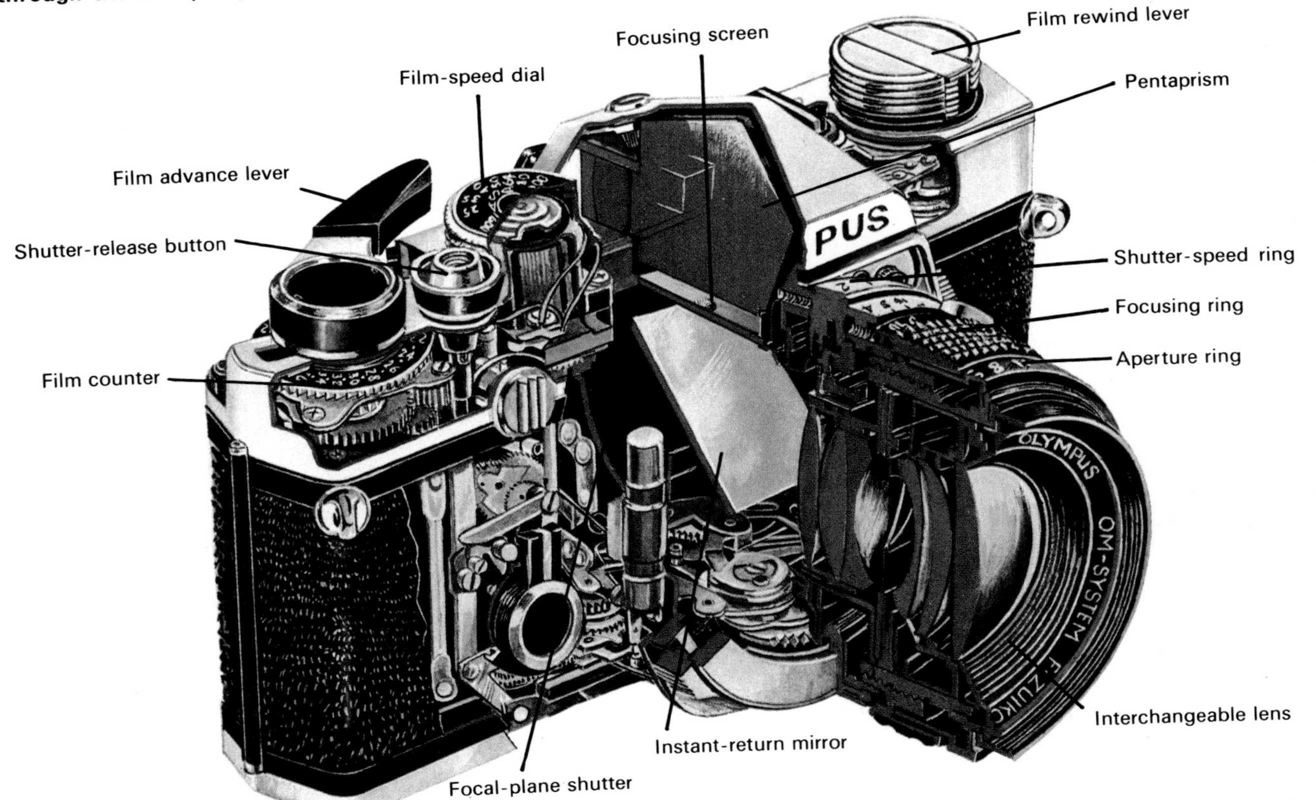

Below: The light path through a different kind of reflex camera, the Polaroid SX-70. Like all the Polaroid cameras, it takes a film pack which contains its own developing chemicals. When you press the shutter-release button, an electric motor propels the picture from the camera, where it proceeds to develop into a finished picture. The film consists of many layers of film, between two of which is a pod containing the developing chemicals. When the motor ejects the film, the pod is broken, and the developing process is triggered off. During development an opaque chemical in the film prevents light getting in from outside. It becomes transparent as development proceeds.

the negative was light. The shades of the original scene are thereby replicated.

The process of colour photography is not as complicated as it first appears. The same kinds of emulsions are used, but they are used in layers made to be sensitive to certain colours. The layers are then developed into metallic silver negatives, which are then dyed. The silver is removed, leaving a coloured piece of film. Two different types of colour film are used, the reversal and negative. The former results in slides, or transparencies, which are held up to the light, or projected to recreate the pictures, while the latter produces colour prints. The make-up of the two types of films is similar, but the layers are dyed differently during processing.

Colour film consists of three layers of emulsion, each sensitive to one of the primary colours blue, green and red. Each layer records the blue-ness, green-ness and red-ness in the original scene. During colour development of a transparency, say, the negative image in the blue-sensitive layer is dyed yellow, that in the green-sensitive layer magenta and that in the red-sensitive layer cyan. These are the so-called complementary colours of the primaries. This has the following effect. Where the original colour was blue, the transparency is dyed magenta and cyan. When light passes through it, the magenta and cyan layers subtract red and green light, leaving only blue – the original. By a similar 'subtractive' process the original red and green are also recreated.

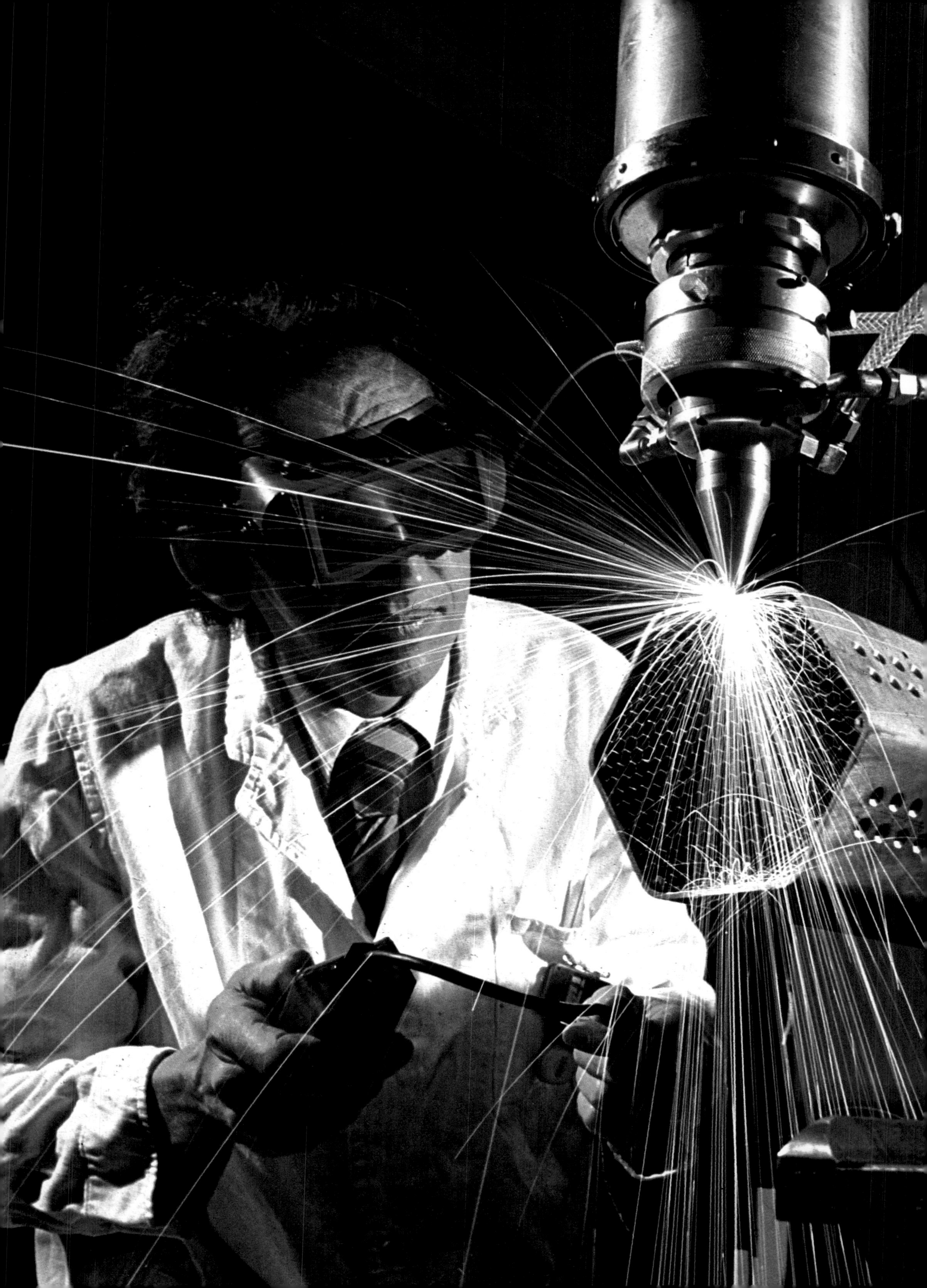

People at Work

Until about 1700 mankind in general pursued a predominantly rural existence in which agriculture was the primary occupation. All things considered the ways of life of the people changed very slowly, if at all, year by year and century by century. People tended to live in small, self-contained communities, producing their own food, clothes and furniture. The only industries to speak of were cottage industries, in which people produced goods in their own homes for a middle man who marketed them. Then suddenly came the Industrial Revolution, and the life of the people began to change at an unprecedented rate. This revolution began first in Britain, but soon spread throughout Europe, the United States and eventually the whole of the civilized world. Within the space of about 150 years Britain was transformed into a heavily industrialized society and because of this had become the dominant world power.

The ramifications of the revolution touched everyone, for good or ill. Communication by road, rail and canal were greatly improved, making long-distance travel feasible to all for the first time. Steam power and plentiful supplies of iron and cheap steel accelerated civil engineering construction. The experimentation that provided the background to, and was a spin-off from the revolution led to the establishment of new branches of science and of the chemical, mechanical and electrical engineering industries. In the home and on the farm too machines began to infiltrate, reducing the drudgery of the housewife on the one hand, and increasing agricultural production on the other.

The spirit of scientific enquiry that prompted the revolution has accelerated over the years, spawning techniques and technologies that touch us all, from housewife and farmer to scientist and industrialist. This chapter describes some of the more important of them.

The nuclear technologist (left) and the farmer (below) use the latest machines (lasers and combines) to speed production.

Spinning a Yarn

Perhaps rather surprisingly the Industrial Revolution was spearheaded by a series of British inventions in textile making, which transformed it from a domestic activity into a factory process. First came John Kay's invention of the flying shuttle (1733), which speeded up weaving so much that the spinners could not produce yarn fast enough. Necessity being the mother of invention, along came John Hargreaves with a spinning machine, or jenny (1764), which could spin thread on several spindles simultaneously. This was improved upon in 1779 by Samuel Crompton's spinning mule.

Meanwhile (1771) Richard Arkwright had begun utilizing water power to drive the spinning frames. To make effective use of this power he had to centralize production. He installed a number of frames in one building and employed people to work them. In so doing he initiated the factory system of manufacturing. In 1785 Edmund Cartwright, a clergyman-turned-inventor, developed a power loom driven by steam, though it was not successful in its early form.

The widespread mechanization of first spinning and then weaving led to a boom in demand for raw cotton, which was imported from the United States. The growers could not produce the cotton fast enough. The main hold-up in production was not the picking of the cotton, but the separation of the lint (fibres) from the seeds. Again, someone came up with a timely solution. He was American artist-turned-inventor Eli Whitney, who in 1792 devised the cotton gin, which could perform the separation of fibres from seeds 50 times faster than a man. It was prodigious increases in production like this that brought about the revolution in working methods.

Iron Masters

The development and exploitation of effective power sources was also a major factor in the Industrial Revolution. Of paramount importance was the steam engine, brought to high efficiency by James Watt (see page 34). The successful harnessing of steam power also led to a reappraisal of the windmill and water wheel as power sources. This led to more efficient exploitation of both, in parallel rather than in competition with the steam engine.

The greater utilization of machines led in turn to a greater demand for iron. A start had already been made in this direction in 1709, when Abraham Darby began smelting iron with coke rather than with the traditional charcoal. The latter had become in short supply as Britain's woodlands dwindled. Since coke is made from coal, there was consequently a great upsurge in demand for that fuel. This led to a rapid expansion in coalmining, which in turn led to a demand for better pumps – steam driven of course. And so it went on; innovation in one field stimulating demand in others.

The new smelting methods produced abundant iron for casting, and cast iron became the major constructional material for bridge and railway construction. More precision machinery required the use of the tougher wrought iron, which was produced by the puddling process from the 1780s on. Steel could also be made in crucibles, but only on a small scale. Not until Henry Bessemer developed his converter process (1856); Siemens and Martin their openhearth process (1863) did steel become cheap.

Precision Tools

The machines developed by Watt and others demanded fine tolerances in their manufacture. This led to the birth of mechanical engineering in general and of the machine-tool industry in particular. Many of the early machine tools were adaptations of the woodworking tools currently used by carpenters and cabinet-makers. The first precision tool was John Wilkinson's cylinder-boring lathe (1775). Among the devices that followed none was more important than Henry Maudslay's all-metal screw-cutting lathe (1800). Two of Maudslay's protégés also subsequently made their mark – James Nasmyth, who invented the steam hammer (1839), and Joseph Whitworth, who began the first large-scale manufacture of machine tools. Whitworth established new standards of accuracy in his machines that ensured complete interchangeability of parts.

Interchangeability of component parts holds the key to mass production, the modus operandi that characterizes the twentieth century manufacturing industry. One of the first practitioners of the method was Eli Whitney, who should have made a fortune from the cotton gin but didn't. In a later venture (1798) he secured a contract from the Army to manufacture 10,000 muskets in two years. Hitherto a host of skilled gunsmiths would have been required for such a task. In the event Whitney employed just a few skilled engineers to operate cutting tools he designed to make musket parts of standard dimensions. Then he employed unskilled and semi-skilled labour to assemble the muskets from the individual parts.

Industrial Chemistry

The rapid expansion in textile production in the 1700s created a demand for chemicals used in processing, cleaning and bleaching the woven cloth. These included sulphuric acid and bleaching powder. By the century's end large-scale processes of manufacturing both were in operation in Britain.

While an embryonic chemical industry was growing up in Britain, France was making similar strides. In 1775 the French Academy of Sciences offered a prize for a process to convert salt into soda ash (sodium carbonate), which was required in very large quantities for the manufacture of soap and glass. In 1790 French chemist Nicolas Leblanc perfected a cheap and effective method of so doing, and claimed the prize. Due to the French Revolution, he never got his just rewards and the process was neglected until the 1820s, when it was revived in Britain.

As the inorganic chemical industry was expand-

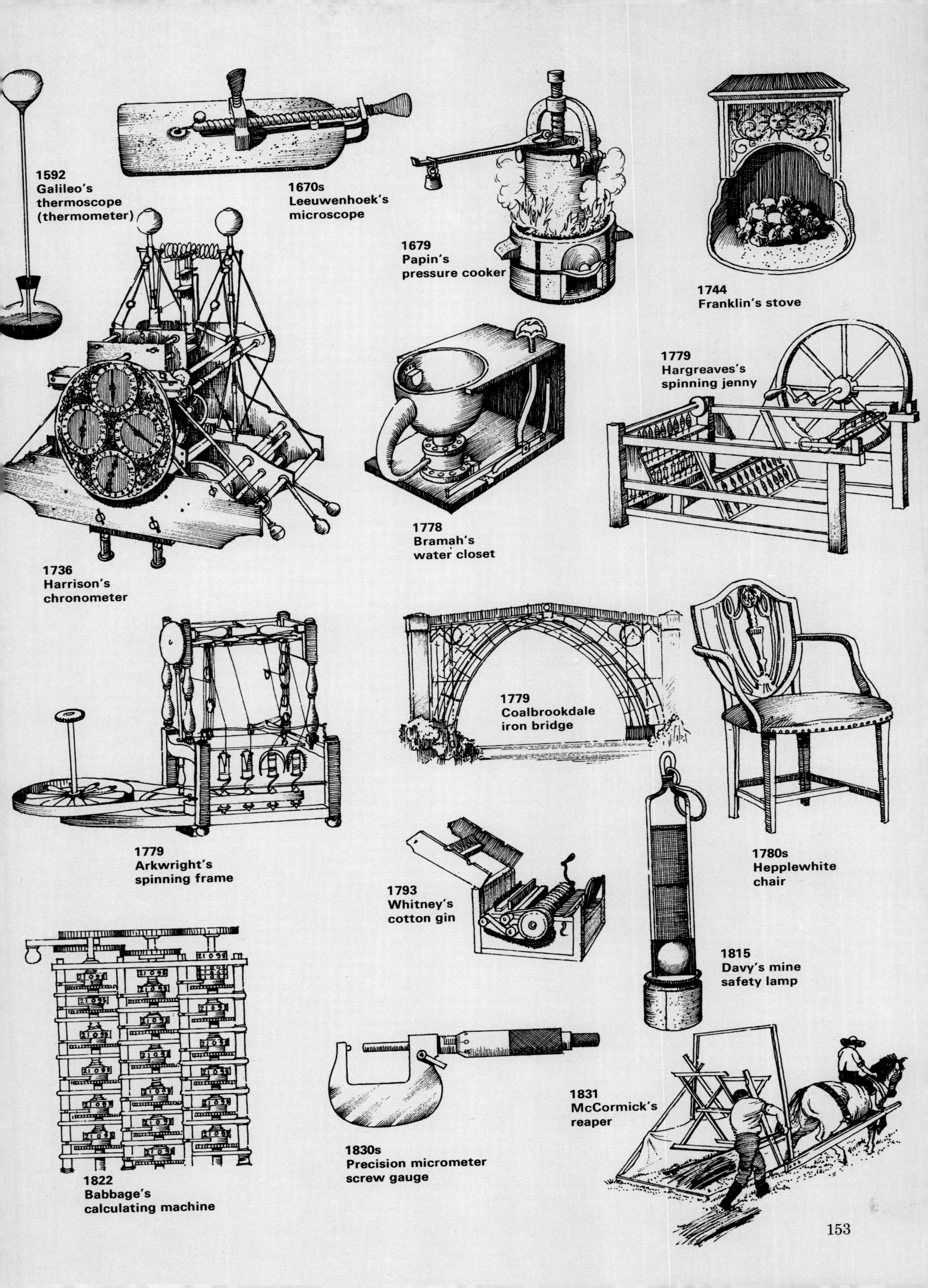
1592
Galileo's
thermoscope
(thermometer)
1670s
Leeuwenhoek's
microscope
1679
Papin's
pressure cooker
1744
Franklin's stove
1736
Harrison's
chronometer
1778
Bramah's
water closet
1779
Hargreaves's
spinning jenny
1779
Arkwright's
spinning frame
1779
Coalbrookdale
iron bridge
1780s
Hepplewhite
chair
1793
Whitney's
cotton gin
1815
Davy's mine
safety lamp
1822
Babbage's
calculating machine
1830s
Precision micrometer
screw gauge
1831
McCormick's
reaper

ing, a new branch of chemistry was being pioneered – organic chemistry. The impetus for this came literally as a by-product of the gas-making industry. Experiments with gas made by destructively distilling coal (heating it strongly in a closed vessel) took place in the 1770s, and in 1802 William Murdock lit one of Watt's factories with coal gas. In 1813 German-born businessman Frederick Winsor established the world's first gas company, in London. Within a few decades most large towns were gas-lit and had their own gasworks. The gas-making process yielded a residue of coke, useful gases such as ammonia and a gooey liquid called coal tar.

Experiments on coal tar revealed that it contained a whole new range of chemicals, called organic chemicals for they had hitherto been associated only with living things. In 1856 William Perkin made the first synthetic dye from coal tar – mauve. In 1899 the effectiveness of the compound now known as aspirin was appreciated. It has been discovered in coal tar by Charles Gerhardt in 1853.

Another product of coal distillation – phenol, or carbolic acid – proved an excellent disinfectant and in 1910 was the starting point for the first synthetic plastic. This was Bakelite, the brain child of Belgian Leo Baekeland, made by reacting phenol with formaldehyde. The man-made plastic known as Celluloid was already in large-scale production, having been developed in 1869 by John Hyatt in the United States as a substitute for ivory. Celluloid is made by treating cellulose (from cotton linters) with nitric and sulphuric acids. It is cellulose nitrate (nitrocellulose) plasticized, or made pliable, with camphor. Cellulose nitrate was also the material used for the first man-made fibre, nitro-rayon, produced in 1884 by Hilaire Chardonnet.

As the twentieth century progressed the chemical industry underwent enormous expansion and diversification, demanding eventually more coal-tar chemicals than were available. This led to the growth of the petro-chemical industry, which obtained organic raw materials by distilling crude oil.

Home on the Range

The revolution in industry was accompanied by a gradual revolution in the home and on the farm. Innovations of the early 1700s included the Franklin stove (1739), made by American statesman Benjamin Franklin to throw more heat into a room than an open fire. The availability of gas from the early 1800s extended the daylight hours inside the home and out, and markedly changed social life.

In about 1840 gas stoves, or ranges, came into use in the United States, while the ubiquitous gas ring had its origins in Britain in the 1860s, by which time primitive gas fires had appeared. From the 1880s gas began to have competition from electricity, first for lighting and then for cooking and heating. But electric stoves and fires did not come into widespread use until after C. R. Belling's invention in Britain of the modern type of heating element. Methods of food preservation also advanced, with the development of canning by Englishman Peter Durand in the 1830s. At much the same time experiments were taking place in refrigeration, resulting by the 1850s in practical machines of the vapour-compression type and the ammonia-absorption type, the descendants of which are still used today.

One of the earliest labour-saving machines to appear in the home was the sewing machine, which was patented in America by Elias Howe (1846) and manufactured by Isaac Singer. A mechanical carpet sweeper next appeared on the scene in 1876, developed by the American M. R. Bissel. Primitive washing machines had also appeared fitfully by then. If housework was not the forte of the lady of the house, there was a growing opportunity for her to break into the business world, hitherto the exclusive preserve of the male, as a secretary-typist. This followed the widespread introduction of the typewriter (invented in 1867) from 1874.

Down on the farm machines had begun to appear as early as 1701, when Jethro Tull built a seed drill. The speed of harvesting was greatly improved as a result of Cyrus McCormick's reaper of 1831. Threshing had already become a mechanized, steam-powered operation, the first machine having been built by inventor of the locomotive Richard Trevithick in 1811. By 1837 a primitive combine harvester had appeared in the United States but took some 80 years to be developed into an effective machine that could harvest the vast wheatfields of the prairies. By then many combines were being pulled by the farmer's best friend, the tractor. The direct ancestor of today's highly efficient machines was the British Ivel tractor of 1902.

Continuing Change

Much more could be written about the background to the many other developments that have shaped today's industrial society, but space does not permit. The accompanying illustrations do, however, build on the themes discussed. They demonstrate effectively that revolution in industry is continuing still. Today's revolution is being spearheaded by automation, just as yesterday's revolution was brought about by mechanization. With automation, machines work automatically with the minimum of human intervention. They do so under the control of the so-called electronic brain, or computer.

Originally conceived in the 1830s by English mathematician Charles Babbage, the first computer was built by Howard Aiken and others at Harvard University in 1944. But only after the invention of the versatile transistor in the United States by William Shockley, John Bardeen, and Walter H. Brattain in 1948 did computer technology advance significantly. Computers became more powerful, smaller and cheaper, a trend that is still continuing, thanks to the miraculous microchip. 'Chip' technology should soon make possible the widespread introduction of the computer into the home.

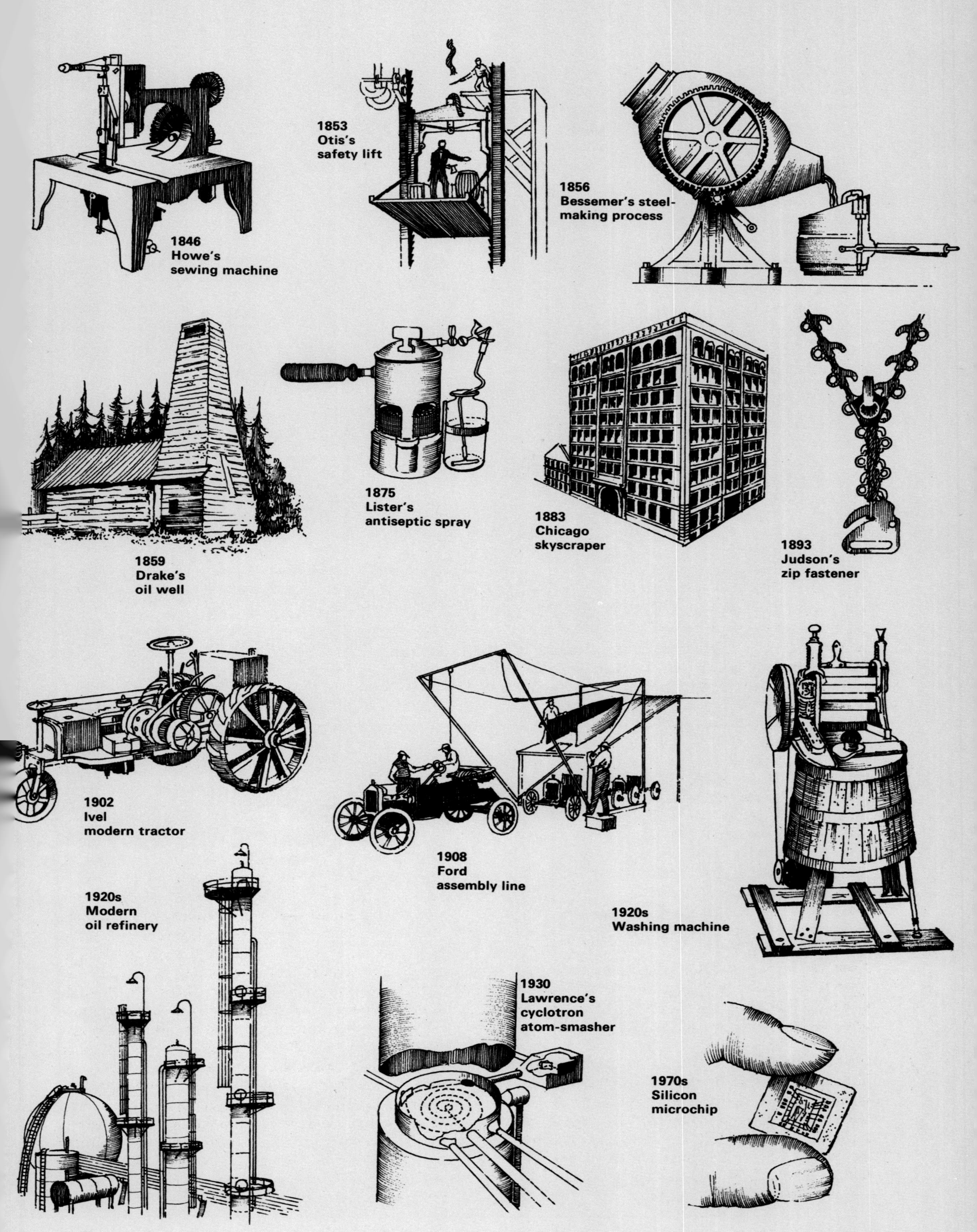
1846
Howe's
sewing machine
1853
Otis's
safety lift
1856
Bessemer's steel-
making process
1859
Drake's
oil well
1875
Lister's
antiseptic spray
1883
Chicago
skyscraper
1893
Judson's
zip fastener
1902
Ivel
modern tractor
1908
Ford
assembly line
1920s
Washing machine
1920s
Modern
oil refinery
1930
Lawrence's
cyclotron
atom-smasher
1970s
Silicon
microchip

Home, Sweet Home

To detail, let alone explain the workings of the paraphernalia or machinery and gadgetry we come into contact with in our work-day lives would need much larger tomes than this. So we have outlined here areas of industrial activity, considered in its broadest sense, where machinery and other products of Man's inventive mind play a significant role. And we must bear in mind throughout that Man not only invents hardware – that is, machines and equipment – but also software – that is, techniques and processes, ways of doing things. It is perhaps in his software that he often displays his greatest ingenuity, reminding us that he is as consummate a philosopher as an engineer.

Work, like charity, begins at home, so it is here that we start. The lot of the housewife these days is a much happier one than it once was, at least if she lives in the advanced industrial societies of the world. She has the pick of a host of labour-saving devices to help her cook, wash, clean and mend. So many of the machines and gadgets she uses are powered by electricity that she almost needs to be as competent an electrician as a housewife!

Among those used for preparing the food are the electric toaster, percolator, can-opener, mixer, liquidizer, grinder, mincer, kettle and carver, and of course electric cooker. They are all relatively simple in operation, containing either simple heating elements or electric motors and gears. Perhaps the most interesting machines relating to food preparation and cooking are the refrigerator and the microwave oven – the one relatively old (1830s), the other ultramodern.

By far the most common type of refrigerator uses the principle of vapour compression. An easily vaporized liquid, called the refrigerant, is allowed to evaporate at low pressure in the refrigerator, absorbing the necessary heat (the latent heat of vaporization) from the surroundings. In a practical refrigerator the refrigerant is a compound known as a fluorocarbon, marketed under the name of Freon. (Freon-12, for example, is the compound dichlorodifluoromethane, CCl_2F_2.)

The refrigerant boils in the coils of the evaporator, around the freezing compartment of the refrigerator, and turns to vapour. The vapour is then compressed by a motor-driven compressor, whereupon it heats up. It then circulates through coils exposed to the air, loses heat, and condenses back into liquid. The liquid then passes through a control device where its pressure is reduced. It circulates back to the evaporator, where it begins to boil and extract heat. And the cycle is endlessly repeated. In the ordinary refrigerator this system usually maintains a temperature a few degrees above freezing, and in the freezing compartment a few degrees below. Deep-freezes use a similar vapour-compression system but are regulated to a temperature of about $-18°C$.

Cooking by Radar and Steam

What can cook a hamburger in half a minute, bake a potato in four and roast a chicken in 15? The answer is the microwave oven, which borrows the technology used in radar. It cooks so fast because the energy it uses penetrates and heats up the food from the inside, not the outside as in normal cooking.

The device that makes microwave cooking possible

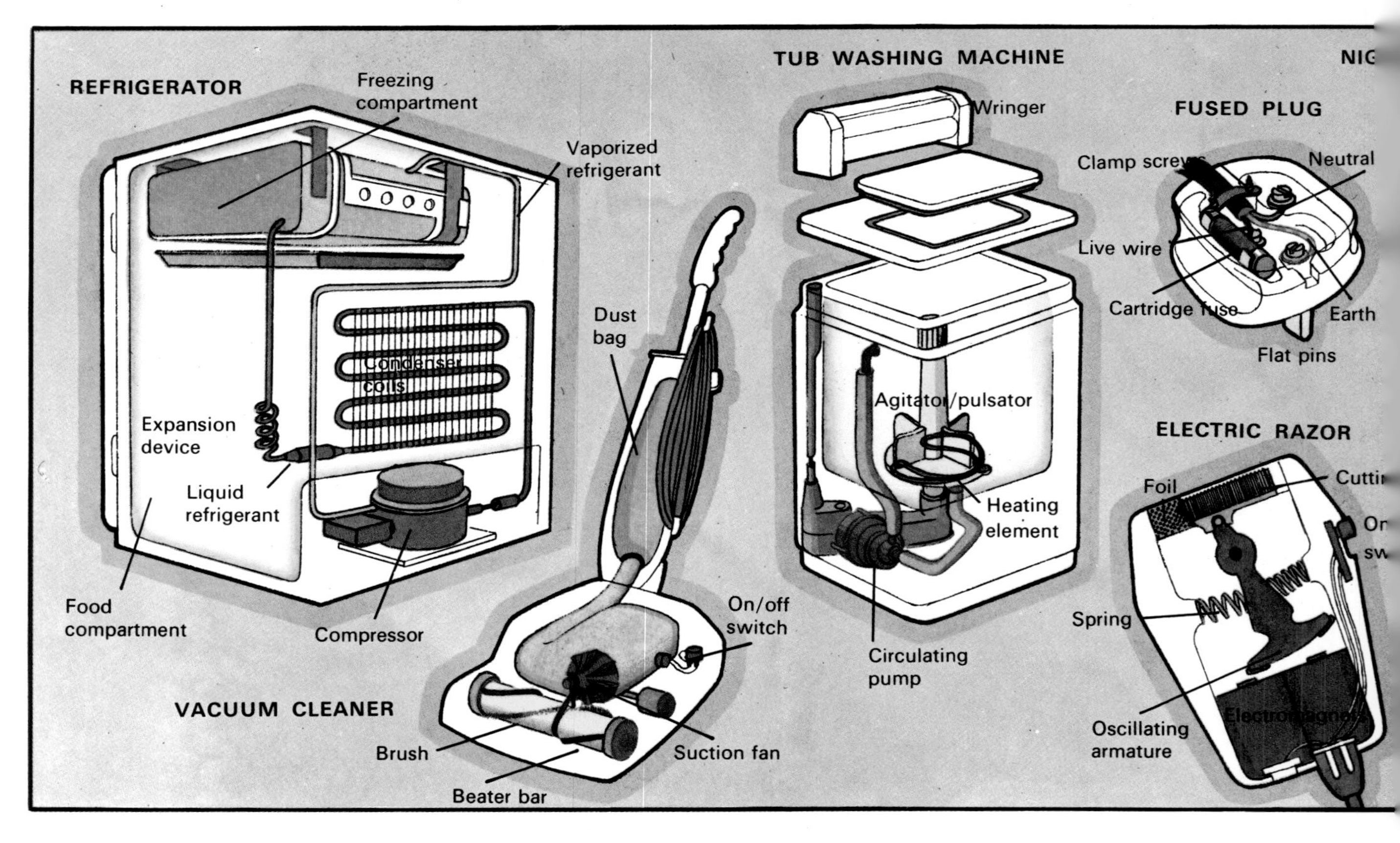

is a vacuum tube known as a magnetron, which generates by means of resonant cavities very high-frequency radio waves called microwaves when current is applied to it. The microwaves have a frequency of some 2000 megahertz (2000 million cycles per second). In the oven the generated waves are directed by a hollow tube known as a wave guide on to a rotating reflector, which projects them directly and via reflection from the oven sides into the food to be cooked.

The rapidly alternating electromagnetic field of the microwaves causes the food molecules to align themselves continually in different directions. This violent agitation of the molecules brings about the heating effect which, since all the molecules are involved, is very rapid.

A rapid-cooking aid that is found in many more kitchens than the microwave oven predates the Industrial Revolution and appropriately enough uses steam technology. It is the pressure cooker, which is not too far behind the microwave oven in speed of cooking but is generally used for different types of food, such as vegetables, casseroles and preserves.

Pressure cooking works because under increased pressure, water, or any liquid for that matter, boils at a higher temperature. And at the higher temperature food cooks faster. The pressure cooker is an airtight vessel with a vent in the lid which allows steam to escape. A weight is inserted in the vent so that the steam can escape only after it has reached a certain pressure (and the food a certain temperature). Under a 1 kg/sq cm (15 lb/sq in) steam pressure water boils at about 122°C.

Washing-Up

The great drawback to cooking a large meal is that it generates a large quantity of washing-up, which is perhaps the most persuasive reason for eating out. But modern technology again has come to the rescue, in the form of the dishwasher. All you need do is stack the dirty utensils on racks in the machine, switch on, and retire to more useful pursuits. In less than an hour those greasy plates and coffee-stained cups will be sparkling clean.

The dishwasher is first filled with cold water, which is then heated by an integral heater to the right temperature, selected from a programming dial. The hot water is pumped through rotating belt-driven spray arms, top and bottom, and emerges as powerful jets which penetrate every nook and cranny of the load. Detergent and water-softening agents are added to speed the cleaning process.

The dishwasher goes through its cycle of washing, rinsing and drying automatically according to the setting selected on the programming dial. Automatic clothes-washing machines also work according to a preset programme. The front-loading automatic consists of a cylindrical tub mounted horizontally, which is supplied with water from a plumbed-in inlet hose and which can be emptied by means of a pump. Inside the tank is a rotating perforated tub, driven by belt from a motor, in which the washing is placed. Detergent is added from a dispenser. After the clothes have gone through their washing and rinsing cycle, the water is drained away. The inner cylinder is then rotated at high speed (typically about 1000 rpm) to 'spin-dry' the clothes, the water being removed from them by centrifugal force.

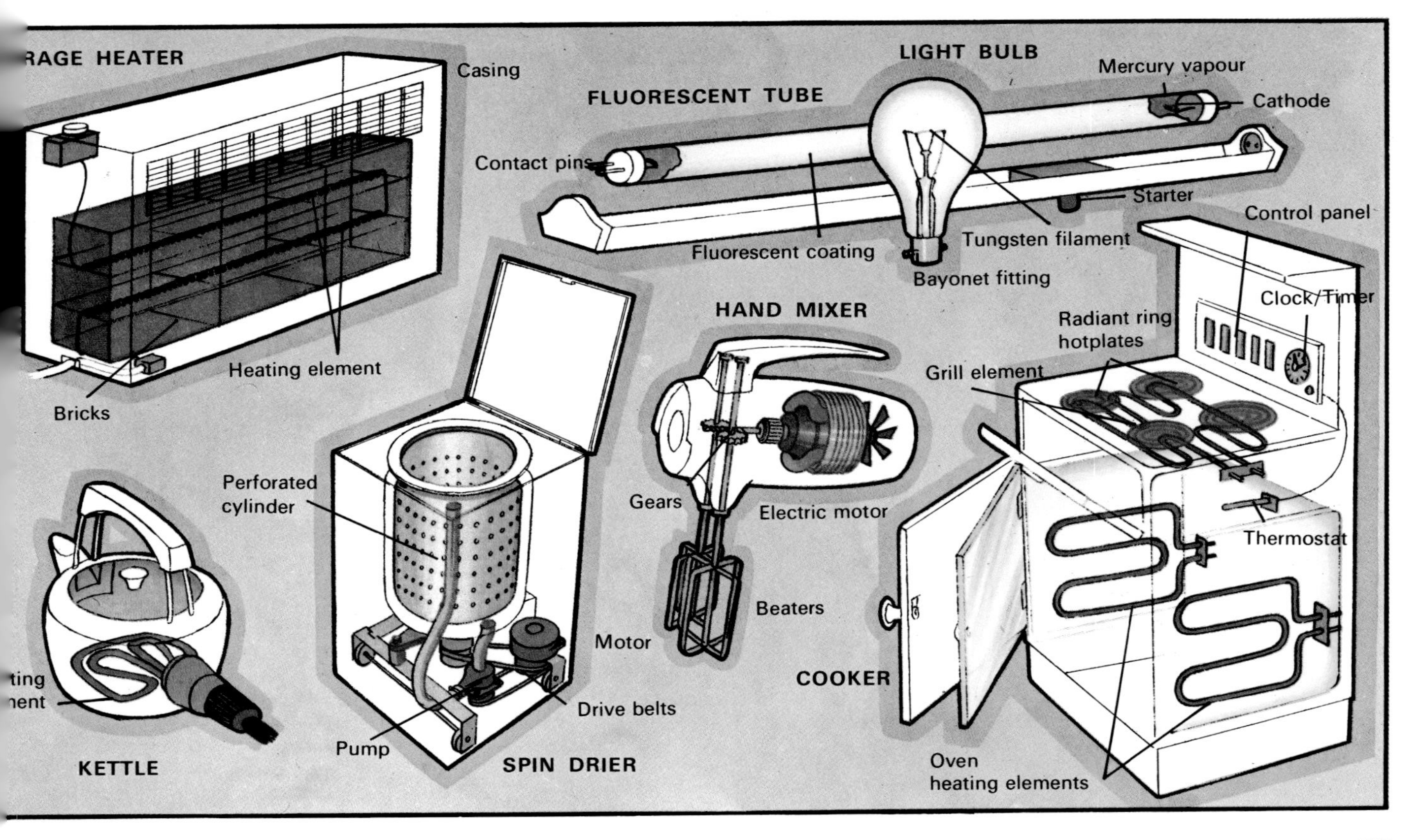

Cheaper and more common than the front-loading automatic is the twin-tub machine, so called because it houses the wash-tub and spin-dryer separately. The wash-tub has a pulsating agitator to keep the clothes moving and a built-in heater. The spin-dryer, like the washing-tub, is mounted vertically and is driven by a separate motor. It can spin at very much higher speeds (2000–3000 rpm) than the automatic.

Plumbing and Heating

Among the household systems we take most for granted are the plumbing and heating, and if they break down we tend to become panic-stricken. Yet they are really relatively simple, and can usually be simply put right. The diagram below shows the main elements of a typical plumbing and heating set-up. The mains water supply enters the house through a stop-cock and is thence piped to wash-basin and sink, storage tanks and boiler, shower and water-closet (WC). In past times the piping was lead, from which the term 'plumbing' got its name, 'plumbum' being the Latin word for lead.

These days the piping is usually iron, copper or, particularly for waste pipes, plastic. Iron pipes are joined by means of screwed iron connections caulked with string and jointing compound. Copper pipes may be joined by screwed brass compression joints. They contain soft copper rings which are squeezed tightly around the pipes being joined to give a water-tight fit. The other method of joining copper pipes is with a capillary joint. The ends to be joined are fitted inside a close-fitting sleeve and heated with a blow torch. While the joints are still hot, solder is applied to them and is sucked inside by capillary action. A solid joint forms when the solder cools.

Central heating has a surprisingly long history, being pioneered by the Romans with their hypocaust underfloor warm-air system. And Matthew Boulton, backer of James Watt, provided steam heating to a friend's house in the 1790s. But the modern hot-water radiator system dates from about 1840, the brainchild of American inventor Robert Briggs. But whereas in Brigg's system natural circulation carried the water round, in modern systems it is pumped round. And the piping is of much smaller bore.

In a typical central heating set-up water is heated in a furnace/boiler burning either solid fuel, oil or

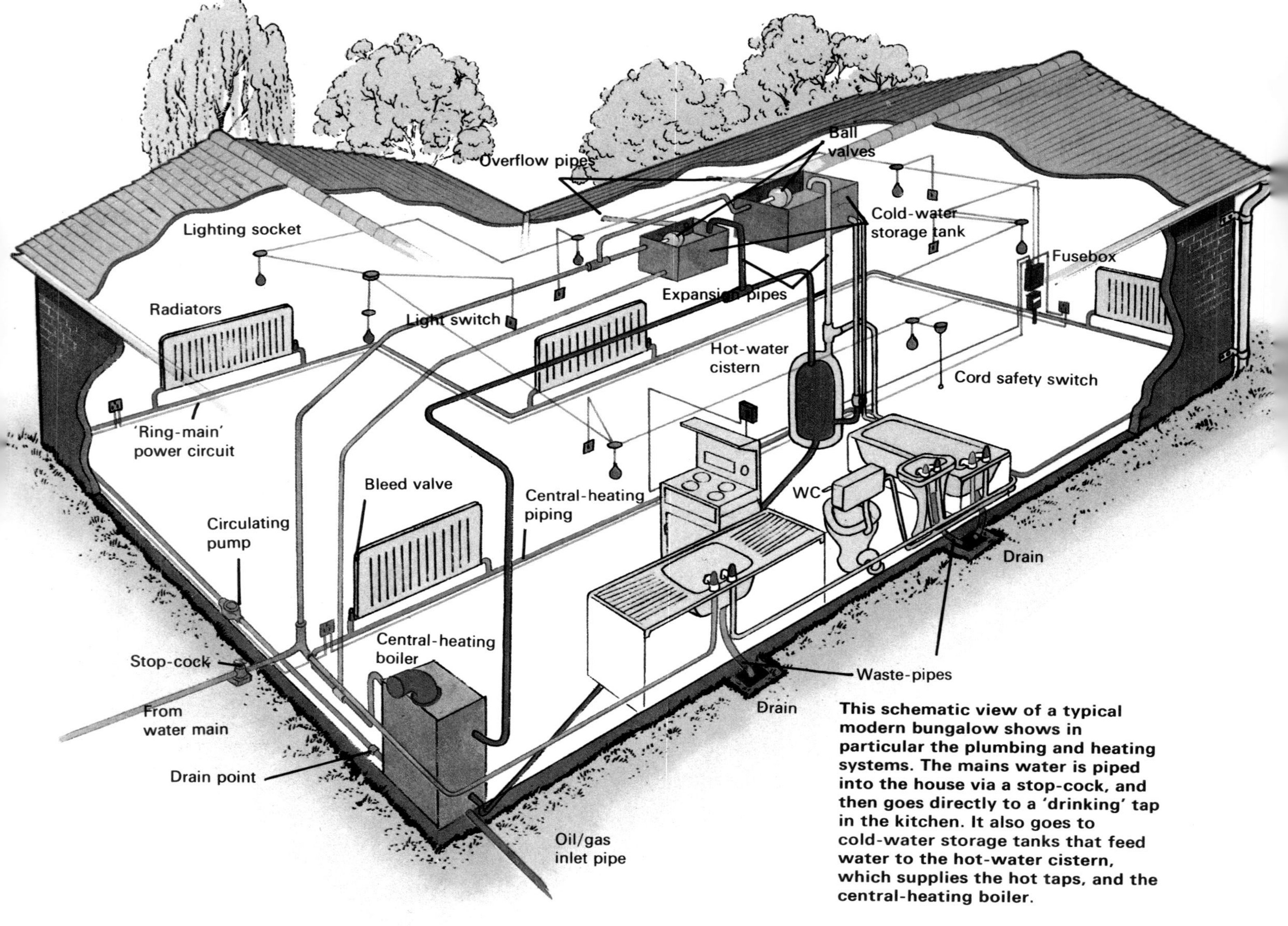

This schematic view of a typical modern bungalow shows in particular the plumbing and heating systems. The mains water is piped into the house via a stop-cock, and then goes directly to a 'drinking' tap in the kitchen. It also goes to cold-water storage tanks that feed water to the hot-water cistern, which supplies the hot taps, and the central-heating boiler.

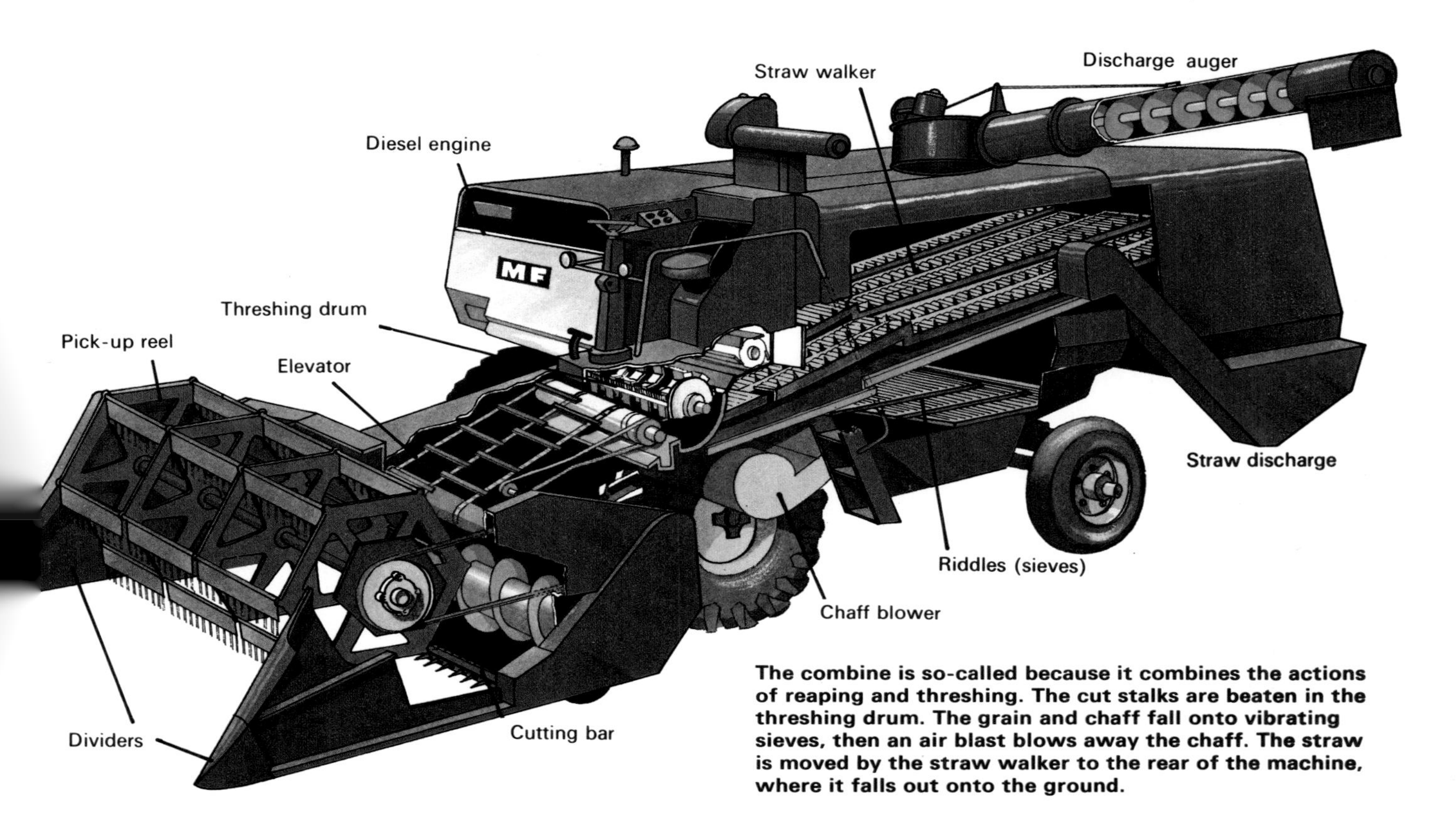

The combine is so-called because it combines the actions of reaping and threshing. The cut stalks are beaten in the threshing drum. The grain and chaff fall onto vibrating sieves, then an air blast blows away the chaff. The straw is moved by the straw walker to the rear of the machine, where it falls out onto the ground.

gas. It is pumped through iron panel radiators, each of which is usually fitted with a control valve, and also a 'bleed' valve. This needs opening periodically to release air which becomes trapped in the radiators. After passing through the radiators it passes back through the boiler to be reheated and recirculated. Another heating circuit heats up water for the hot-water taps. A pipe carries hot water from the boiler up to a heat-exchanger in the hot-water cistern (tank). There it heats up the water in the tank before returning to the boiler. The heat-exchanger may consist of a coil inside the cistern or a jacket around it. An immersion heater – electric heating element – is also incorporated in the cistern to provide hot water when the heating boiler is not operating.

Machines on the Land

Mechanization was relatively slow in coming to agriculture owing to the ready availability of abundant cheap labour. But over the last few decades people have been increasingly attracted to the cities by the promise of better wages and better prospects for less arduous toil. So today, except in under-developed regions of the world farming by machine has become the norm. A diversity of machines is used to help prepare the soil, nurture and harvest the crops, and look after the livestock.

Undoubtedly the farmer's best friend today is his tractor, whereas half a century ago it would have been his carthorse. The standard tractor has either a petrol or diesel engine with a power output up to 100 horsepower or more, though a few tractors have engines which use kerosene or liquefied petroleum gas (LPG) as fuel. Drive to the rear wheels is via a propeller shaft and differential, and these wheels are massively tyred and of large diameter for high-torque traction. Usually the front wheels are of small diameter, though in four-wheel drive machines they are also large. Four-wheel drive tractors can remain going in conditions that immobilize both ordinary tractors and even those with caterpillar tracks.

The tractor is equipped with a variety of devices to pull and operate other machines. It has a strong and rigid drawbar, and units to power drawn and stationary machines. One is an engine-driven pulley, which drives machinery such as circular saws by means of a belt. Another is the power-take-off (PTO), which is used to drive such implements as forage harvesters and mowers, hay balers, and spray pumps. The PTO mechanism usually consists of a telescopic shaft, flexibly linked by universal joints so that it allows relatively independent movement of tractor and implement, while still maintaining drive. Most tractors also have a hydraulic-ram device for adjusting the height of the plough, harrow, or whatever it is pulling. The hydraulic system also operates shovel and digging attachments if they are fitted.

Of the tillage machines used to prepare ground for planting crops, the plough is the most important. It turns over the soil and buries the weeds. Following the plough comes the harrow, which breaks up the ploughed furrows and leaves a level surface. Finally comes the rollers which crush the large clods of soil, leaving a suitable tilth for sowing. The sowing is done by drilling, the seed being placed underground at the correct depth in parallel rows and in carefully metered amounts. Fertilizer is usually applied at the same time, and later as the crop grows. Later too, fungicides and selective herbicides may be sprayed over the crop to control diseases and weeds.

Many ingenious machines have been devised to

Above: A forage harvester at work. The cut grass is gathered and blown through the overhead chute into a wagon drawn alongside. Grass and other forage crops, such as clover and lucerne (alfalfa), are now usually harvested while they are still green and juicy to be made into silage for winter feed.

Below: Where the coal seams are suitable, the coal is cut with coal shearers like this. The water jets keep the cutting head cool and help lay the dust.

Right: The world's biggest excavator, which works West Germany's lignite deposits. Over 200 metres (650 ft) long and 80 metres (260 ft) tall, it weighs 13,000 tonnes and can shift 200,000 cubic metres (260,000 cubic yards) of lignite per day.

gather the mature crops, such as potato lifters, fruit-tree shakers and giant cotton pickers. But by far the most useful and widely used harvester of all is the combine. Without the combine the vast wheat-fields of the North American prairies and the Steppes of Russia could not be cropped, with disastrous consequences for mankind. These days all kinds of cereal crops, including rice, may be combined, as may legumes such as peas and beans.

Farming is still the world's greatest industry and arguably the most important. It is what is called a primary, or extractive industry, which directly exploits and extracts natural resources and raw materials. Forestry and fishing also come into this category but are less important globally speaking. Vying in importance with farming is the other great primary industry, mining. This provides, with minerals and ores, the raw materials for all the other, manufacturing industries; and, with coal, oil and gas, the means of powering those industries. Though both farming and mining are primary industries they differ in one vital respect. Farming is effectively a self-propagating process, in which resources are continuously recycled. Contrast mining, in which resources are tapped and will never be replenished, at least in this geological era.

Mining Methods

To many the word 'mining' is synonomous with underground working, but in terms of productive output most mining is done from the surface. It is known as open-cast or strip mining. Most coal, iron ore and bauxite (aluminium ore) are mined on the

surface. Strip mining is done when the coal or ore deposit lies on or just beneath the surface. Then giant excavators strip off the intervening soil, or overburden, and power shovels or other excavators load the deposit into wagons, trucks or conveyors as the case may be.

The excavators used in strip mining are some of the world's largest land machines. The biggest excavator is the bucket-wheel excavator operating in West Germany's 'Brown Coal Triangle', which contains one of the world's largest deposits of lignite (brown coal). It is shown in the picture below.

Other kinds of surface mining include placer mining, which is practised for cassiterite (tin ore). This ore is heavy for a mineral, and it tends to concentrate in stream beds. These are mined by large floating dredges, which also house equipment to separate the ore from the worthless material dredged up with it. Diamonds and gold, being heavy, can also be found in stream-bed placer deposits. They can be extracted using a method derived from the traditional gold-mining method of panning. Stream-bed gravel is washed through sluice-boxes, and the gold tends to settle out while the lighter matter is washed away. Where the gold-bearing gravels are dry, they are broken up by powerful water jets and then directed through the sluice-boxes.

A great deal of gold and other precious metals are today extracted as by-products when other ores are refined. Many copper, lead and nickel ores contain minute traces of precious metals, which are nevertheless worth extracting because of their scarcity and value. The majority of gold, however, is obtained by underground mining. The gold mines in South Africa are the world's deepest mines, going down over 3 km (2 miles).

The method of underground mining for gold and other ores is broadly similar. Shafts are sunk down to the level of the ore vein and tunnels are struck from them to the vein. The ore is removed from the rock by explosives, loaded into railway wagons for transport to the shafts and then lifted by skip to the surface. The surrounding rock may be strong enough unsupported to allow tunnelling, otherwise the tunnels have to be shored up with props. Where large ore bodies are found, pillars of ore are left here and there to act as supports. This room-and-pillar method is also practised for underground coal-mining.

Coal differs from other mineral deposits in several respects. First, it is not a true mineral, because minerals are inorganic and have a definite composition. Coal is of organic origin, being the fossilized remains of ancient giant ferns and horsetails, and varies in composition. Also coal is very soft, and can readily be cut, allowing machines to be used for its extraction.

In modern underground mines coal is cut by machines called shearers and trepanners. They have a rotating cutting head covered in steel spikes, controlled by probes so that they stay within the coal seam. The cut coal drops onto a belt or chain conveyor and is removed. The machines work under the protection of hydraulic props, which hold up the roof of the tunnel while the coal is extracted. The props are lowered and advanced in turn as the coal

is cut, which is why they are termed 'walking props'. In mines which are not so mechanized, the coal may be undercut by coal cutters and brought down by explosives or drilled by hand-operated pneumatic picks.

Drilling of another kind is used to mine that other precious energy resource – oil, or petroleum. The oil is usually trapped deep down inside the rocks, in folds in the rock layers, in faults, near salt domes and elsewhere. Painstaking geological detective work is needed to trace new oil deposits, involving aerial and ground surveys and seismic soundings. In seismic exploration explosions are set off in the ground, and the sound waves reflected from underground rock strata are detected by geophones (ground microphones). The seismic traces show up the underlying strata, whose formations might suggest that oil – or natural gas for that matter – is present.

Drilling on land is a laborious enough process, but is easy compared with drilling offshore. As progressively deeper and deeper waters are being drilled for oil, as in the North Sea for example, new technologies have had to be developed to win the oil. These have included new designs of drilling vessels, such as the semisubmersible, and new production platforms, such as the massive reinforced concrete 'Condeep' type, 250 metres (800 ft) high and weighing 350,000 tonnes.

Above: A Condeep production rig being towed from its construction site in Norway, bound for the North Sea oil fields.

Left: In many regions of the United States oil-well pumps like this are a common sight on the roadside.

Right: Magnificent structures like this, the famous Golden Gate Bridge in San Francisco, could not be built without the use of metals – in particular steel. Each of the 227-metre (745-ft) high towers of the bridge alone contain nearly 30,000 tonnes of steel.

Working with Metals

Ninety-two chemical elements are found in appreciable quantities in the Earth's crust, and 70 of these are metals. The two most abundant metals are aluminium (8%) and iron (5%). They also happen to be the two most important metals, though in the reverse order, as far as our civilization is concerned. They are the prime structural metals, both of which are used most extensively in the form of alloys.

Iron, relatively weak and soft by itself, is transformed into the hard, strong alloy steel by the addition of minute fractions of carbon and other elements. When aluminium is alloyed with other metals, such as magnesium, copper and zinc, it too is made harder and stronger. Though dearer than steel, it has the advantage over steel of being very light and corrosion-resistant (rust-proof). This gives it widespread application in airframe construction,

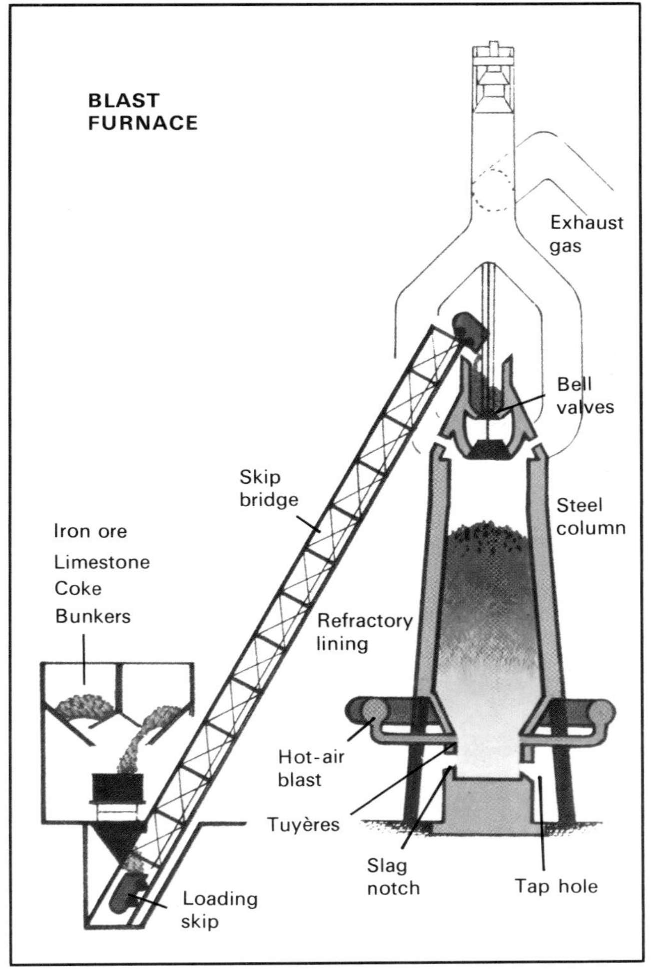

for example, where weight is at a premium. Aluminium is also an excellent conductor of heat and electricity, which is why it is used for cooking utensils and electricity transmission lines.

After aluminium, copper is the next most important metal. It is too soft and weak to be a structural metal, but it finds its application in the electrical industry, for it is a superlative conductor of electricity. It is also widely used in the form of alloys, for example, in cupronickel ('silver') and bronze ('copper') coinage.

The ways of extracting and purifying these three metals are quite different and representative of the types of processes used throughout the metallurgical industries.

Blast-Furnace Smelting

The ores from which iron is extracted are, fortunately, in plentiful supply. They are usually oxide ores, such as haematite and magnetite, and are often so concentrated that they can be processed virtually in the same state as they leave the ground. Smelting – the conversion of ore into metal by heat – takes place in the blast furnace. This is a tall steel structure lined with firebricks in which the iron ore is fiercely heated (up to 1500°C), together with coke and limestone.

The coke, which provides the fuel for the furnace, is burned in a blast of hot air. It also takes part in the chemical reactions that reduce the iron ore to metallic iron. The limestone takes no part in these reactions, but is added to act as a flux. It absorbs the bulk of the impurities in the ore charge to form a molten slag. The iron and slag fall to the bottom of the furnace and are periodically and separately removed.

Refining Furnaces

The molten iron is often poured directly into huge travelling ladles for further processing, or into moulds called pigs, which give it its name – pig iron. Pig iron is too impure to be useful – in particular it contains too much carbon, which makes it very brittle. With slight refining, however, it becomes cast iron, still widely used to cast machinery parts, such as car engine blocks. But more extensive refining of pig iron is necessary before it becomes the strong tough metal we call steel.

The biggest tonnage of steel is now made by the basic-oxygen, or Linz-Donawitz (L-D) process; the electric-arc process; and the open-hearth process. The basic-oxygen process has taken over from the latter as the major steel-making method. It is a development of the Bessemer process, pioneered by Henry Bessemer in 1856. In the process a conical steel vessel lined with refractory bricks is charged with 350 tonnes or more of molten pig iron, together with steel scrap. A water-cooled tube, or lance, is

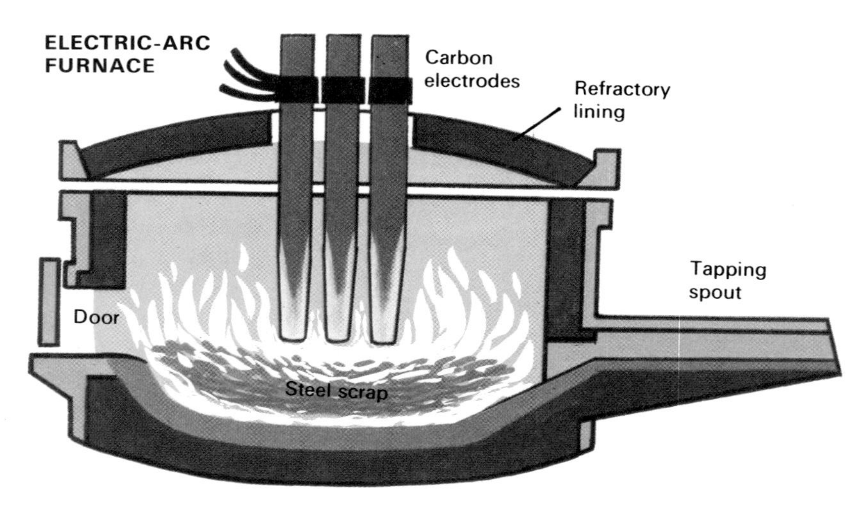

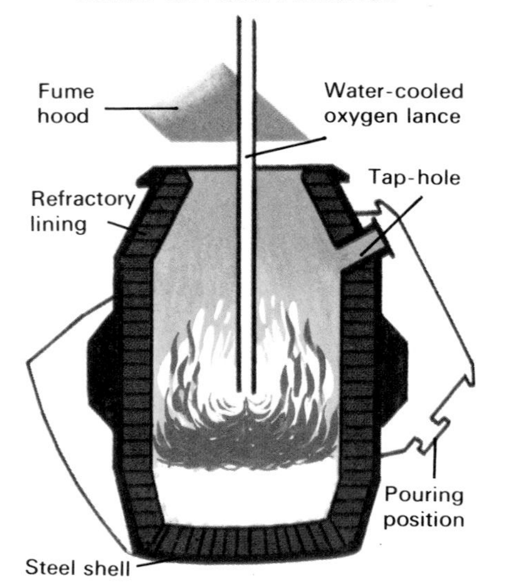

Opposite left: Molten pig-iron cascades from a blast furnace as it is tapped.

Opposite right: 'Poling', one stage in the smelting of copper from its ores.

Below: Molten pig-iron being charged into a basic-oxygen furnace from a travelling ladle.

ARC-FURNACE OPERATING CYCLE

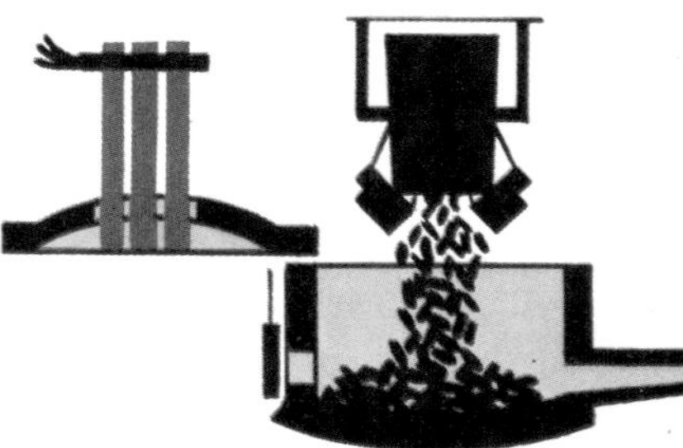

Charging scrap

Melting

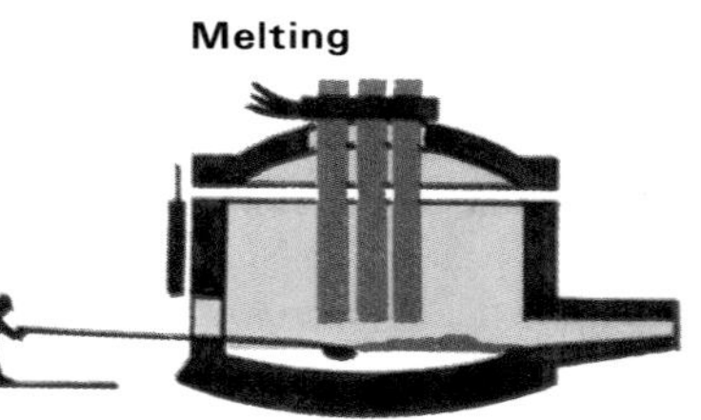

Sampling

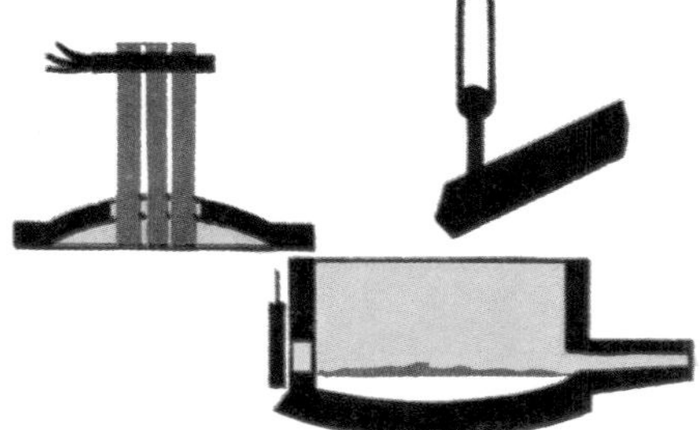

Additions

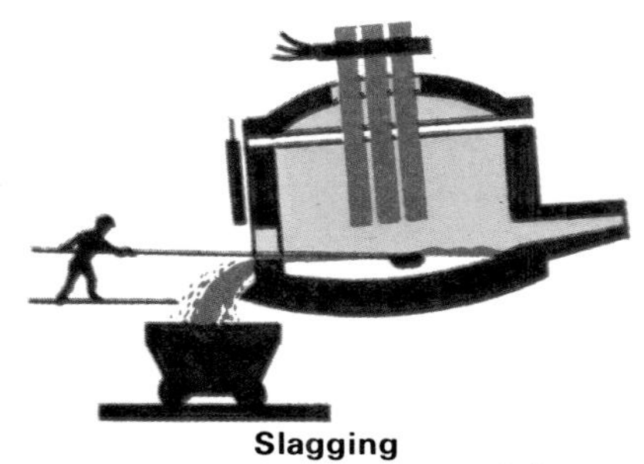

Slagging

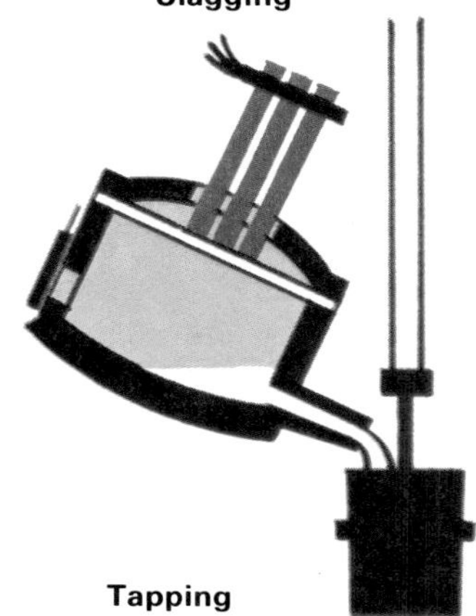

Tapping

lowered into the furnace and through it high-purity oxygen is blown onto the metal at supersonic speed. Lime is also added at this stage. The oxygen combines with the excess carbon in the charge, burning it away in a spectacular pyrotechnic display. Other impurities are oxidized and pass into the lime to form a slag. After about 40 minutes the steel is ready, and is poured into a ladle car. In the ladle alloying elements may be added to bring it to the desired composition. Manganese is almost always added at this stage.

In the open-hearth process, a charge of molten metal, steel scrap, iron ore and lime is melted on a shallow-hearth furnace open to the flames from overhead burners. The hot exhaust gases exit via one of two brick chequerwork chambers and heat it. Air for combustion is blown through the other chamber, which had previously been so heated. Flow through the chambers, or regenerators, is periodically reversed, ensuring that the ingoing air is always preheated. As in the basic-oxygen process, the excess carbon is burned out of the charge and impurities pass into a slag. But the open-hearth process is slow, taking up to 10 hours, which is why it is being superseded.

The third main steel-making process uses the heat from an electric arc to refine scrap metal only. The design of the electric-arc furnace is outlined opposite, while the diagrams alongside show the various stages of operation. Scrap metal is charged into the furnace and an arc is struck between it and huge carbon electrodes in the roof of the furnace. The heat produced melts the charge. The molten steel is then sampled and its composition adjusted by additives of iron ore. Lime and fluorspar are also added as fluxes to aid slag formation. At the end of the process, which takes about 4 hours for a 150-tonne furnace, the slag is raked off and then the whole furnace is tilted to pour off the molten steel. The steel produced in an electric furnace is of high quality because the temperature of the furnace can be carefully controlled. It also does not suffer contamination from the combustion products associated with orthodox furnaces.

Aluminium Cells

Although aluminium is the most plentiful metal in the Earth's crust, it is widely dispersed in the many minerals in which it occurs. Its only profitable ore is bauxite, which contains hydrated forms of aluminium oxide (alumina), together with other metallic oxides. Treatment with hot caustic soda by the Bayer process (page 180) yields pure alumina.

The pure alumina is then separated into its elemental constituents – aluminium and oxygen – by means of electrolysis. It is not practical to electrolyse molten alumina by itself since it has too high a melting point (over 2000°C, it is one of the best refractories). So it is first dissolved in a mixture of molten cryolite (the mineral sodium aluminium fluoride) and calcium fluoride, at a temperature of about 1000°C. This process was pioneered independently in 1886 by Charles Hall in the United States and Paul Héroult in France and is now called the Hall-Héroult process.

The aluminium reduction cell, or pot, is a shallow steel vessel with a carbon lining, which forms the cathode (negative electrode). Above it is suspended a consumable carbon anode. The molten electrolyte lies in between. When heavy current is passed through it, aluminium collects at the cathode, while oxygen collects at the anode. The molten aluminium is periodically siphoned off; the oxygen combines with the carbon anodes to form carbon dioxide gas, which is removed with other gases by a fume hood. The aluminium produced is over 99% pure.

Copper Extraction

Copper occurs occasionally in the Earth's crust as nuggets of pure metal, but these specimens are of geological rather than economic importance. Commercially copper is extracted from a variety of ore minerals, which are usually either oxides (such as cuprite), carbonates (azurite, malachite) or sulphides (chalcocite, chalcopyrite, bornite). The method of extraction used depends on the type of ore.

Oxide and carbonate ores are processed by leaching, or hydrometallurgy. They are treated with a suitable solvent, for example sulphuric acid, and pass into solution. The associated earthy, or gangue minerals are little affected. After chemical treatment to remove dissolved impurities, the leach solution – copper sulphate – is subjected to electrolysis, using an insoluble anode such as carbon and a copper cathode. The electric current splits up the copper sulphate and causes the copper to be deposited on the cathode.

Sulphide ores require more extensive processing, especially when they occur intermixed with other metallic sulphides, as in the famous Sudbury ore of Ontario. This ore consists of mixed copper, nickel and iron sulphides. Even deposits of similar sulphide ores, however, are contaminated with earthy impurities and iron ore. The first step in the extraction process is to concentrate the ore. This is done by froth flotation, a common form of what is called mineral dressing.

In the flotation process finely ground ore is mixed with water and selected surface-active, or wetting agents, and violently stirred. This produces a mass of froth, which preferentially attracts the fine ore particles, while earthy matter sinks. The froth is then skimmed off and dried into what is called copper concentrate. This is next roasted in air to expel volatile impurities and to drive off some sulphur.

The roasted ore is then smelted in a reverberatory furnace – one in which heat is radiated from the roof. Silica or lime is added to extract impurities and form a slag, which is removed. The resulting mass – called matte – is then transferred to a converter and air is blown through it. This burns out some of the sulphur from the copper and iron in the matte. The iron oxide and copper sulphide remaining react to yield metallic copper. The yield, called blister copper, can be up to 99% pure. But it is usually refined to higher purity by electrolysis, being made the anode in an electrolytic cell. During electrolysis copper from the anode is deposited, pure, on the cathode.

Getting into Shape

Metals are required in a diversity of shapes and sizes, and many different types of machines have evolved to produce them. Depending on the metal concerned and its application, it may be shaped molten, red-hot or cold. For simplicity most of the following description refers to steel-shaping methods. But analogous processes and machinery are used for other metals.

When the steel comes from the refining furnaces, it is usually poured directly into moulds and cools into ingots. The process of moulding molten metal is known as casting. The casting of ingots puts the steel into a form in which it can be handled by other shaping machinery. So does the more modern method of continuous casting, or 'strand' casting. In this type of casting molten metal is poured from the ladle

Below: Common methods of shaping metals. Some methods adversely affect the structure of the steel, which must be heat treated to restore its strength.

Right: Forging a generator rotor shaft on Britain's largest press-forge at British Steel's River Don works. It can exert a force of over 10,000 tonnes.

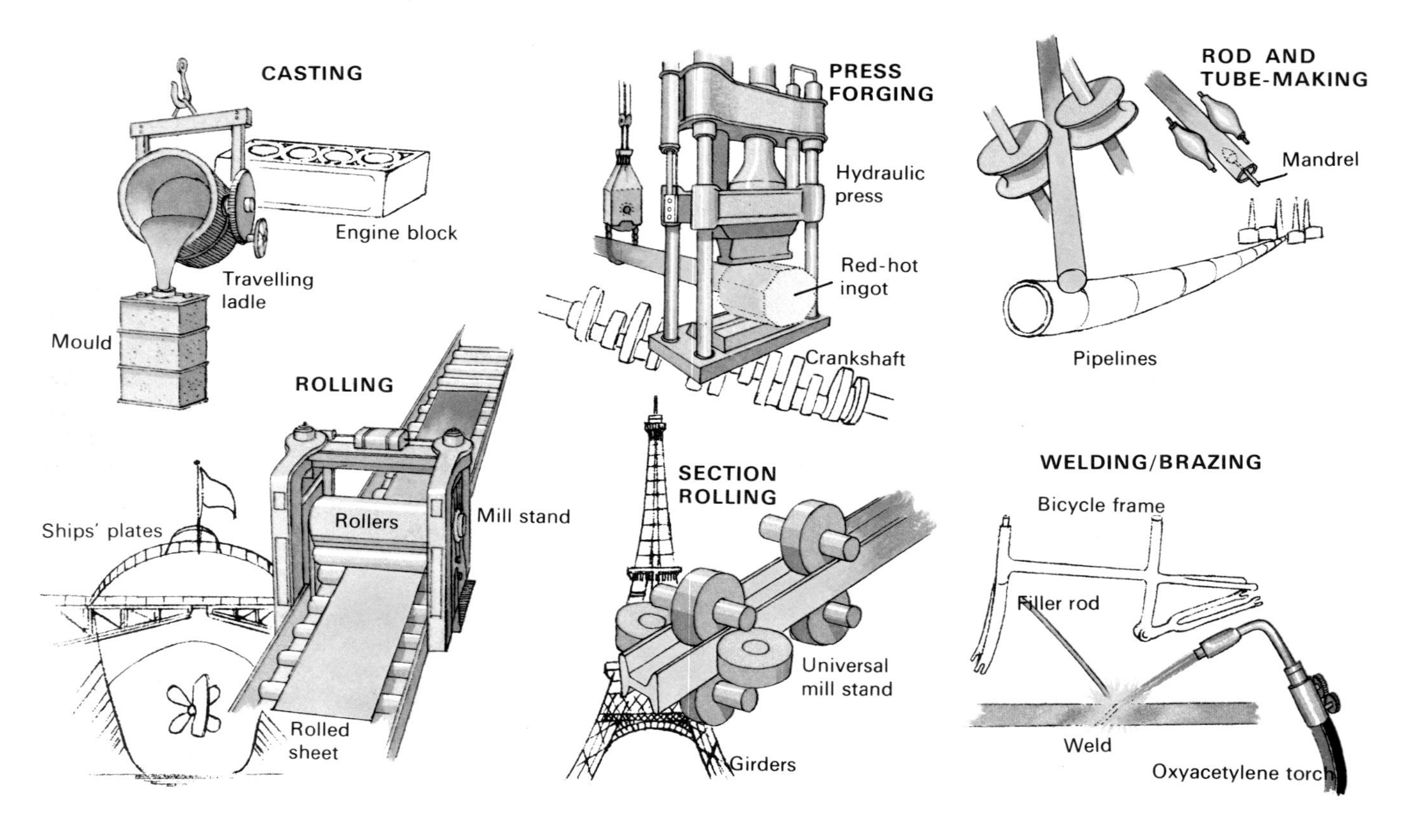

into a basin (tundish), from where it flows downwards into an open-ended, water-cooled mould. It emerges from the mould semi-solid and is then sprayed with water jets to accelerate the cooling. Rollers then grip it and withdraw it at a controlled rate. A cutting torch or saw cuts it to the required length.

More intricate casting is practised to form objects of complex shape. This can be done with refined steel and also with slightly modified pig iron (cast iron), as mentioned earlier. This kind of casting is usually done in a sand mould. The shaped cavity in the mould is produced by packing sand (actually a moist mixture of sand and clay) around a pattern of the object to be made. The mould is made in two halves, the upper half of which has risers (channels) to allow the molten metal to be poured in and the air inside to escape. The mould is dismantled when the metal inside has cooled and set.

For some applications permanent moulds, or dies, are made in metal and used repeatedly. This method is particularly applicable to the mass production of small metal parts – for example, toy cars – and the process is called die-casting. Very fluid, low melting point zinc alloys are widely used for die-casting. They are injected into closed water-cooled moulds, and quickly solidify. The two halves of the moulds spring apart and the cast objects are ejected. The moulds then close ready to receive the next injections.

Keep Rolling

Rolling is perhaps the commonest shaping method. It is used to manufacture products directly, such as steel sheets, plates, rail, bars and rods. It is also used to reduce the size of the ingots to smaller slabs and billets so that they can be more readily handled in other shaping processes. In rolling, red-hot metal blocks are passed repeatedly back and forth between pairs of heavy rotating rollers. The separation of the rollers becomes smaller with each pass so that the metal becomes thinner and longer.

The rollers are arranged in heavy rugged housings called mill stands and powered by electric motors. The stands may consist of only two rollers or of several rollers on top of one another, the upper and lower 'back-up' rollers helping to minimize bending of the smaller working rollers in between. The rollers can vary in size from a few centimetres to over two metres in diameter and may weigh 150 tonnes or more. Many rolling mills are now highly automated, requiring just one or two operators. They also operate very rapidly, the emergent strip of metal merging from the final rolls at speeds of 60 km/h (40 mph) or more.

The oldest method of shaping metal is forging, practised since the dawn of civilization. It shapes hot metal by means of repeated hammer blows or by progressive squeezing. In drop forging a heavy hammer or ram is lifted by steam or hydraulic pres-

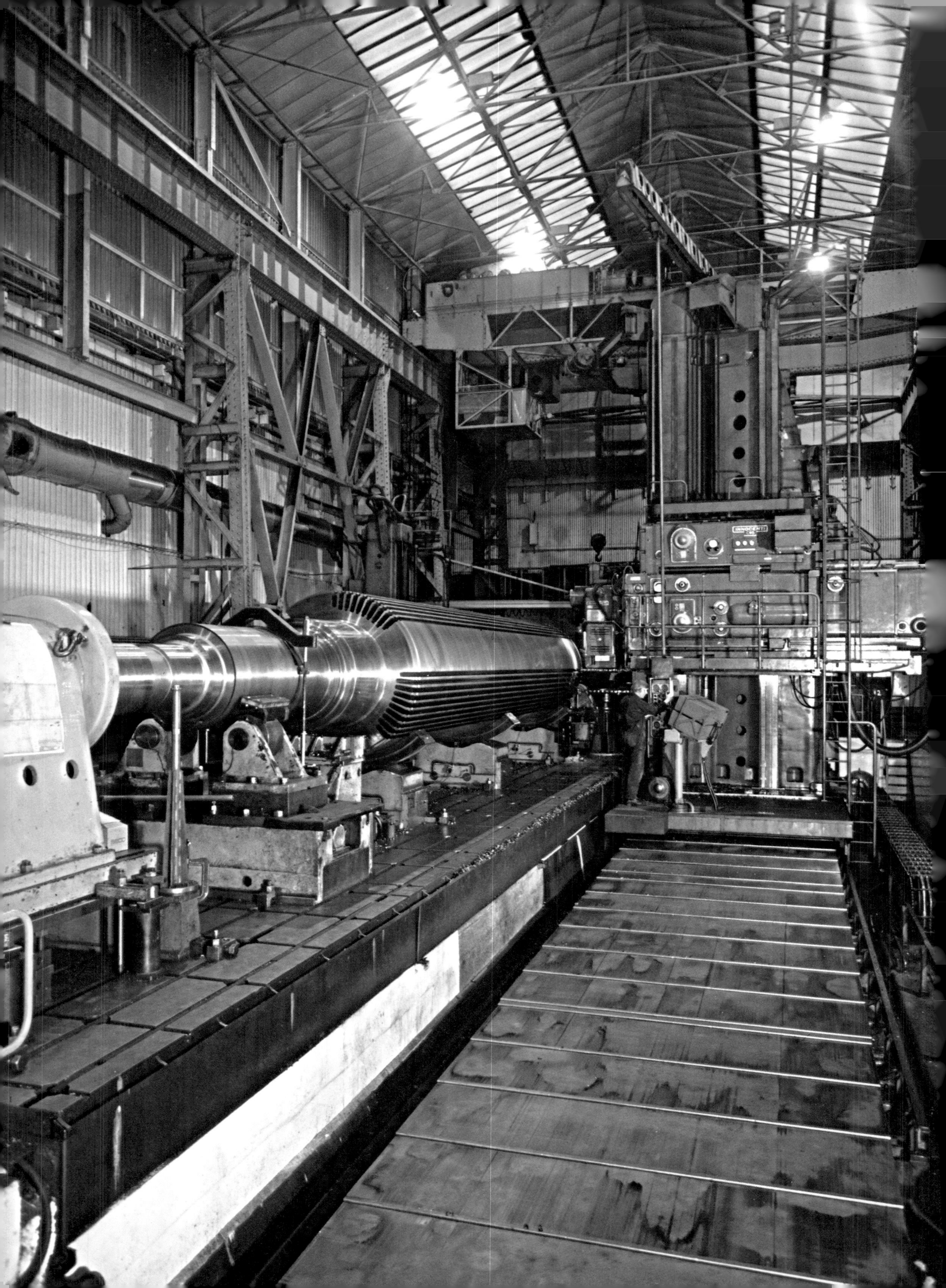
INNOCENTI

sure and is allowed to drop onto an anvil on which rests the metal to be shaped. Usually shaped moulds, or dies, are mounted on the underside of the hammer and on top of the anvil. The force of the impact between hammer and anvil forces the metal into the die cavity. Typical products of the drop forge are engine crankshafts, axles and tools.

In press forging the workpiece is slowly squeezed into shape by a hydraulic press, which may exert a pressure of 10,000 tonnes or more. Products like marine propeller shafts and turbine rotors are forged directly from steel ingots on such massive machines.

Among the other methods employed to shape metal are extrusion, drawing, powder metallurgy and explosive forming. In extrusion hot metal is squeezed through an orifice, acquiring the cross-sectional shape of that orifice. More complicated shapes can be produced in this way than with rolling. This technique is used widely to shape aluminium sections – for greenhouses, for example. Drawing is used to make wire, rod being drawn cold through tapered holes in hardened steel dies of progressively smaller diameter. Cold working increase the strength and hardness of the wire.

In powder metallurgy an object is shaped from metal powder. It is particularly suited to metals and other materials of high melting point which are difficult to shape in the usual way, such as tungsten. In the process the power is compacted under pressure into the desired shape and sintered, or heated strongly in a furnace. Another modern method, explosive forming, uses the force of an explosion, often transmitted through water, to press metal into a die. Large and complicated pieces can be so shaped without the need for massive forging presses. Explosive techniques can also be used to make metal laminates and compact, shape and bond metal powders.

Machining Operations

The metal-shaping processes discussed above often take place in workshops attached to the massive steel-making plants. The products are then transported to the factories in which they are to be used. There they will often require slight or extensive attention to bring them to the correct shape or size. This is done by machining, which is carried out by power-driven machine tools. These machines bring a variety of cutting tools to bear on the workpiece and thereby remove metal from it. They have provision to grip the workpiece and to move it relative to the tool or vice versa.

Since they have to cut metal, the tools must be hard and retain this hardness and their cutting edge at elevated temperatures, which will be reached because of friction during cutting. So-called high-speed steel is extensively used for cutting tools. It is an alloy containing tungsten, chromium and vanadium. For very high cutting speeds tungsten carbide tools are used, which are almost as hard as diamond and which retain their cutting edge even when red hot. To lower operating temperatures, most machines have the facility to direct a coolant fluid over the cutting tool. An emulsion of oil and water is often used, this also acting as a lubricant.

The most important machine tool by far is the lathe, which rotates the workpiece and brings cutting tools to bear on it. In a standard lathe the workpiece is held between a headstock and tail-

Left: A massive milling machine being used to cut slots lengthwise in a power-station generator rotor to take the armature windings.

Below: A standard lathe of a size widely used in light industry. It has a working bed length of 125 cm (50 inches) and can be driven at speeds up to 2000 rpm.

stock, and rotated by the headstock at a variety of speeds. A carriage called the saddle can move along the lathe bed parallel to the workpiece. The saddle carries a cross-slide, appropriately named because it can slide crossways at right-angles to the workpiece. It carries the tool posts on which are mounted the cutting tools. This arrangement allows the lathe operator to apply the tools to any part of the workpiece.

Other common machining operations include planing, milling, drilling and grinding. In planing the workpiece is mounted on a reciprocating table. A stationary cutting tool removes a slither of metal as the workpiece passes beneath it. In milling the workpiece is fed against a rotating toothed cutting wheel. Drilling bores holes by pressing a rotating drill bit into the workpiece. The bit has a cutting edge at the base and spiral fluting along its length to allow the cut metal (swarf) to escape. Grinding is done with a rotating abrasive wheel or a moving abrasive belt. In the operation metal is removed very gradually, making it a very precise metal-finishing process. It can machine parts to an accuracy of ± 0·0025 mm (± 0·0001 inch).

Assembly Lines

Consistent accuracy is, in fact, the hallmark of the modern machine tool, which is capable of turning out identical-sized components in quantity. The realization of the advantages that would accrue from the use of standard-sized components led to the birth of the machine tool in the late eighteenth century (see page 152). These advantages include the rapid assembly of the components into a finished product, often by semi-skilled labour. This is the basis of many modern methods of manufacturing.

Where a simple product is concerned, it may be assembled from its standardized parts by a single worker. But where the product is large or complicated, assembly usually takes place in various stages. One worker or group of workers put together one subassembly, which is then married to a number of other subassemblies to make the final product. This operation may take place in a single manufacturing complex, as it does in shipbuilding, for example. Or it may take place in different plants, as it does in car assembly.

Car assembly incorporates the two other elements required to streamline the assembly process into a method of mass production – the assembly line and the moving conveyor. It is appropriate that car manufacture be mentioned here, for it was in this industry that the assembly line and conveyor were first introduced (the former by Ransome E. Olds, the latter by Henry Ford, see page 68).

On a typical assembly line workers are positioned at different points (stations) along a moving conveyor. Each is assigned a specific, limited task and is supplied with the materials he requires to perform it, usually by other conveyors. At the beginning of the assembly line the main conveyor carries, as it were,

Above: the traditional car assembly line with workers at each station adding to the vehicle as it moves along. Components arrive via chain conveyor.

Top right: A new type of spinning machine that is coming into use, the rotospin, which spins directly from sliver. In the machine the sliver is opened out and the fibres fed to a rapidly rotating rotor. There they are compacted into continuous thread by centrifugal action.

Below: Workers are noticeably absent from this car assembly line at a Fiat factory in Italy. Most of the assembly operations are carried out by computer-controlled robots.

the embryo of the product to be manufactured. As it passes each station in turn, workers attach to it in sequence the appropriate parts. And the embryo grows progressively. But not until the final act of assembly does the finished product emerge, as it were, newly born after labour.

Different kinds of conveyors are used in assembly lines. The main conveyors, both on the floor and overhead, are usually chain operated, while components are usually delivered to operator stations by conveyor belt. In assembly processes involving heavy components, overhead travelling cranes may be employed to move the subassemblies. In others – for example, airframe manufacture – air cushions, or hoverpads may be used. The assembly is mounted on a pallet, and compressed air is forced beneath it to form a low-friction air cushion. The pallet can then be readily moved by tractor.

Processing Raw Materials

Assembly of components into a finished product forms a major branch of the manufacturing industry. Other major branches are concerned with the processing of raw materials into finished products or into intermediate products that are sold to someone else for making into finished products. As contrasting examples of raw-material processing let us consider ancient and modern sectors of the textile industry – the conversion of cotton into cloth and the production of man-made fibres; and oil refining.

In the cotton industry a natural raw material is physically processed, or manipulated, but does not basically change. Man-made fibres are produced by chemically processing a natural raw material or are created entirely from man-made chemicals. Many of these chemicals are obtained by refining crude oil, today's primary source of organic chemicals. In oil refining we see in action many processes common to the rest of the vast chemical industry.

Spinning a Yarn

The cotton used to make textiles comes from the lint, or long fibres around the seed in the cotton boll. (The shorter fibres, or linters, are too short for textiles but form a valuable raw material for the cellulosic man-made-fibres industry.) At the cotton mill machines open up the bales of cotton and untangle the fluffy mass of fibres into a loose blanket, or lap.

The lap is then fed to a carding engine, which removes the short fibres and straightens the long ones remaining, in a kind of combing operation. This is done by means of wire-toothed cylinders rotating at different speeds. The cotton emerges from the carding engine as a loose rope called a sliver. Several slivers are then combined and passed through a draw frame, where drawing, or drafting takes place. The loose ropes of fibres are passed through several pairs of rollers rotating at ever-increasing speeds.

There they are drawn out to longer lengths, lightly twisted, and combined to form what is called roving. This is now in a suitable state for spinning. It is wound onto a bobbin which is mounted on the spinning machine, the ring-spinning frame.

On the spinning frame the roving is further drawn out by drafting rollers into a finer yarn and then wound onto a bobbin on a rotating spindle. It is threaded onto the bobbin through a wire clip, or traveller, which is free to travel around a ring circling the bobbin. As the bobbin rotates, the clip moves around the ring and imparts twist to the yarn.

Cotton used to be spun on another type of spinning frame, called the mule, but this method is most often used these days for spinning wool. Wool goes through similar opening, carding and drafting operations prior to spinning as cotton, but for worsted yarn additional carding, or combing stages take place. Worsted yarn is made from only the longest wool fibres, and is therefore much stronger than ordinary woollen yarn, which is made from short as well as long fibres.

On the spinning mule, roving is drawn out by drafting rollers into finer yarn and then wound onto a rotating spindle mounted on a carriage, which moves alternately back and forth. As the carriage moves forward, the yarn slips off the top of the spindle and is twisted because the spindle is rotating. Then the carriage moves back, and the twisted yarn is wound onto the spindle bobbin. A third common type of spinning frame, the flyer, is used mainly to spin linen yarn, linen being made from the bast (stem) fibres of the flax plant.

Get Weaving

The process of weaving – the interlacing of threads at right-angles to one another to form cloth is the oldest and is still the commonest means of making textiles. Most cloth today is woven on high-speed power looms, though hand looms are still used in craft centres and in regions where traditional cottage industries still thrive. But both power looms and hand looms work on the same basic principles.

One set of threads, the warp, passes lengthwise through the loom. Certain of these threads are raised or lowered to form a gap, called the shed. Another thread, the weft, is then passed through the shed by a shuttle to form a line of weave. The warp threads are then reversed – lowered and raised – to form another shed, and the shuttle is passed through it again to form another line of weave. To ensure a tight weave, a kind of comb called a reed pushes the newly woven line against the previous line. The endlessly repeated operations of making a shed, passing through the shuttle, and tightening the weave, are termed respectively shedding, picking and beating-up.

The diagram below shows the basic features of a hand loom. Most power looms have similar components, but shedding, picking and beating-up take place automatically and very speedily with the minimum of operator intervention. Highly patterned fabrics can be woven on specially designed Jacquard looms. The Jacquard mechanism, introduced by Joseph-Marie Jacquard in 1805, is important historically in that it was the first automatic control device used in industry. It uses a system of needles and punched cards to control the raising and lowering of the warp threads. The lines of weave are represented by lines of holes punched in the card. Needles which control the raising of each warp thread press against the card. Where there are holes, they pass through and cause the appropriate threads to be raised. Where there are no holes, the thread-lifting mechanism does not operate.

The greatest innovation in weaving in recent years has been the introduction of shuttleless looms. In the ordinary loom the shuttle carries a bobbin of weft thread. In shuttleless weaving only a loop or line of weft is carried through the shed each time. The commonest method uses a dummy shuttle, a 'missile' that shoots through the shed, dragging the weft with it. In the rapier loom a metal rod or tape (rapier) grips a line of weft, thrusts it through the shed and then withdraws. But the most interesting looms of all carry the weft thread through the shed by a jet of water or air.

Man-Made Fibres

These days a great proportion of the fibres used in textiles are not natural fibres, like cotton, wool and linen, but are man-made. Man-made fibres include those, like rayon, which are derived from natural materials and processes by man. It also includes synthetic fibres, like nylon, which are derived totally from man-made chemicals.

Man-made fibres are produced by processes known as spinning, though they bear no relation to the

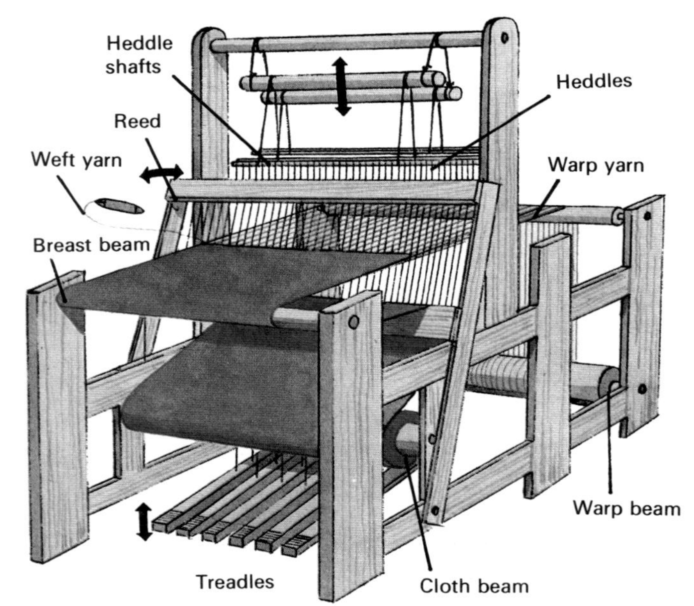

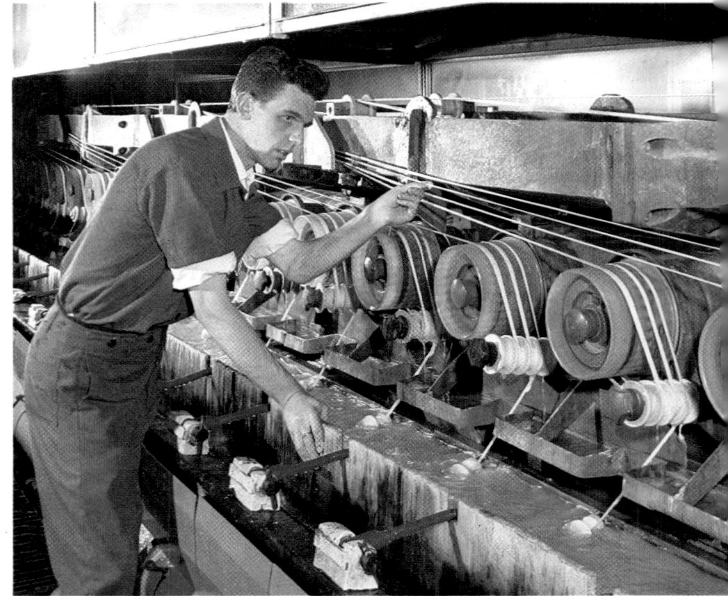
Continuous filaments of rayon emerging from the acid bath during the 'wet' spinning process.

Patterned fabric being woven on a Jacquard loom. The punched cards of the Jacquard mechanism can be seen at the top.

spinning processes just described. They are usually produced initially in the form of long continuous strands, or filaments, rather like silk is. These filaments can be woven as they are, or they can be chopped up into shorter, staple fibres and then spun into yarn on one of the spinning frames. The spun yarn is then woven in the usual way.

Refining the Crude

Crude oil, or petroleum ('rock oil') is the main source not only of the raw materials used in the synthetic fibres industry, but of those used in the manufacture of plastics, dyes, insecticides, explosives, adhesives, antifreeze, drugs, detergents, paints, rubber, innumerable solvents and a host of other synthetic products. It is a viscous greenish-black liquid which is composed almost entirely of compounds containing hydrogen and carbon only, called hydrocarbons. It becomes useful only after it has been refined. The processes in the oil refinery aim at separating the hydrocarbon compounds into groups with similar boiling points. It is not practical to isolate individual hydrocarbons, since there are hundreds upon hundreds of them.

The main refinery operation is fractional distillation, or fractionation, a physical process widely used throughout the chemical industry to separate liquids of different boiling points. The crude oil is first passed through tubes in a furnace ('tube still'), where it is heated to a temperature of about 300°C and partly vaporizes. Liquid and vapour are fed into the middle of a fractionating column, which is a tall tower with perforated trays spaced at regular intervals up and down it. They are maintained at certain temperatures, which decrease going up the column.

Each tray contains liquid just below its boiling point. Vapour rising up the tower bubbles through the liquid via mushroom-shaped caps. The liquid absorbs from the vapour heavier (lower boiling point) components while itself losing lighter (higher boiling point) components to the vapour. It continuously overflows to the next tray down, bringing about further contact and exchange with the vapour. This happens on each level, and when equilibrium is reached, the tray at the top of the column holds the lightest fraction in the oil, while the bottom tray holds the heaviest. Going downwards from the top, the main fractions obtained are petrol, kerosene, gas oil and fuel oil. The uncondensed vapour emerging from the top of the column is a mixture of hydrocarbon gases such as propane, butane, methane, ethane and ethylene.

The heavier fractions in the lower part of the column are usually further distilled in a second fractionation unit under a vacuum, which results in various grades of machinery lubricating oil, greases and a residue of coke and asphalt. The fractions obtained from both primary and secondary (vacuum) distillation usually undergo further purification and blending before they are fit for use.

SPINNING WITH A TWIST

There are several kinds of man-made fibre 'spinning' processes, which can be illustrated by reference to the common fibres viscose rayon, triacetate and nylon. Rayon, the most widely used of all fibres, natural or man-made, consists of pure cellulose, the main substance in woody tissue. It is called a regenerated fibre because the cellulose is first extracted and dissolved and then reformed in the spinning process.

In viscose-rayon manufacture cotton linters or wood-pulp (both nearly pure cellulose) is cooked with caustic soda, whereupon the cellulose dissolves. The solution is then treated with carbon disulphide and afterwards dilute caustic soda solution. This results in a viscous, orange-brown 'spinning solution' of sodium cellulose xanthate. The spinning solution is then pumped through the fine holes of a metal spinneret and into an acid bath. The fine streams emerging from the spinneret react with the acid and change into filaments of pure cellulose.

Rayon manufacture is an example of 'wet' spinning. The production of triacetate is an example of 'dry' spinning. In the process cotton linters or wood-pulp is treated with acetic acid, whereupon the cellulose is converted to cellulose triacetate. This cellulose ester is dissolved in a mixture of alcohol and methylene chloride, both of which are very volatile. The solution is next pumped through a spinneret into a stream of hot air. The solvents evaporate to leave delicate filaments of cellulose triacetate.

Many wholly synthetic fibres are formed by melt spinning. They are kinds of heat-deformable plastics known as thermoplastics, and include nylon (the original) and polyesters like Terylene. Nylon is a polymer, or plastic, formed by polymerizing together the chemicals hexamethylene-diamine and adipic acid. The molecules of these chemicals join together in long chains, which are responsible for their fibre-forming characteristics. To make fibres, the nylon is remelted and pumped through a spinneret into cold air. The emergent streams solidify as long filaments, and are then drawn out to give a strong, lustrous fibre.

Cat Crackers

It is the lighter petroleum fractions which are the most valuable, particularly petrol, but the yield of petrol from straight fractionation is only about 30% (it varies with the quality of the crude). However, the petrol yield can be substantially increased by further processing of the heavier and less useful fractions. The main method employed is cracking, a chemical process in which the large molecules of the heavy hydrocarbons are split into smaller ones.

Cracking may be achieved by heating the heavy oil fraction, usually gas oil, under pressure, temperatures of some 450°C and pressures up to 30 atmospheres being necessary. This method, known as thermal cracking, is bettered by the method of catalytic cracking, which uses a catalyst to help bring about molecular breakdown, under only slightly elevated pressure. A typical 'cat cracking' complex will include a reactor, a regenerator, and a fractionating column.

Cracking takes place, usually with a fluidized catalyst, in the reactor. The catalyst is finely divided clay made 'fluid' by the passage of the oil-feed vapour through it. As cracking takes place, carbon is deposited on the catalyst particles. They are therefore continuously recycled through the regenerator, where the carbon is burned off, and back again. The cracked oil vapour rises through the reactor and is led off to a fractionating column for separation into its fractions. Petrol yield from gas-oil feedstock can be as high as 60%. And it is of the high-octane rating required by modern engines.

A number of other refinery processes aim at improving the straight-run yield. They include catalytic reforming, in which the hydrocarbon molecules

Above: Operator at the control console of an automatic steel bar mill. In the modern steel mill rolling is a highly automated operation, masterminded by a control computer. It reacts instantly to changes in operating conditions and instructs the machines to make the necessary adjustments to ensure a consistent product.

Left: Automation has reached its zenith in the oil refinery, where a variety of physical and chemical manipulations of the crude oil are carried out. The modern refinery comprises a huge complex of tall stainless steel distillation towers and reactor units covering an area of some some 250 hectares (over 600 acres). It can handle up to 20 million tonnes of crude a year.

are rearranged. For example, straight-chain molecules are converted into branched-chain and ring molecules (aromatics), which have higher octane rating. The light gases taken from the top of the catalytic cracking fractionating towers may be polymerized into heavier, petrol-type fractions. Treatment of certain fractions with hydrogen, itself a refinery by-product, also results in more desirable fractions.

There is virtually no limit to the number of hydrocarbon compounds that can be obtained from crude oil by intricate chemical manipulation. Yet most oil is converted into fuel to be burned in cars (petrol), jet planes (kerosene), lorries (diesel, gas oil) and central-heating boilers (fuel oil), all of which are consuming it at a prodigious rate (over 50 million barrels, 8000 million litres, 1750 million gallons per day worldwide). What a waste this is of such a precious commodity, that has been accumulating in the rocks for maybe 200 million years or more.

Automatic Control

An oil refinery covers a vast area with tall gleaming steel distillation towers and cracking units, interconnected by kilometres of snaking pipes. There is obviously a lot going on, for there is a continuous hum, hiss, rumble and click as liquids flow, heaters switch on and off and valves open and close. Yet for all this apparent activity there is scarcely anyone to be seen. The few operating personnel present are not scattered around the plant but concentrated in a central control room. However, it is not they who man the switches, press the buttons, and issue the orders. The 'boss' is a computer programmed so that it can react to and correct any irregularities in operating conditions throughout the refinery. It can act much quicker than a human controller and carry out a multitude of operations simultaneously, which a human being, however dextrous, could never do.

The computer-controlled refinery still provides the best illustration of automatic, self-regulating machinery control, or automation. Mechanization ushered in the first Industrial Revolution; automation is ushering in a second.

The control computer at a refinery contains in its 'memory' details of the ideal operating conditions in the various parts of the plant required for optimum output. These include such things as the temperature and pressure levels in the distillation towers and cracking units and the flow rates of feedstocks and products through the labyrinth of pipes. The computer is linked to sensors (thermocouples, pressure gauges, flow meters) in various locations throughout the refinery which monitor the actual conditions prevailing during operation. It is programmed to compare details of these actual conditions with the details of the ideal conditions stored in its memory, and take remedial action should they differ. If they do differ, the computer sends 'error' signals to mechanisms in the appropriate locations, ordering them to switch heaters or pumps on or off, or to open and close valves.

Many other industries, though not as suited to such complete automation as oil refining, have introduced at least partial automation. Modern steel rolling mills are now highly automated, the separation of each set of rollers being determined by the temperature and thickness of the steel coming from the previous set, the appropriate roller separations being calculated and implemented by the control computer. Many machinery shops now use automated numerically controlled (NC) machine tools. Operating instructions are fed into the machine in the form of punched-paper tape or magnetic tape. A scanning unit senses and interprets the instructions on the tape, sending signals to the machine to advance the appropriate tool, rotate the cutting head, position the workpiece, and so on. Sensors monitor the actual performance of the machine and feed the information to a control computer, which can take remedial action if necessary. All these examples illustrate the difference between automatic and automated machines. The former work automatically but need human operators to adjust and regulate them. The latter work automatically but adjust and regulate themselves.

Electronic Brainpower

The computer has become an invaluable tool not only in industry but in business, commerce, transportation and, in the guise of the pocket calculator, in the home. It is itself in essence an electronic calculating machine, which can perform mathematical operations with lightning rapidity. It works with figures, or digits, which is why it is called a 'digital' computer. (Another more specialist kind of computer is known as an 'analog' computer, for it represents a variable quantity in the form of a variable electric voltage or current so that the one is analogous to the other.)

The digital computer can not only perform mathematical calculations, but by extension can sort out, compare and analyse figures and any information that can be represented by figures, which means virtually everything. In other words it is a superlative data processor, which is why it has found such widespread use in such areas as banking and stock control.

The computer handles figures not in the usual form of the digits 0, 1, 2 . . . 9, but in the form of the two binary digits 0 and 1. In the binary, or base-2 number system, place numbers indicate powers of two, just as in the decimal, or base-10 system place numbers indicate powers of 10. In binary 1 = 1, 2 = 10, 3 = 11, 4 = 100, 5 = 101, and so on. Inside the computer the two binary digits can be represented by a current being switched 'On' (1) or 'Off' (0), or a piece

Above: Laser and computer join forces at the latest supermarket checkout counter. The laser scans a visual bar code printed on each item, and the linked computer retrieves the item's name and price, displays it, prints it out on a list, which it finally totals.

Below: A silicon chip in situ, compared in size with a needle and thread. Fine wires connect the chip to contacts on its mounting, which lead to the external connections at the sides.

Left: A digital watch with liquid crystal display (LCD). Battery powered, it is regulated by a wafer of pure quartz which vibrates exactly 32,768 times a second. A micro-circuit etched on a silicon chip converts this 'time base' into the digital display. The display may be provided by minute light-emitting diodes (LEDs) instead of liquid crystals. Depending on their sophistication, digital watches can be programmed not only to tell the time but to give the date, act as a stop watch, issue alarm signals and do mathematical calculations.

Right: Silicon-chip technology has brought the computer into the small business, where it can be used for accounting and stock control.

of iron being magnetized in one direction or another.

Instructions informing the computer what it is to do with the data are also coded in binary. But the computer operator need not always feed his instructions into the computer in digital form, he can do so using a specially developed computer 'language', which the computer then translates into digital form. Such languages include COBOL (Common Business Oriented Language) and FORTRAN (Formula Translation). The instructions are given in the form of a 'program', and each stage of the program involves a simple comparison or question that demands the answer 'Yes' or 'No'. The computer can interpret this again as a 1 or a 0 – a current switched 'On' or 'Off'.

The data fed into and manipulated by the computer, and the program the computer follows, form what are called the 'software'. The equipment that feeds the data in and processes it forms the 'hardware'. In a typical computer set-up the hardware consists of an input device, data store and memory, control unit, arithmetic unit and output device.

The input device feeds in data in the form of a pattern of holes on punched paper tape or cards; a magnetic pattern on magnetic tape or disc; or directly via a keyboard. The control unit commands the other parts of the computer to carry out the instructions in the program. The arithmetic unit carries out mathematical calculations and processes data retrieved from the memory or data store. The memory, or internal data store, contains data that is repeatedly needed, while the external data store contains data that is needed for the specific operation in progress. When the computations have been carried out, the results are presented by the output device. This is usually a line printer similar to a teleprinter but very much faster, but it can also be a video screen or punched or magnetic tape.

The computing power of the modern computer (as long as it is correctly programmed, of course!) is prodigious. The most powerful are capable of performing 100 million calculations per second. They can store in their memory information equivalent to tens of million of binary digits (also called 'bits' and 'bytes'). They can locate in this voluminous data bank specific items in a ten-millionth of a second. And they can print out results at the rate of 100 lines per second. It is its incredible speed that makes the computer such a marvellous tool. But it is worth remembering that it is only as good as the person who programs it.

In recent years there has been a revolution in computers brought about by the introduction of microprocessors – the so-called miracle chips – for data storage devices. Ten of thousands of storage circuits can be etched in a wafer-thin 'chip' of silicon a few millimetres square (see page 186). This has brought about the powerful 'desk-top' computer and on a smaller scale the digital watch and pocket calculator.

Behind the Scenes

Industry puts to work machines and processes that have been developed over long periods of time. The original impetus may have come from a lucky discovery (like the vulcanization of rubber by Charles Goodyear), but chances are that it resulted from years of painstaking research by individuals or groups of workers.

Research scientists work by observing, probing and measuring, as a result of which they acquire relevant data about the topic they are researching. Having acquired this data, they then have to process and interpret it and draw conclusions from it. These stages form part of what is called the scientific method.

No matter how perceptive we might be, we have limited powers of sight, hearing, touch, taste and smell which severely restricts our ability to observe. And our observations will only tend to give a qualitative view of the world – what things are – rather than a quantitative view – how big, how long, how heavy things are. To extend the power and range of our perception and our ability to probe and measure, we have therefore created a host of scientific instruments.

Our vision, for example, is vastly extended by the microscope and telescope. Using the electron microscope we can see deep into the cells that make up living tissue and view the very genes that determine our inheritance. Using the telescope we can see worlds so far away that their light takes thousands of millions of years to reach us. And with photometers and bolometers we can measure the brightness and heat of the faint starlight reaching us. With Geiger counters, we can detect radiation we cannot even see or feel – emanating from the heart of unstable radioactive atoms. By using particle accelerators we can actually split atoms into fragments, and with bubble chambers witness the event. Deep down inside a mountain in South Dakota a tank of dry-cleaning fluid is used to detect elusive particles called neutrinos born in the interior of the Sun, which are invisible, have no charge or mass, only the energy of spin.

Today the hardware and software – the tools and techniques – of the men and women engaged in science and industry are becoming ever more sophisticated as time passes. And the ordinary layman can be forgiven if he feels uncomfortably out of his depth at times. To help him assimilate the new culture we include the following glossary which outlines the principles behind a variety of instruments, processes and techniques in widespread current use, and which do not appear elsewhere in this book. While the reader may find some of them familiar, he will not necessarily be familiar with how they work.

Left: The 400,000-litre (88,000-gallon) tank full of perchloroethylene, which forms the sensor in apparatus designed to detect the presence of neutrinos emanating from the Sun's interior.

GLOSSARY OF TOOLS AND TECHNIQUES

aerosol Strictly, aerosol container; a can which contains a substance which is sprayed into the air. The substance to be sprayed – for example, paint, hair lacquer or insecticide – is dissolved in a volatile liquid in the can. Above the liquid is a propellant gas under pressure. Often the dissolving agent and the propellant are one and the same – usually a fluorocarbon such as Freon. When the release button of the aerosol can is pressed, the propellant forces the carrier liquid out through a nozzle into the air. Being highly volatile, this liquid rapidly evaporates, leaving an aerosol, or fine mist of droplets of paint, lacquer or whatever.

altimeter An instrument that measures altitude – the height above sea level or the ground. One kind of altimeter relies on measurement of air pressure, for the air pressure varies with altitude. It decreases with increasing altitude in a known way. The basis of this kind of altimeter is an aneroid capsule (see **barometer**), which responds to changes in air pressure. Its movements are magnified by cogs and levers to drive pointers over a graduated dial. An altimeter normally measures height above sea level, and must be zeroed frequently as the sea-level pressure changes with varying weather conditions.

A pilot of an aircraft has to subtract the elevation of the ground beneath him to find his ground clearance with this kind of altimeter. Most aircraft these days, however, are fitted with a radio altimeter which measures the height above ground directly. It works by timing the passage of radio pulses down to the ground and back (in an analogous way to a boat's echo-sounder, which uses pulses of sound waves to measure depth).

ammeter An instrument that measures the flow of electric current, in units of amperes (amps). The most common type, the moving-coil ammeter, has its pointer attached to a wire coil which is free to rotate. It is located between the poles of a permanent magnet. The electric current to be measured is passed through the coil, and by the so-called motor effect (page 60) makes the coil move. The amount of movement is proportional to the strength of the current passing.

anemometer An instrument that measures wind speed. The type most widely used at weather stations is the cup anemometer which has three or four cups mounted horizontally on spokes. They catch the air and spin when the wind blows, and their rate of revolution is recorded on a dial. Alternatively the spinning shaft turns a tiny generator which operates an electric meter calibrated directly in wind speed. Another type sensitive enough for low air speeds is the hot-wire anemometer, measurement of speed being based on the cooling effect of the air stream on a fine, electrically heated wire.

Above: One of the world's most powerful atom-smashers, the Stanford Linear Accelerator at Stanford University, California. It is 3050 metres (10,000 ft) long.

atom-smasher Properly called particle accelerator; a machine for accelerating charged atomic particles to high energy. The particles are then used to bombard atoms, or rather the nucleus of atoms. The bombardment often results in the splitting of the nucleus, with the ejection of new atomic particles. Study of such collisions helps physicists understand the make-up of matter. There are two basic types of accelerator – the linear and the circular – which describe the path the particles follow.

In the linear accelerator an electric field is applied so that it pushes and pulls (repels behind and attracts in front) the charged particles along a so-called drift tube. In one type of circular accelerator, the cyclotron, charged particles are accelerated in a spiral as they pass between two D-shaped plates, charged alternatively positive and negative and located in a strong magnetic field. The most powerful kind of circular accelerator, called a synchrotron, accelerates particles within a great circular tube. Powerful magnets constrain the particles in the tube while it is accelerated by electric pulses applied in synchronization with the passage of the particles. The synchrotron at CERN, the Centre for European Nuclear Research, near Geneva, has an accelerating ring over 2 km ($1\frac{1}{4}$ miles) across. It has a power consumption of 135 megawatts of electricity.

balance A sensitive pair of scales used in laboratories. The traditional balance uses a pair of scale pans. They are suspended from knife edges at the ends of a balance arm, which also balances on a knife edge during weighing. The object to be weighed is placed in one pan and weights are added to the other pan until the two are in equilibrium. Modern balances, however, have a single weighing pan and suspended above it a variety of weights. With nothing on it, the pan exactly balances a counterweight. The object to be weighed is placed in the pan and then weights are *removed* until the arm is in equilibrium again. Very sensitive precision balances, using the twisting of a quartz fibre or electromagnetism to restore equilibrium, can measure to a millionth of a gram (3·5 hundred-millionths of an ounce).

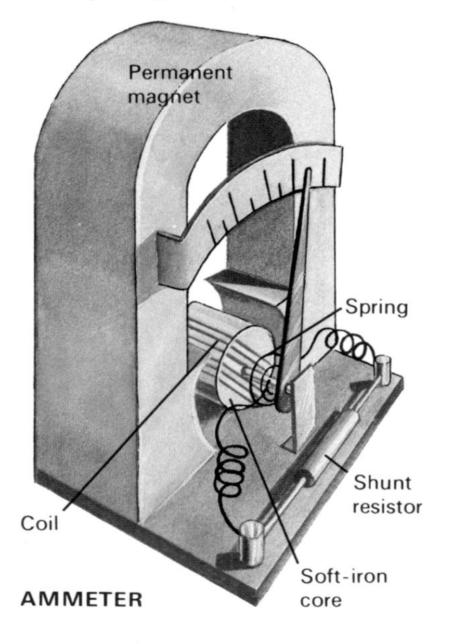

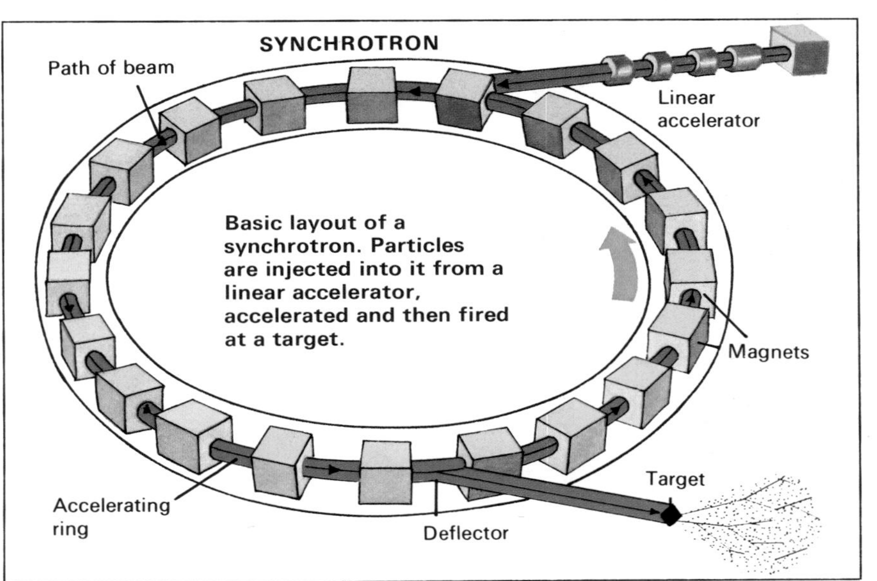

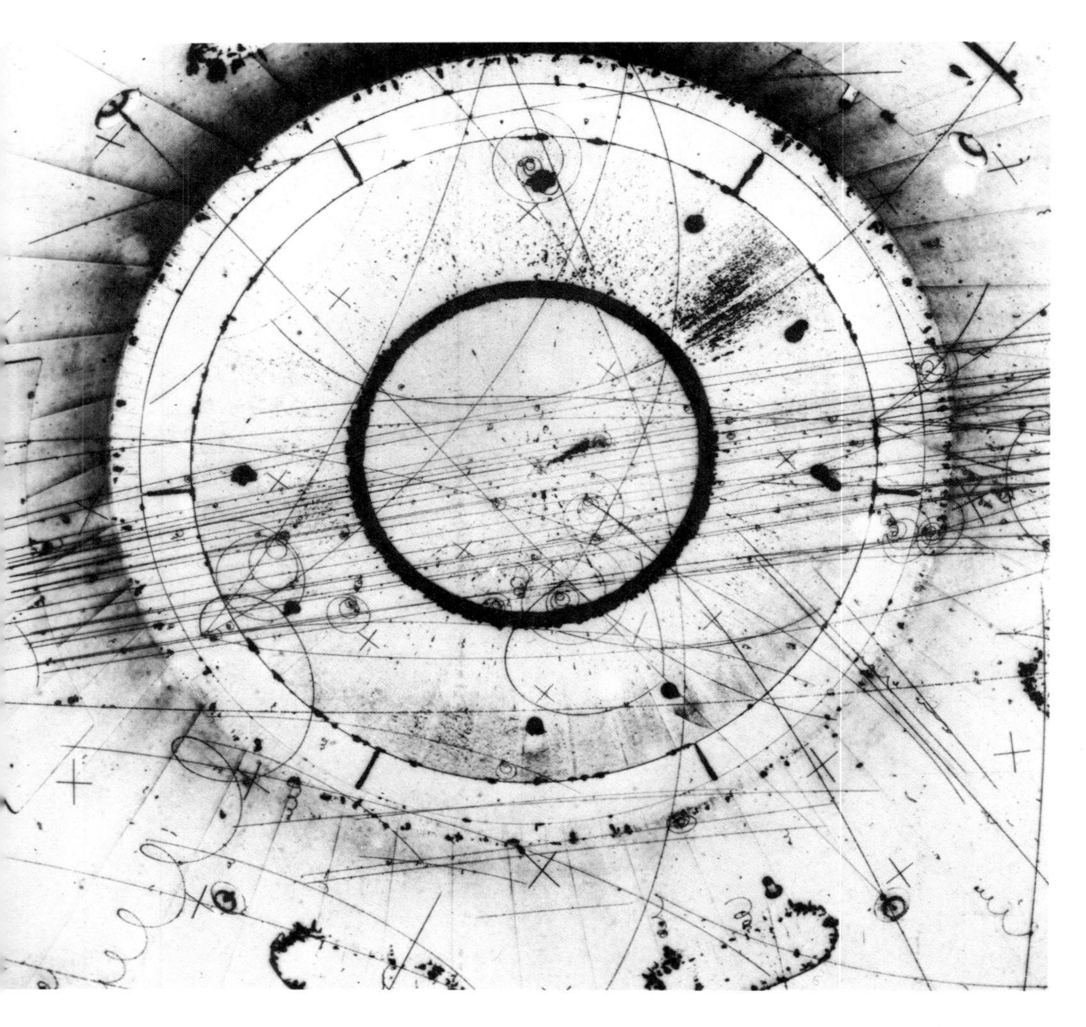

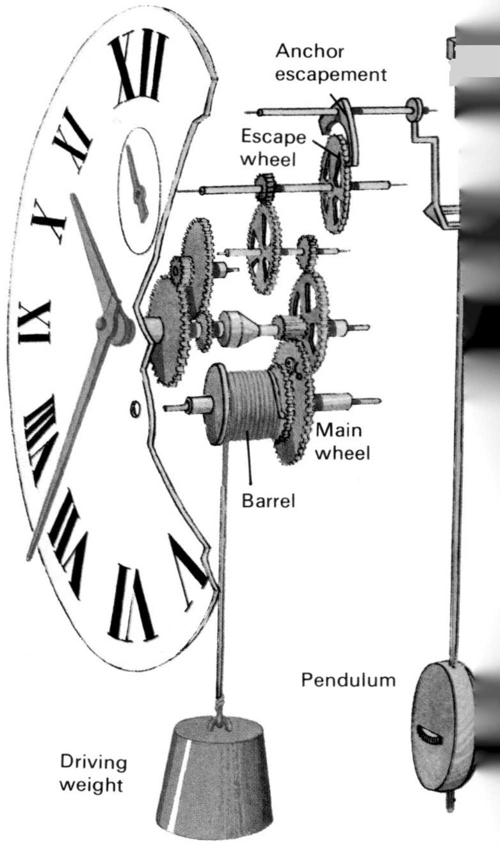

PENDULUM CLOCK

barometer A device for measuring air pressure, dating to Torricelli's experiments in 1643. The mercury barometer consists of a long glass tube filled with mercury, up-ended in a bath of mercury. The column of mercury in the tube falls to a point at which it is supported by the pressure of the ambient air. The space in the tube above the mercury column is a partial vacuum. The height of the mercury column at sea level is approximately 760 mm (30 inches). It changes according to the state of the atmosphere, and provides warning of imminent changes in the weather.

The home barometer, however, is an aneroid type. It contains a partially evacuated bellows, which expands or contracts as the ambient pressure decreases or increases. The movement is exaggerated by levers to move a pointer over a calibrated scale.

Bayer process The means by which the oxide of aluminium, alumina, is extracted from the aluminium ore bauxite prior to smelting. The process involves treating the bauxite with hot caustic soda, whereupon the oxide dissolves. The solid that separates out on cooling is converted to alumina by heating.

Bosch process A method of making hydrogen industrially by reacting water gas (hydrogen and carbon monoxide) with steam in the presence of metallic catalysts. Carbon dioxide is also produced, which can easily be removed by scrubbing, or absorption in limewater. This process could become commercially much more important in the years ahead.

bubble chamber A device used to detect atomic radiation. It uses a superheated liquid – one heated above its boiling point – as the detection medium, often liquid ether or liquid hydrogen. Pressure is applied to prevent it boiling, until just before operation. Then the pressure is released, and any particles passing through the liquid will ionize molecules along their path. Bubbles will form around the ions as the liquid beings to boil, and so the passage of the atomic particle can be followed.

Bunsen burner The standard gas burner used in laboratories. It is a metal tube into which gas is introduced at the base through a narrow orifice. Sculpted from the base of the tube are two holes, which can be opened or closed by a collar. This regulates the amount of air mixing with the gas to form an inflammable mixture that burns at the top of the tube. By opening the air hole the flame becomes blue and very hot (1500°C). by closing it, the flame becomes luminous yellow (because of unburned carbon particles) and cool.

cement-making Ordinary Portland cement is a fine grey powder made by roasting a mixture of limestone, iron ore, clay and gypsum in a rotating kiln. Other substances, such as alumina, may be added. Cement is mixed with sand, gravel and water to make concrete. The setting of concrete, which can take place under water, is a chemical reaction. It gives out heat, which must be allowed for in massive concrete constructions. Concrete is relatively weak when it is under tension, and is often reinforced with steel.

centrifuge A revolving device used to separate solids from liquids, or liquids of different densities. It makes use of centrifugal force. As the centrifuge spins the solid particles or heavier liquid drift to the bottom, leaving the lighter substance on top. Ordinary centrifuges rotate at speeds up to several thousand revolutions per minute (rpm). Ultracentrifuges used in medicine and biological research can rotate at speeds up to 100,000 rpm.

chromatography A method of separating chemicals from one another, now widely used in chemical analysis, particularly when minute quantities of material are available. Simplest is paper chromatography, carried out with highly porous paper like blotting paper. A drop of the substance to be analysed is placed on the paper, the end of which is then dipped into a solvent. As the solvent is absorbed (by capillary action), it travels through the sample, differentially absorbing the component chemicals within it. When these components are coloured, they separate out as coloured bands at differing distances up the paper. In column chromatography a vertical column of absorptive material is used instead of porous paper. In gas chromatography a stream of inert gas, such as nitrogen, is used to carry vaporized samples through an absorptive column. The components in the sample travel through the column at different rates and emerge at different times. Trace substances present in quantities as low as one part per million can be detected in this way. They may then be identified by other sensitive techniques, such as **mass spectroscopy**.

clocks and watches The most accurate and precisely engineered common machines. The latest digital watches, for example, are accurate to minute fractions of a second per day. The way they operate is a far cry from the grandfather clock deliberately ticking the time away second by second. But the latter provides the simplest illustration of the function of a timepiece. All clocks have two main basic elements. One is a power source to drive the hands around a calibrated dial (clock face). The second is a regulator to allow the power to escape, and drive the hands, at a steady rate. Most clocks also have a train of gears – interlinking toothed gearwheels – to transmit the power from source to the hands.

In the grandfather clock the power source is a weight, suspended by a cord from a cylinder, or barrel, which has a gearwheel at the end. A gear train connects the barrel with the hands and also with the regulator, which is a pendulum. The pendulum – simply a long rod with a weight, or bob, on the end – makes a good regulator because it always has the same time a swing. An anchor-shaped lever rocks back and forth with the pendulum, alternately engaging and disengaging a gearwheel ('escape wheel') at the end of the gear train, allowing it to move round tooth by tooth. It also 'nudges' the pendulum each time to keep that swinging.

Falling weight and pendulum are of little use in a watch, and in this a coil spring (mainspring) provides the power and the regulator is an oscillating wheel ('balance wheel'). The balance wheel is made to oscillate by means of a fine hairspring. Most ordinary watches you have to wind

WATCH

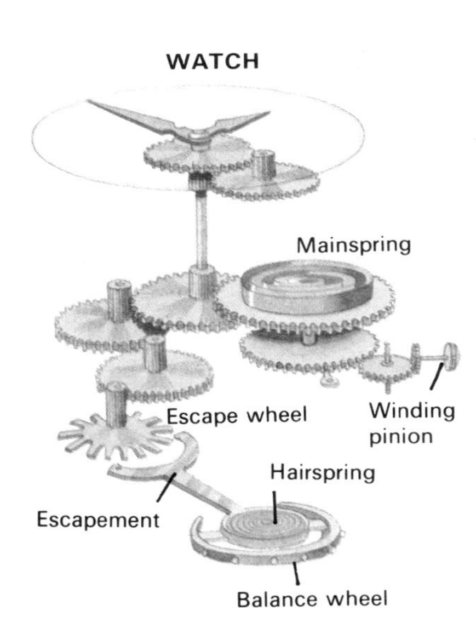

up yourself, but in self-winding watches an oscillating weight does the winding for you.

In recent years electronic watches have come to the fore. The earliest type uses a battery-operated electronic circuit to vibrate a tuning fork at a very precise frequency (usually about 360 vibrations a second). It does this by means of tiny electromagnets. The movement of the fork is passed through a reducing gear train to the hands. The latest electronic digital watch is the product of semiconductor, microprocessor technology (see page 176).

compass A device for indicating direction. The simplest compass is a magnetized needle balanced on a pivot – or suspended from a thread or floated on water. It works because the Earth has a magnetic field, behaving as if there were magnetic poles near the geographical north and south poles. A magnetized needle aligns itself parallel with the invisible lines of magnetic force that girdle the Earth and run in a north–south direction. In practice the lines do not run precisely north-south because the magnetic poles are not coincident with the geographic poles but are several degrees off. The discrepancy between magnetic and geographic ('true') north varies regularly and must be allowed for when map-reading. (See also **gyrocompass**.)

contact process The main method of making sulphuric acid, the most important industrial chemical, of which some 80 million tonnes are produced each year. In the process sulphur dioxide (SO_2), made by burning sulphur or roasting iron pyrites, is converted into sulphur trioxide (SO_3) by being passed over trays of heated catalyst (finely divided platinum or vanadium pentoxide). The sulphur trioxide is then absorbed in water (actually dilute acid), with which it combines to form sulphuric acid (H_2SO_4).

cryogenics The study of materials at very low temperatures and methods of reaching such temperatures. It began with the liquefaction of air (at −190°C) by progressive compression and rapid expansion (see **Linde process**). These days liquid hydrogen and liquid oxygen, obtained by distilling liquid air, are extensively used. Their most exotic application is as space rocket propellants. The most difficult gas to liquefy, helium, becomes liquid at −269°C, only about 4°C above absolute zero, which is the lowest temperature that can ever be obtained. It exhibits incredible behaviour (superfluidity) in that it will in some situations seemingly defy gravity and flow upwards. Another phenomenom exhibited by some metals at very low temperatures is superconductivity, in which they lose virtually all their electrical resistance.

detergent Strictly speaking any substance that cleans, including **soap**. In common parlance, however, it means a synthetic cleansing agent made from petroleum chemicals. Any detergent works in the following way. Its molecules have ends with different properties. One end is water-loving (hydrophilic), the other water-hating (hydrophobic). When a detergent encounters grease and dirt, the hydrophobic ends of its molecules attach themselves to the grease particles. They effectively become covered with a water-loving layer, and float away in the water.

electrocardiograph An instrument used in medicine which displays electronically on a cathode-ray tube or graphically on paper, the beating of the heart. A type of galvanometer, it registers the tiny electrical impulses associated with the contractions of the heart muscles. The electrical pulses are picked up by probes attached to various parts of the body. Electrocardiograms (ECGs) can supply information about possible heart disorders and other diseases. A related but more sensitive instrument, the electroencephalograph (EEG) produces traces displaying the electrical activity of the brain.

Above: Electrical engineers carrying out tests using cryogenic conductors – conductors cooled by liquid nitrogen (−196°C).

Below: The electroencephalograph produces a trace of the brain's electrical activity – literally brainwaves.

electron microscope A microscope that manipulates beams of electrons rather than light rays to magnify objects. In high powers it can magnify objects hundreds of thousands of times, and is even capable of viewing large molecules. The 'lenses' used in the electron microscope are electric or magnetic fields, which have the property of bending beams of electrons rather as optical lenses bend (refract) light rays. In an electron microscope a beam of electrons produced by an electron 'gun' is accelerated by means of a high-voltage (50,000–100,000 volts) anode and concentrated by a condenser lens onto the specimen. This is mounted in the transmission electron microscope on a very thin slide. The specimen scatters the electrons as they pass through it and an objective lens focuses them into an enlarged image. A projector lens then magnifies the image further and throws it onto a fluorescent screen, which the microscope operator views through binoculars. The whole microscope is maintained under high vacuum to prevent the electron beam being scattered by air molecules. In the scanning electron microscope solid objects can be viewed. A beam of focused electrons is scanned over the object, and the electrons triggered off from the surface of the object are used to form an image. It operates in a manner not unlike the TV camera and receiver.

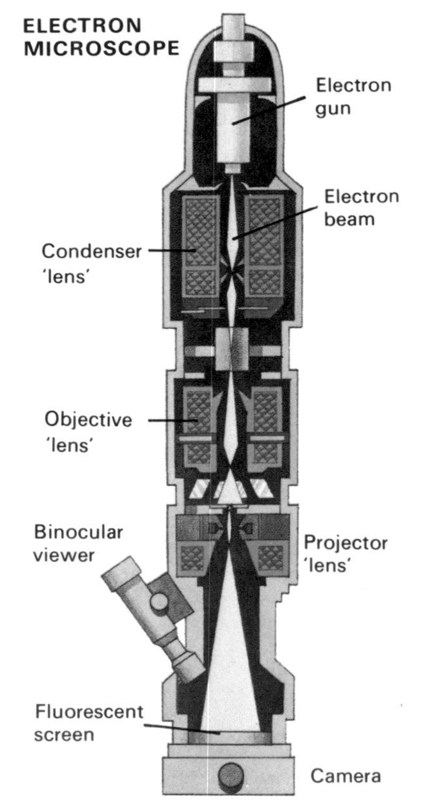

electroplating The coating of one metal on another by means of electricity. It is a means of giving metals a hard, rust-proof attractive finish. Car bumpers, for example, are chromium-plated steel. To coat a metal object with, say, copper, it is placed in a bath containing a solution of a copper salt, for example, copper sulphate. A plate of pure copper is also placed in the bath, and object and plate are wired to a battery or other direct-current supply. The object is wired to the negative terminal, becoming the cathode, the plate to the positive terminal becoming the anode. When current is switched on, copper ions (positive) migrate to the cathode. At the cathode the copper comes out of solution and is deposited. Simultaneously at the anode copper goes into solution; the anode thus dwindles away.

electroscope An instrument that can detect electric charges and radiation. In the classic gold-leaf electroscope, a pair of gold leaves hangs from the bottom of a rod, which is contained in, but insulated from a glass jar. When an electric charge is brought near the top of the rod, both ends of the rod themselves become electrically charged. Since both leaves acquire a like charge, they repel one another and move apart. When the charge is neutralized, or leaks away, the leaves collapse. Radiation can be detected by a charged electroscope. The radiation ionizes the gas inside it, permitting the charge on the leaves to leak away. The rate of collapse of the leaves is a measure of the intensity of radiation.

feedback A concept used in many branches of science and technology, often as a means of control. A thermostat works on the principle of feedback, information about the temperature reached (heat output) being fed back to control the heat input. A governor works by feedback too. So does the fantail on a windmill, which keeps the sails pointing into the wind.

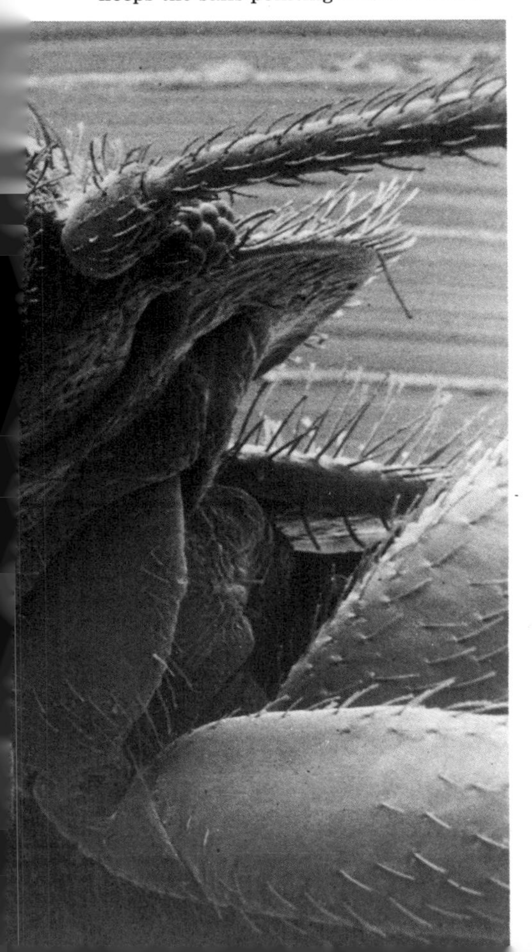

flame test A method used in chemical analysis to identify elements. It consists of dipping the end of a platinum wire in the sample under test, and then heating it in a Bunsen flame. Certain elements colour the flame in a characteristic way. For example, when sodium is present, it colours the flame a deep yellow; calcium, apple-green; strontium, crimson.

fluidics A recent branch of science that makes use of tiny jets of fluids – gases and liquids – as signals in control devices. Fluidic components can be used as switches and amplifiers in conditions where electronic devices break down – at high temperatures and under high levels of radiation.

Frasch process The main method of extracting sulphur from deep underground deposits. It uses a column consisting of three concentric tubes. Superheated water is forced down the outer tube, compressed air down the inner tube. The superheated water melts the underground deposit of sulphur, and the compressed air forces the molten sulphur to the surface through the middle tube as a frothy mass.

galvanizing Coating iron and steel with zinc, by means of electroplating or hot-dipping in the molten metal. The zinc coating protects when it is intact because zinc does not corrode in damp air. It even protects when the coating is scratched and the steel underneath exposed. It does so by sacrificial protection. This means that it becomes the anode in an electrolytic cell and as such dissolves away in preference to the exposed steel, which is the cathode (compare **electroplating**).

galvanometer An instrument that detects electric current. It is like a very sensitive ammeter. One type (D'Arsonval) has a fine coil suspended from a wire in the magnetic field of a permanent magnet. Attached to the coil is a small mirror, onto which a beam of light is shone. When electric current flows, the coil twists, and deflects the light beam over a scale some distance away. A slight twist will cause the light beam to be deflected a lot, making possible great sensitivity.

Geiger counter Or Geiger-Müller tube; an instrument that detects radiation. It is one type of ionization counter. The detector consists of a gas-filled tube containing two electrodes, across which there is a large potential difference, or voltage. One electrode is the tube itself, the other a fine wire in the centre. When a charged particle or other radiation enters the detector, it ionizes the gas therein. The gas ions (atoms with an electric charge) become electrically conducting and pass current between the electrodes. The passage of current is shown by a pulse of light, the kick of a needle, or a click in headphones, depending on the type of Geiger counter in use. The counter is sensitive enough to detect the passage of a single atomic particle and can be connected with a counting meter. For field survey work – for example, when prospecting for uranium – geologists often prefer to use scintillation counters to detect radiation. These rely on the scintillation (light-emission) by phosphors when struck by radiation.

glass-making Glass is a type of substance known as a super-cooled liquid. It is amorphous in form – has no regular internal structure – unlike the crystal of a mineral. Ordinary glass, called soda-lime glass, is made by fusing together in a high-temperature furnace (about 1700°C) a mixture of sand, soda ash (sodium carbonate) and limestone (calcium carbonate). Lead-crystal glass, which has a high refractive index and consequent 'sparkle', includes in the glass-making recipe red lead. Heat-resistant borosilicate glass, such as 'Pyrex', includes boric oxide (from borax) in its ingredients. It is appreciably heat-proof because it expands scarcely at all when heated. Coloured glass is made by including traces of metallic oxides – copper, cobalt, cadmum and gold – in the glass-making recipe. 'Float glass' is glass sheet that has been made by floating a ribbon of molten glass on molten tin.

governor A device that regulates the speed of a machine, extensively used in industry. The classic engine governor invented by James Watt consists of a set of weighted balls fixed to rods that are hinged at the top to a shaft that is spun by the engine. The faster the engine and shaft spin, the higher the balls are flung out by centrifugal force. If the balls rise over a preset level – engine too fast – then a simple linkage closes a valve allowing fuel or steam into the engine. And the engine slows down. When the balls fall below the preset level – engine too slow – the linked valve opens to allow in more fuel or steam. And the engine speeds up. The action of a governor is an example of feedback.

Opposite: The electron microscope has become one of the most invaluable tools in pure and applied research.

Bottom: A horrifying close-up of the fangs of the common bed bug, revealed by the scanning electron microscope.

Below: The fantail of a windmill is an early feedback device. It is mounted with the sails on a revolving turret, and automatically keeps the sails pointing into the wind as the wind direction changes.

gyrocompass The trouble with the magnetic compass is its magnetism, for it is affected by the proximity of other iron objects. It is also useless in high latitudes near the geographic poles. The former shortcoming can be compensated for in the ships and aeroplanes in which it is used, but the latter cannot. Modern navigation therefore relies rather on the gyrocompass for directional information. This contains a gyroscope which is kept spinning constantly by electricity (becoming itself an electric motor). Once set spinning and pointing in the direction of true north, the gyro rotor will continue to point north no matter how the craft it is housed in twists and turns.

gyroscope A device containing a spinning wheel, whose inertia, gives it many interesting properties. A practical gyroscope has its spinning wheel, or rotor, attached to an axle that is mounted in gimbals. It is so mounted that the supporting frame can be tilted in any direction without disturbing the rotor. Once set spinning a gyroscope tends to remain in the same position in space making it useful as a compass (see **gyrocompass**).

The same property gives it application in self-levelling devices, such as a ship's stabilizers. The stabilizers are movable fins fitted to the lower part of the hull. When the ship rolls, it moves relative to the fixed-pointing gyroscope, and the relative movement causes signals to be sent to move the fins in such a direction as to counter the rolling.

GYROSCOPE

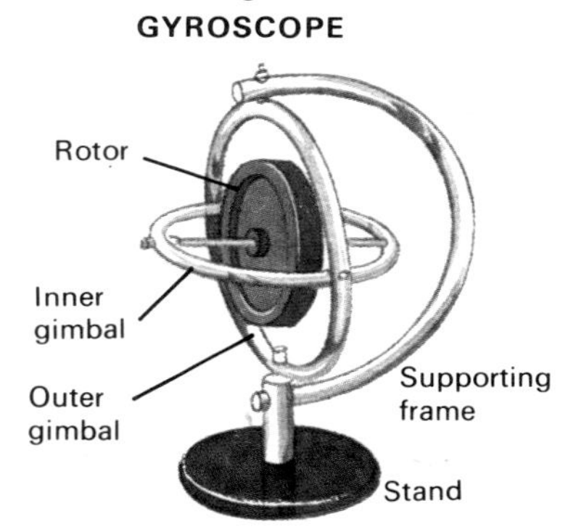

Haber process A major industrial process which 'fixes' the nitrogen of the air to make ammonia, one of the most important raw materials of the chemical industry. The Haber process combines nitrogen, obtained from the distillation of liquid air; and hydrogen, obtained from water gas. The process takes place at a temperature of about 500°C, at a pressure of some 200 atmospheres, and in the presence of an iron catalyst.

heat treatment Subjecting metal alloys to varying combinations of heating and cooling in order to improve their physical properties. It modifies the internal structure of the metal, for example, by altering the size or composition of its crystal grains. In quenching, metal is heated and then plunged into cold oil or water. This treatment makes the metal hard but brittle. In tempering the metal is first quenched and then reheated slightly and allowed to cool slowly. This preserves the metal's hardness but makes it less brittle. In annealing the metal is heated and allowed to cool slowly. This strengthens it by relieving internal stresses.

holography A fascinating technique by which realistic three-dimensional pictures of an object can be made, using a laser beam. In holography a laser beam is split in two – one part is directed onto the object and the reflections from the object are allowed to fall on a photographic film. Simultaneously the film is illuminated directly by the other part of the laser beam. Interference takes place between the direct and the reflected beams, creating patterns that show on the developed film as a mosaic of light and dark bands, or fringes. The film is called a hologram. To reconstruct an image laser light is shone through the hologram, the lines on which scatter the light in exactly the same way that the real object did. And an image of the object appears floating in the air. It is a perfect three-dimensional image, which uncannily changes as you move your viewpoint.

hydraulic press Also called a Bramah press; a device that works by the transmission of liquid, or hydraulic pressure. It relies on the basic properties of liquids that they are incompressible and will transmit pressures equally in all directions. A simple press has two pistons able to move in interconnected cylinders. The cylinders are of different diameter. Force applied to the smaller piston is magnified by the larger piston according to their relative sizes. A 1 tonne load pressing on a piston 1 square cm in area creates a pressure of 1 tonne/sq cm in the liquid. When it presses against a piston 1000 square cm in area, it will exert 1000 tonnes.

hydrometer A device for measuring a liquid's specific gravity, or relative density (density in relation to that of water). It consists in simplest form of a weighted glass bulb, which floats upright. To float (by Archimedes' principle) it must displace a weight of water equal to its own weight. In a liquid of high density, it needs to displace less liquid and therefore floats high. In a liquid of low density, it needs to displace more liquid and therefore floats low.

hygrometer An instrument used in meteorology for measuring humidity. The simplest hygrometer uses a bundle of human hair, the length of which (test for yourself) varies according to the humidity, or amount of moisture in the atmosphere. In the hair hygrometer one end of the hair is fixed while the other is kept in tension and is attached to a pivoted pointer. This moves over a graduated scale as the length of the hair alters. Another common type is the wet-and-dry bulb thermometer, or psychrometer. It comprises two identical thermometers, one of which has its bulb covered by a piece of fabric that dips into water. Because water evaporates from the 'wet' bulb, it is cooled and reads lower than the dry bulb. The extent of evaporation – and cooling – depends on the humidity. So the difference in temperature between the two bulbs is a measure of the humidity.

Kipp's apparatus A common means of generating gases in the laboratory, used for the production of carbon dioxide, hydrogen sulphide and hydrogen, for example. It consists of three linked glass bulbs, the middle one of which is the reaction vessel. To make carbon dioxide, the reaction vessel contains chalk or limestone (calcium carbonate). The upper and lower bulbs contain dilute sulphuric or hydrochloric acid. A tube, sealed by a tap, leads from the reaction vessel.

When the tap is opened the acid in the upper bulb runs into the lower one, forcing the acid there into the middle bulb and onto the chalk. The acid reacts with the chalk to produce carbon dioxide gas, which discharges through the tube. When the tap is closed, pressure builds up in the middle bulb and forces the acid back into the bottom bulb and thence into the top one. The reaction ceases until the tap is again opened.

KIPPS APPARATUS

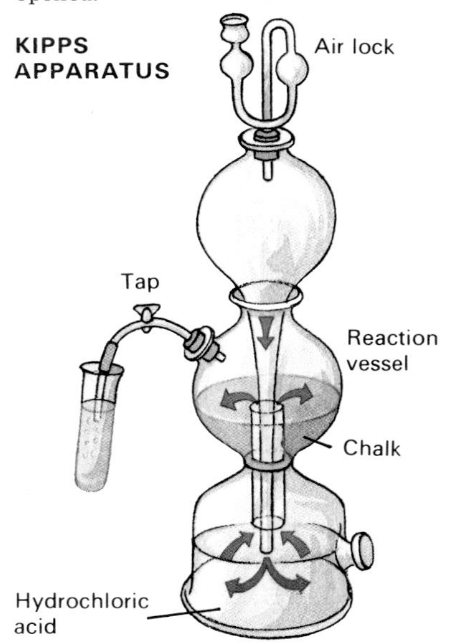

laser An optical device which produces an intense, parallel beam of very pure light. The word laser is an acronym for 'light amplification by stimulated emission of radiation'. There are three main types of laser, the original ruby laser, the gas laser and the semiconductor laser. The ruby laser is the most powerful of the three, capable of producing pulses of such high power that they can be reflected from the Moon. The gas lasers, although they are less powerful, can produce that power continuously. So can the feebler semiconductor lasers.

The function of the laser can be illustrated by reference to the ruby laser. It consists of a rod of synthetic ruby, whose ends have been machined exactly parallel. One end is silvered, making it into a mirror, while the other end is partly silvered. It is surrounded by a high-intensity light source. When the light is switched on, trace atoms of chromium in the rod (which is mainly aluminium oxide) become excited and emit radiation. This radiation in turn stimulates other atoms into emitting radiation of the same wavelength. Much radiation escapes from the sides, but radiation travelling parallel to the axis of the rod is reflected by the ends, returning to stimulate still more atoms into emitting still more radiation. This process builds up into a powerful surge, or pulse of radiation which eventually escapes through the partly silvered end of the rod. The pulse is very pure light because it consists of radiation of exactly the same wavelength and it is accurately parallel because of the way it is built up.

Right: This spectacular picture shows lasers being used to measure air flow rate in a wind tunnel.

lie detector Also called a polygraph, a device which may be able to determine whether a person answers questions truthfully or not. The person under interrogation is fitted with numerous probes which monitor a variety of physiological conditions, including blood pressure, pulse rate, breathing and perspiration. These activities are recorded on a moving paper roll or a videoscreen. In most people the telling of a lie is usually accompanied by minute fluctuations in bodily activity. The machine is calibrated by getting the person to tell the truth and then lie in answer to certain lead questions.

Linde process The method by which air is liquefied, at a temperature of about −190°C. By distilling liquid air we get liquid oxygen and liquid nitrogen. In the Linde process atmospheric air is compressed to a pressure of some 200 atmospheres, whereupon it heats up. It is first water cooled and then allowed to expand suddenly through a valve. The sudden expansion causes the gas to lose energy, and it therefore cools. (This is called the Joule-Thomson effect.) The process is repeated until the air has cooled sufficiently for it to condense into liquid.

manometer A simple and widely used device for measuring the pressure of a gas. It consists of a U-tube containing liquid, often water. One end is connected to the gas pipe, or tank, while the other is open to the air. The pressure of the gas can be gauged from the difference between the water levels in the arms of the tube.

mass spectrometer A sophisticated instrument for analysing mixtures of substances and measuring their mass and abundance. The sample to be analysed is first vaporized and then passed into the instrument. A beam of electrons ionizes the gas – its atoms become electrically charged. The ions are then accelerated by an electric field and then are deflected by a magnetic field. They are deflected by differing amounts, according to their mass and electric charge, and can be separately detected.

microprocessor A wafer-thin slice or 'chip' of silicon a few millimetres square which contains many complex electrical circuits. These circuits can be tailor-made to perform a variety of functions – for example, measure time, as in digital watches; and carry out conversions and calculations, as in pocket calculators. The silicon chip is the product of what is called large scale integration (LSI). And the electronic circuits it contains are called 'integrated circuits'. They are made up of the same components as ordinary electronic circuits – resistors, capacitors, transistors and diodes – but these components are fashioned out of the same piece of silicon and are connected by thin metal film, not wires. Complex stage-by-stage masking, etching and 'doping' processes are required to make the chips. In doping, controlled amounts of impurities are allowed to diffuse into certain regions of the pure silicon, turning them into semiconductors (see page 188). These regions are then interconnected in various ways to form the required electronic components. The final chip consists of the components, insulated from one another by oxidized silicon and with aluminium film contacts, this being termed MOS (metal-oxide-silicon) technology.

microscope An instrument that magnifies objects. The standard optical instrument uses two lenses or lens combinations and is called a compound microscope. The front lens (objective) forms a magnified image of the object, which a second lens (eyepiece) magnifies further. The best optical microscopes gives a maximum magnification of about 2500 times. They are usually equipped with several objectives to provide different magnifications. The specimen to be viewed is placed on a platform, or stage, and is illuminated by light from a lamp, concentrated by a lens called a condenser. Coarse and fine adjustments are provided for moving the body tube up and down for focusing.

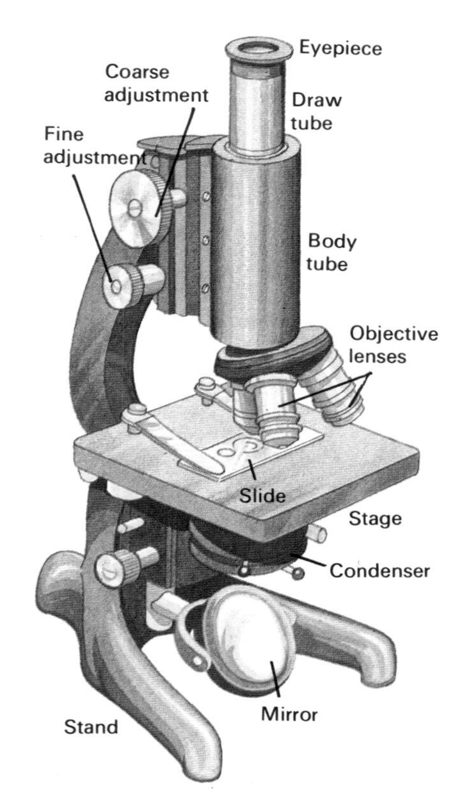

Above: The standard compound microscope widely used in laboratory work. The nosepiece revolves to bring different strength objectives into operation.

Left: Photomicrographs of a thin slice of dunite (top) and ascorbic acid crystal.

Below: An advanced phase-contrast microscope, which brings out details in transparent specimens without staining.

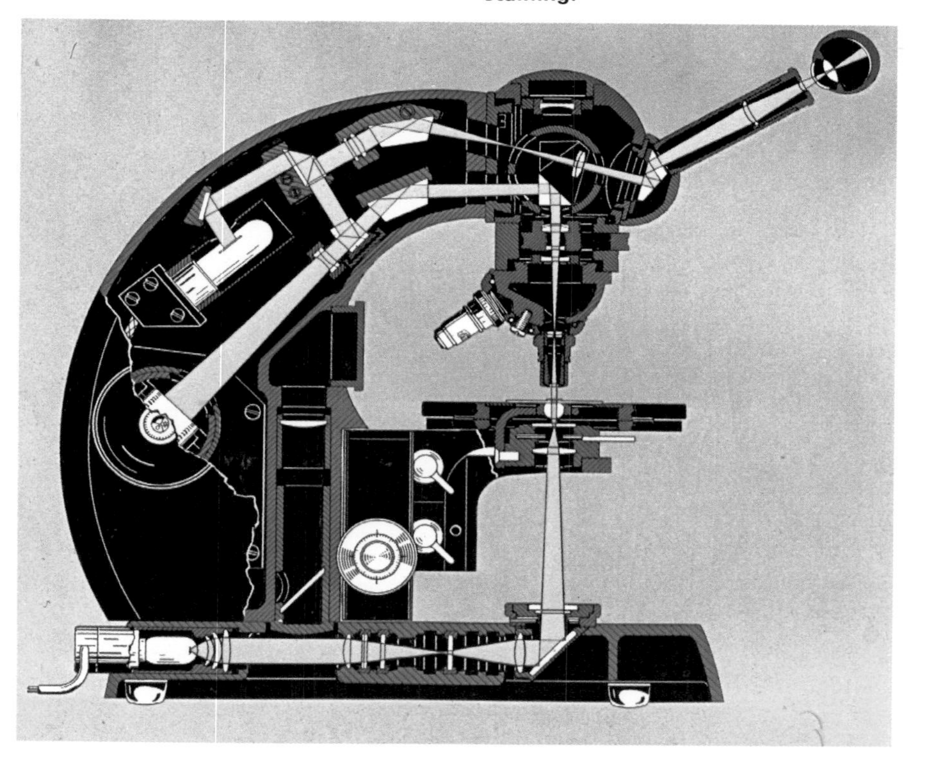

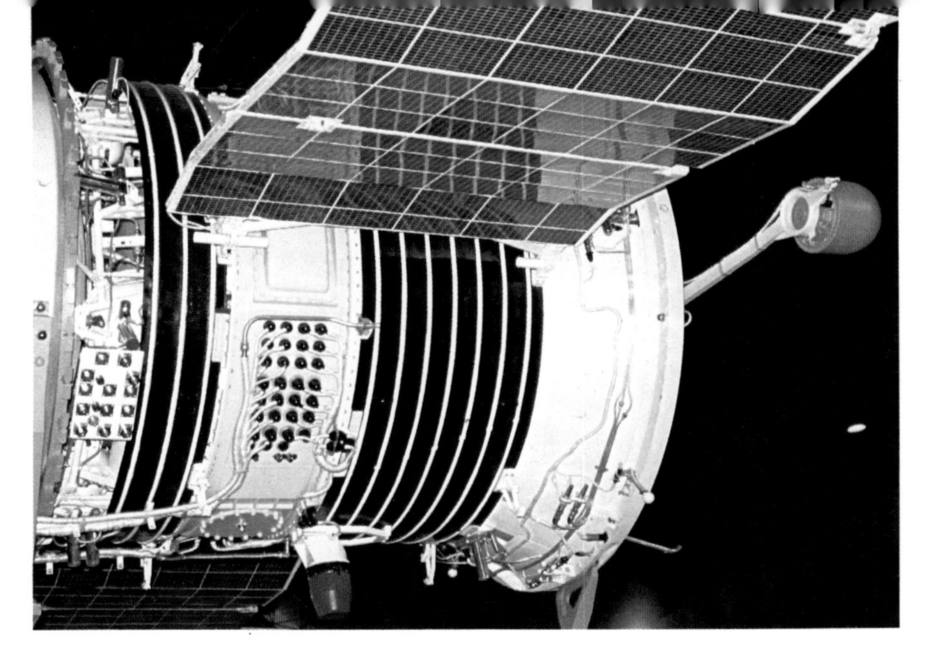

photocell Or 'electric eye', properly called photoelectric cell. A device whose electrical state changes in one way or another when exposed to light. Photocells form the basis of such things as photographic exposure meters, automatic door-opening devices and burglar alarms, and the solar cells used in spacecraft. Three distinct types of photoelectric effect (first explained by Einstein) exist. One is 'photo-emission', a phenomenon displayed by caesium and selenium, for example. When light falls on selenium, electrons are given off from it. One type of photocell consists of a two-electrode electron tube (valve) in which the cathode is selenium coated. It is incorporated in an electrical circuit. When light falls on the cathode, electrons are emitted and flow to the anode, and the tube as a whole passes electric current. Photocell burglar alarms and automatic door-openers are triggered to work when a beam of light falling on the photocell is interrupted.

Photographic exposure meters generally rely on the 'photoconductive' effect in which the electrical resistance of a substance changes when light falls on it. The extent of the change is proportional to the intensity of light falling on it. The traditional photosensitive material in exposure meters is cadmium sulphide. Other exposure meters utilize the 'photovoltaic' effect, in which a substance actually generates an electric current when light falls on it. Many semiconductors display this effect, including gallium and silicon (used in solar cells).

photometer An instrument that measures light intensity. Most photometers work by means of the photoelectric effect and are really sophisticated exposure meters. Photometers used in astronomy, which deals with the very feeble light from the stars, use a photographic method. It relies on the degree of darkening of photographic film when exposed to the light from a star. Spectrophotometers measure the intensity of light at a specific wavelength (colour).

pitot tube A device used to measure the velocity of flow in a fluid. It forms the basis of one type of anemometer and the air-speed indicator in aircraft. It exploits the difference in pressure between the air in a tube pointing into the direction of the air stream and the surrounding air (sampled at the side of the tube). The faster the air stream is flowing relative to the tube (or vice versa) the greater will be the pressure build-up inside the tube. The pressures inside and outside the tube are transmitted to a diaphragm device which moves back and forth as the pressure differential changes. Movement of the diaphragm moves a pointer over a calibrated dial. A similar device is used to measure a ship's speed through the water.

plaster of Paris An interesting material used to form plaster casts around broken bones. It is made by roasting gypsum, calcium sulphate dihydrate ($CaSO_4 2H_2O$), which turns into calcium sulphate hemihydrate ($CaSO_4 \frac{1}{2} H_2O$), a fine white powder. When this powder is mixed with water, the water combines chemically with it to re-form the chalk-hard gypsum.

polymerization The process by which synthetic plastics are made. Plastics derive their inimitable properties from the nature of their molecules, which consist of long 'chains' of repeated units. Their alternative name, polymer, means 'many parts'. The basic units from which these polymers are built up are called monomers ('one part'). Two basic types of polymerization occur: addition polymerization, in which monomers simply link together by themselves; and condensation polymerization, in which a molecule of water is eliminated for each repeated link in the chain. Polymerization may result in linear polymers – long-chain molecules which are unconnected with one another. These constitute the so-called thermoplastics (like polythene), which can be shaped readily and remelted with ease. Polymerization may also bring about cross-linking between long-chain molecules, resulting in a rigid, three-dimensional macromolecule. These constitute the so-called thermosetting plastics (like Bakelite), which can be shaped and melted only once and are thereafter heat resistant.

Top: Solar cells exploit the photovoltaic effect.

Right: Extruding PVC pipe.

Below: Plaster casts have a tendency to collect autographs.

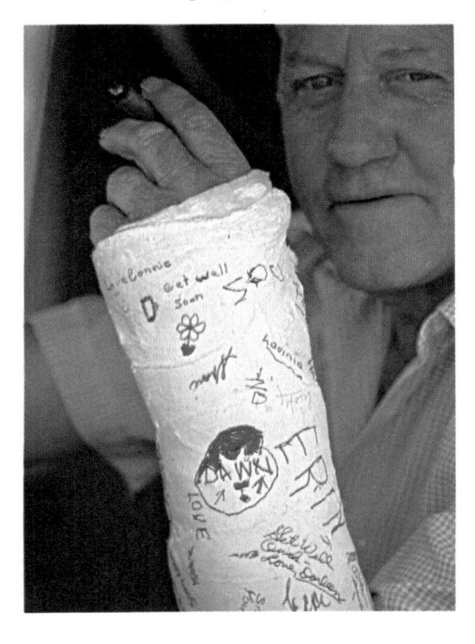

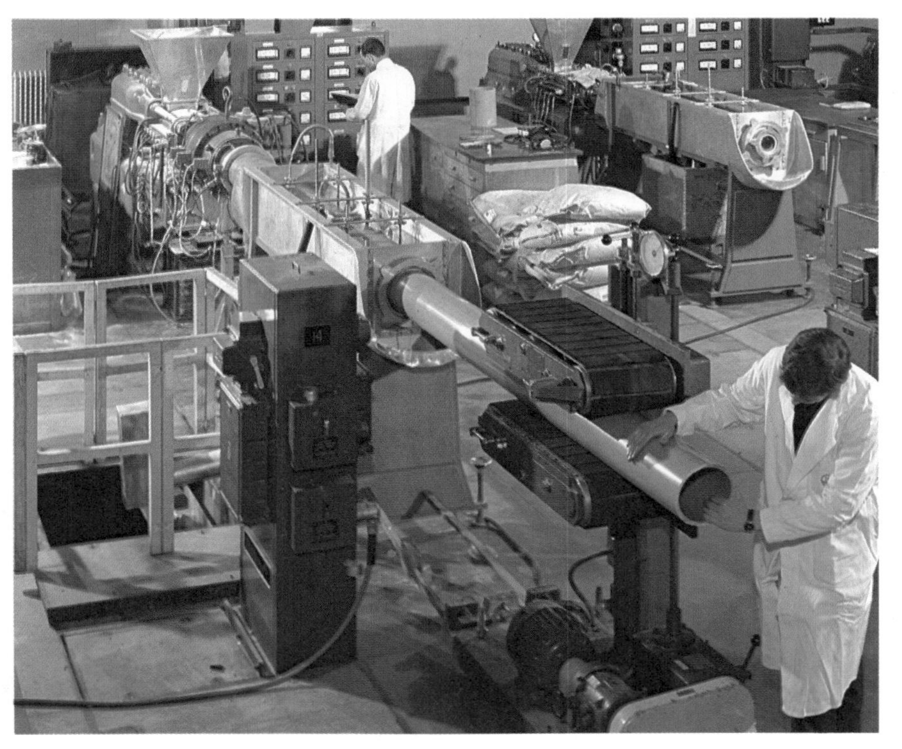

radar An electronic navigation aid that uses high-frequency radio waves (microwaves) to detect distant objects. It works by sending out microwave pulses and timing the reflections, or echoes, from objects in their path. The word radar is an acronym of 'radio direction and ranging'. In a typical radar set up a transmitter generates microwave pulses by means of electron tubes such as magnetrons and klystrons. The pulses are then fed to a rotating antenna, which transmits them. Because the transmitter is 'Off' for part of the time (in between pulses), the same antenna can be used to receive the pulse echoes. These are fed to the receiver, which typically displays them as points of light, or 'blips' on the fluorescent screen of a cathode-ray tube. The position of the blip indicates the direction and range of the objects that originated the reflections.

seismograph An instrument that detects and measures the strength of ground tremors, resulting from earthquakes or man-made explosions. The standard instrument consists of a weight suspended from a frame by a spring. Attached to it is a thin coil of wire, located in a magnetic field and linked to a sensitive galvanometer. When the ground on which the instrument is resting shakes, the frame moves up or down. The weight, by its inertia, tends to remain in its former position. The coil moves relative to the magnetic field, and a small current is thus generated inside it. This is detected by the galvanometer, which deflects a beam of light onto a moving roll of photographic paper. Alternatively the generated signals are recorded on magnetic tape.

semiconductor A substance that is neither a conductor (like a metal) nor an insulator (like glass) but something in between. Semiconductors form the basis of transistors and other so-called solid-state devices that have revolutionized electronics in recent years. Best known is silicon, used to make the miraculous microchip. The difference between a conductor and an insulator is that the former contains many 'free' electrons, which can move readily from atom to atom; while the latter contains no free electrons at all. In some materials that are normally insulators, a few free electrons can be introduced by 'doping' – treatment with controlled amounts of impurities. By suitably doping other materials a few electrons can be taken away from some atoms. The lack of an electron is called a 'hole'. When included in an electric circuit, the free electrons in the one (n-) type of semiconductor and the 'holes' in the other (p-type) can move from atom to atom, thus permitting the flow of current. The movement of electrons and holes can be readily manipulated by altering the polarity (positive or negative) and voltage of the electrical supply. A wide variety of effects (amplification, detection, and so on) can be obtained by using different combinations of n-type and p-type material.

Above: Air-traffic controllers in the radar control cabin at a small airport keep constant watch over the display screens.

sextant An instrument used in navigation to measure the angle of a heavenly body above the horizon. It is so-called because the arc of its frame covers a sixth (Latin 'sex') part of a circle. The arc is marked off in degrees. A movable arm, the index bar, pivots at the centre of the circle. A small mirror (index glass) is fixed to the top of the index bar, while a half-clear, half-silvered mirror (horizon glass) is fixed to the frame. An observer looks through a small telescope on the frame and views the horizon through the clear half of the horizon glass. He moves the index bar until an image of the Sun or a certain star appears in the horizon glass to be on the horizon. He then takes the reading on the scale.

soap-making Soap, the detergent used by Man since Roman times, is made by treating animal and plant fats and oils with alkalis, usually caustic soda and caustic potash. Chemically soaps are the sodium and potasium salts of the fatty acids present in the fats and oils. In the soap-making process fat and alkali are boiled together, whereupon saponification takes place, and a mixture of soap and glycerine results. The mixture is then treated with brine so that the soap and glycerine separate out into layers. The crude soap is removed and purified.

Solvay process Also called ammonia-soda process; the commercial method of producing sodium carbonate (washing soda), widely used in the manufacture of soap, glass and other important materials. In the process, brine (salt water) is saturated first with ammonia gas and then with carbon dioxide. This results in sodium bicarbonate, which is converted by heating into sodium carbonate, carbon dioxide being given off. The ammonium chloride that is also formed in the process is treated with slaked lime (calcium hydroxide), whereupon ammonia is released and recycled, while calcium chloride remains as a waste product. The Solvay process provides a classic example of recycling, of both ammonia and carbon dioxide, to achieve more economical production.

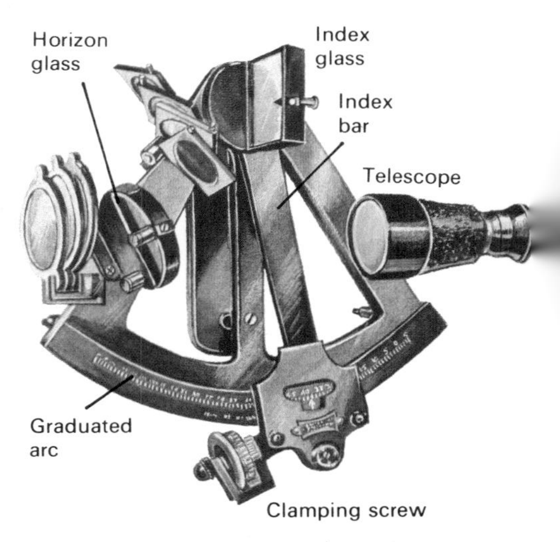

Above: The marine sextant has been a major navigational instrument since its invention by John Hadley in about 1730.

Below: Cross-sections through two common types of semiconductors, showing the regions that have been made electrically different by doping.

SEMICONDUCTORS

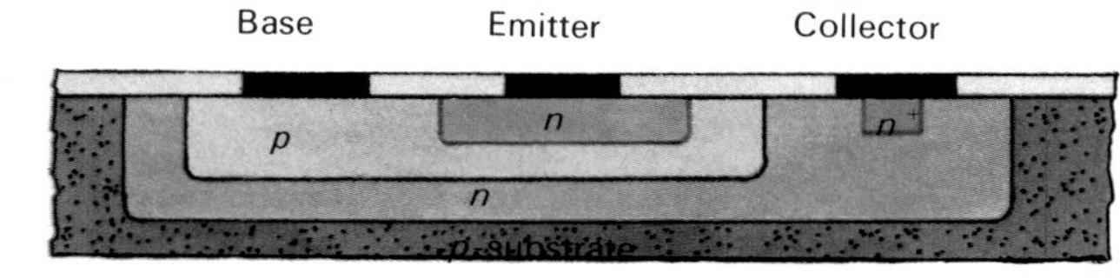

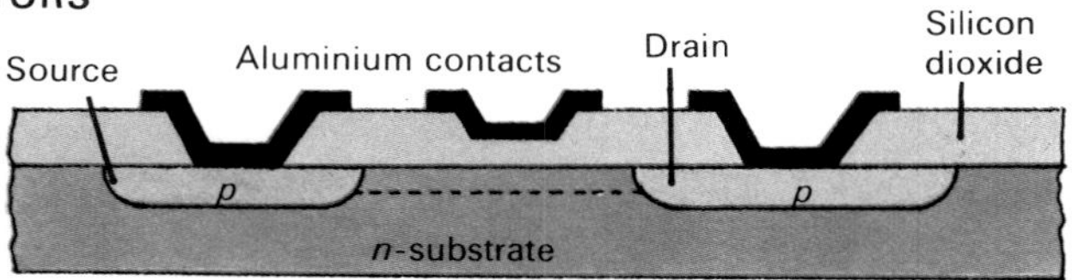

spectacles Lenses used to improve eyesight. They correct defects inherent in the eyes by converging (with convex lenses) or diverging (with concave lenses) the incoming light rays. Classic myopia, or short-sightedness, results from the eyeball being too long. Images are brought to a focus in front of the light-receiving retina rather than on it, resulting in blurred vision. This defect is corrected by having spectacles with concave lenses. Long-sightedness, or hypermetropia, results from the eyeball being too short. Images are brought to a focus behind the retina, again causing blurred vision. This defect is corrected by wearing convex lenses.
sphygmomanometer The instrument doctors used to measure a patient's blood pressure. It consists of a mercury manometer, one leg of which is set against a scale while the other is connected to an inflatable armband, or cuff. To take a patient's blood pressure, the doctor winds the cuff around the arm above the elbow and pumps air into it. He continues pumping until the air pressure causes constriction of the artery so that the blood no longer flows. The doctor puts his stethoscope on the arm just below the cuff and listens while he gradually lets out the air from it. The pressure reading on the scale when he hears the blood begin to flow again represents the systolic pressure, when the heart is fully contracted. Still listening through his stethoscope, he continues to release the air. When the noise he hears changes, he again notes the pressure, which represents the diastolic pressure, or the pressure between heart beats. In a healthy person the systolic pressure is about 120 mm (of mercury) and the diastolic pressure 70 mm, the blood pressure being expressed as 120 over 70 (120/70).
stethoscope The device doctors use to listen to noises made by a patient's body. It consists of two ear-pieces and a chest-piece or contact piece connected by rubber tubing. The ear-pieces fit snugly, thereby excluding unwanted external noise. The chest-piece includes a small cone, or bell, and a flatter cup, or diaphragm. The former registers low notes best, the latter high notes. Doctors use a stethoscope to listen to heart, lungs, intestines, veins and arteries, the listening technique being known as auscultation.
stroboscope An instrument that emits a rapidly flashing light, used for example to examine rotating machinery and measure the speed of rotation. The modern stroboscope produces its light electronically from a gas-discharge lamp (like an electronic flash-gun in photography). The rate of flash can be varied. To examine rotating machinery – say a spinning shaft – the rate of flash is adjusted until it coincides with the shaft's rate of rotation. Then the shaft is always illuminated by the stroboscope at exactly the same point in the rotational cycle. As far as your eyes are concerned, the shaft appears stationary.
theodolite An instrument used in surveying to measure horizontal and vertical angles. It is essentially a telescope mounted so that it can be swivelled horizontally and vertically. It is carried on a tripod with adjustable legs and incorporates a spirit level so that it can be accurately levelled. A form of theodolite called a transit is used in astonomical timekeeping.
thermostat A device that automatically regulates temperature by controlling the heat supply. A simple room thermostat, which regulates a central-heating boiler, consists of a bimetallic strip. This is a strip made of two different metals, for example, steel and brass, bonded together. These metals have different coefficients of expansion, brass expanding more than steel for a given temperature rise. When the temperature changes, the strip bends one way or the other. In the thermostat this movement is utilized to switch electric current (to the boiler pump) on and off. Another common type of thermostat uses the expansion of a liquid in a bellows to operate valves to restrict the flow of liquid in a heating or cooling system (as in a car's cooling system, see page 40).
ultrasonics The science concerned with the generation and use of sound waves of very high frequency, which are so high-pitched that they are inaudible to human beings. Human beings cannot hear sounds with frequencies beyond about 20,000 hertz (cycles per second). Ultrasonic devices use frequencies of 100,000 hertz or more. Bats rely on ultrasonics to navigate and find their prey – they emit ultrasonic pulses and listen for echoes that indicate obstacles in their path or food. This sound echo location is the basis of sonar, used by Man for underwater detection, navigation and depth sounding. Similar use of ultrasonic waves is made in medicine to scan internal organs and foetuses in the womb. This is much safer than using X-rays. The remote control devices used to switch channels in television sets also work by ultrasonics.
vacuum flask Also called Dewar flask or Thermos flask; a vessel for keeping liquids hot or cold. The flask consists of a double-walled glass bottle, housed in an outer container. Glass is used because it is a good insulator, minimizing heat transfer by conduction. Between the double wall is a vacuum (or rather partial vacuum). This minimizes convection since there is no air to circulate. The inner surfaces of the double wall are silvered, this minimizing heat transfer by radiation.
xerography A dry printing process used in photocopying, or 'xerox' machines. In the machine the document to be copied is scanned by a strong light. Lenses and mirrors focus an image of the document on to a rotating drum. This is coated with selenium and electrically charged. Where light falls on it, the drum loses the charge; in dark areas the charge remains. 'Toner' powder is sprinkled over the drum and adheres to the charged areas only. It is transferred to the charged copy paper, which is then heated to fuse the powder into a permanent image.
X-ray tube A device for producing the penetrating X-rays used in medical radiography. Using X-rays, photographs can be taken of the human body, for instance, which show the bones and sometimes other organs. This is possible because X-rays can penetrate body tissue more readily than bones, which therefore show up. The X-ray tube is a vacuum tube in which a stream of electrons from a cathode is accelerated by a high voltage and directed at an anode, or target, usually made of tungsten. The impact of the electron stream triggers the tungsten atoms to give off X-rays. The target is rotated to keep it cool.

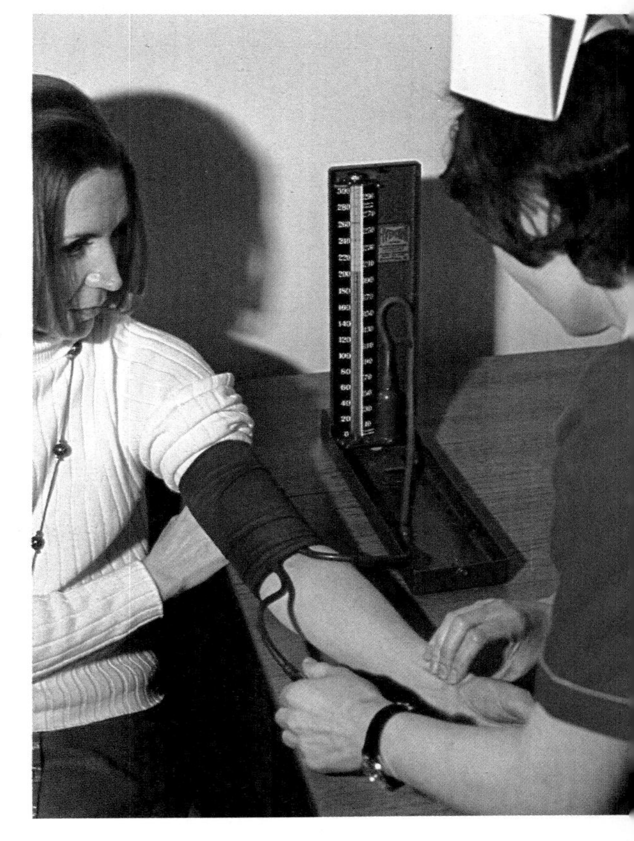

Above: A nurse taking a patient's blood pressure with a sphygmomanometer.

Below: An X-ray scan of the brain, colour coded to retrieve more information. It was taken by an X-ray scanner machine, which investigates a thin 'slice' of the brain in situ.

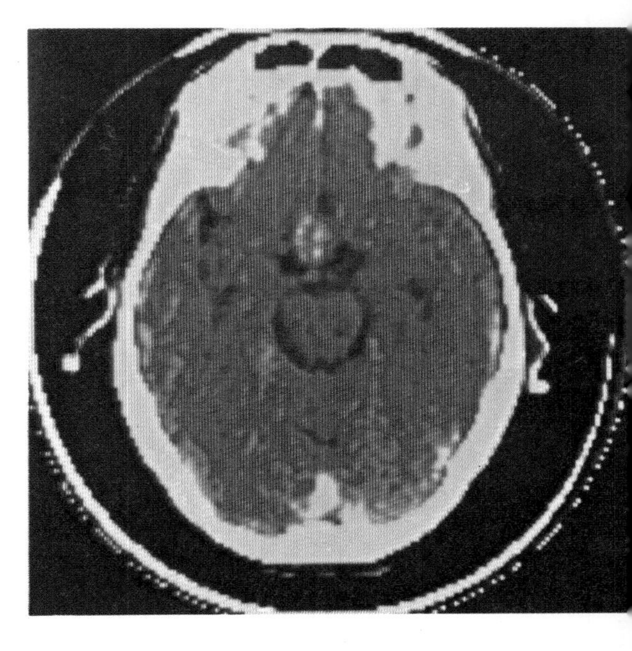

Journey into Space

The introduction of the American space shuttle ushers in a new era in space exploration. It represents a quantum leap forwards in Man's progressive mastery of things out of this world.

The shuttle system makes space launching a much safer, cheaper and more routine operation than hitherto. For the first time ordinary men and women and not only highly trained astronauts, will be able to venture into space. Though the non-astronauts will initially be skilled scientists and engineers, one wonders how long it will be before an enterprising travel company offers the ultimate in package tours: 'Seven days in orbit, with optional excursions to space station Salyut'.

For many years, though, the shuttle's usual cargo will be satellites, space laboratories and deep space probes. The benefits of satellite technology have been with us for some time. They include speedier and more reliable round-the-world communications; more accurate weather forecasts; and greater appreciation of the Earth's resources.

Other satellites view the universe from above the Earth's distorting atmosphere. Deep space probes venture to the planets and even beyond the solar system, some carrying messages to other beings should they exist elsewhere among the stars.

Such satellites and probes have immeasurably increased our knowledge of the universe, supplementing the information acquired gradually over the years by the more traditional astronomers' tools – optical and radio telescopes. The big new telescopes are themselves becoming more and more sophisticated, looking deeper into the universe than ever before. Both probe and telescope are revealing stranger goings-on in the universe than anyone could possibly have suspected. They tell of rapidly rotating balls of solid nuclear matter a few kilometres across (pulsars); starlike objects brighter than galaxies (quasars); and dark abysmal whirlpools that swallow all matter and radiation (black holes).

Left: An artist's impression of the first true space vehicle, the space shuttle, returning to Earth after a successful mission.
Below: The 150-cm (60-inch) McMath solar telescope at Kitt Peak Observatory in Arizona.

New Look at the Universe

While astronomers only had their eyes to rely on, astronomy advanced but slowly. To be sure, successive observers greatly refined the art of observation and constructed more accurate instruments – astrolabes, armillary spheres, quadrants, octants and so on. In the famous observatory he established at Hven (1576), Tycho Brahe engineered instruments with which he mapped the movements of the stars and planets with great precision.

Though the evidence was there before him, Tycho never supported Copernicus's solar system (1543), in which the Sun, not the Earth is the centre of our corner of the universe. His protégé, Johannes Kepler, saw how Tycho's meticulous observations fitted in with Copernicus's theory. In 1609 he published his first two laws of planetary motion, in which he stated that the planets move in elliptical orbits around the Sun.

In the same year Galileo built his first telescope and trained it on the Moon, Venus and Jupiter. In the phases of Venus and the moons of Jupiter he also found support for a solar system. Soon astronomers the world over were grinding their own lenses to make bigger and better telescopes. But all these telescopes suffered from a great defect – the image was colour-blurred. This defect, chromatic aberration, is caused by the lenses acting as prisms and splitting white light into its component colours, violet, indigo, blue, green, yellow, orange and red, each of which is brought to a separate focus.

On Reflection

Perceiving this inherent fault of the lens-type, or refracting telescope, Isaac Newton designed a mirror-type, or reflecting telescope (1668), which was free from chromatic aberration. At much the same time he was pondering why the heavenly bodies move as they do. He found the answer in his universal law of gravitation, which he announced in the monumental work known as the *Principia* (1687). He also worked out how a body could become a satellite of the Earth.

One solution to the chromatic aberration of refractors was discovered in 1758, when John Dolland produced the first achromatic lenses. He made them from a combination of two different types of glass, which act together to prevent the spread of colour.

But any large telescope has to be a reflector. It is difficult to support large lenses without them distorting, whereas mirrors can be supported easily from behind. In 1789 William Herschel, recent discoverer of the planet Uranus (1781), built a 122-cm (48-inch) reflector; in 1845 the Earl of Rosse completed a 183-cm (72-inch) reflector. With this he was able to discover the spiral nature of galaxies.

A few years later the spectroscope came into widespread use in its classic form for producing and viewing the spectrum of starlight. At much the same time William Bond turned his camera on the Moon to pioneer astrophotography. Though the photographic plates of the time were relatively insensitive, they demonstrated their superiority over the eye as a means of recording faint stars and nebulae.

With such new tools at their disposal nineteenth century astronomers began to discover that the universe was far bigger than they had imagined, for they could now measure distances to some of the nearer stars. The Moon and the planets came into clearer focus. This undoubtedly helped to popularize a new form of literature – science fiction. The great pioneer of this genre was Jules Verne, with *From the Earth to the Moon* (1865). In it he fired his heroes to the Moon from a giant cannon, from a site in Florida, uncannily close to Cape Kennedy, from where the Apollo astronauts would be launched a century later to the same destination.

The Rocket Revolution

But a cannon, however powerful, cannot launch anything into space. Only one machine is able to – the rocket. Rockets were around in Verne's day, as fireworks, as military missiles, and as life-saving rockets for emergency use at sea. But in Russia a school teacher saw that rockets could be used to launch bodies into space.

This man, Konstantin Eduardovich Tsiolkovsky, developed his ideas on rocketry and space travel over many years and not until 1903 did he publish them in a famous article 'Exploring Space with Reactive Devices'. In this and subsequent articles he laid the foundations for the science of astronautics. He realized that rockets alone could work in space; that liquid-fuelled rockets would be most suitable; and that a step rocket would be needed to carry something into orbit. Prophetically he was later to say: 'Man will not remain on the Earth forever. In his pursuit of light and space he will at first timidly probe beyond the atmosphere, then conquer all of circumsolar space'.

Whereas Tsiolkovsky was exclusively a theorist, American rocket enthusiast Robert H. Goddard put his ideas into practice. In 1926 he launched the first liquid-fuelled rocket. His interest was echoed in Germany by a group who formed themselves into The Society for Space Travel (Verein für Raumschiffart, VFR). A member of that society was to feature prominently in the history of space travel. His name was Wernher von Braun.

Von Braun headed the team at Peenemünde in the Baltic which developed the notorious V-2 rocket bomb used to bomb London in 1944. After World War 2 von Braun and some of his V-2s went to the United States, while other V-2s found their way to Russia. The V-2 was the direct ancestor of all the launching vehicles that thrust the world into the Space Age.

That Age dawned on 4 October 1957, when a tiny aluminium cylinder, Sputnik 1, bleeped its way around the Earth in orbit. Astonishingly, less than four years later, Man himself was orbiting the Earth. And before 1970 he had twice landed on the Moon.

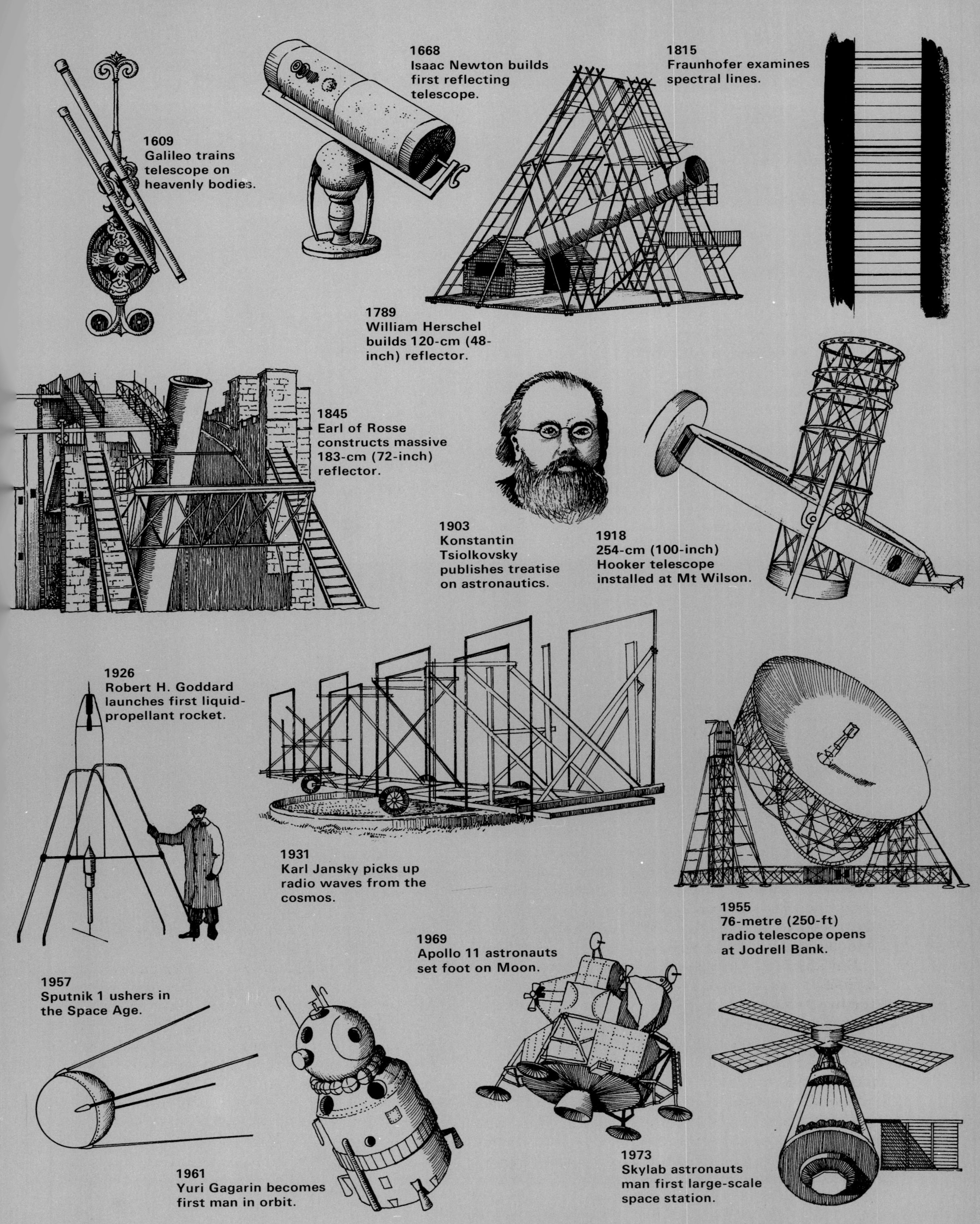
1609
Galileo trains telescope on heavenly bodies.
1668
Isaac Newton builds first reflecting telescope.
1815
Fraunhofer examines spectral lines.
1789
William Herschel builds 120-cm (48-inch) reflector.
1845
Earl of Rosse constructs massive 183-cm (72-inch) reflector.
1903
Konstantin Tsiolkovsky publishes treatise on astronautics.
1918
254-cm (100-inch) Hooker telescope installed at Mt Wilson.
1926
Robert H. Goddard launches first liquid-propellant rocket.
1931
Karl Jansky picks up radio waves from the cosmos.
1955
76-metre (250-ft) radio telescope opens at Jodrell Bank.
1969
Apollo 11 astronauts set foot on Moon.
1957
Sputnik 1 ushers in the Space Age.
1961
Yuri Gagarin becomes first man in orbit.
1973
Skylab astronauts man first large-scale space station.

Stargazing

The eye, as an optical instrument for viewing the heavens, suffers from one big disadvantage. The pupil – the part that lets light in through the lens – is very small. This limits the eye's ability to gather light and its power to separate, or resolve stars that are close together. Telescopes, however, can be made with a much larger aperture and hence greatly increased light-gathering ability and resolving power.

Astronomical telescopes use a combination of lenses or mirrors to collect and focus the light from the stars. A large objective lens or mirror gathers the starlight and brings it to a focus. Then a smaller eyepiece is used to view the image formed. The image is inverted, or upside-down, but this does not really matter for astronomical work.

The light-gathering ability of a telescope, as already mentioned, is dependent on aperture – in other words on the diameter of the objective. This ability increases as the square of the diameter, so a 200-mm (8-in) objective will gather four times as much light as a 100-mm (4-in) one. Light-gathering and resolving power are far more important than magnifying power, which is given by the ratio of focal lengths of the objective and eyepiece. For a given objective magnifying power can be varied by changing the eyepiece. But by increasing magnification, you also increase distortion and reduce the field of view and brightness of the object.

The Refractor

The refractor uses lenses to gather and focus starlight. Both objective and eyepiece are convex, or converging lenses systems. This arrangement, known as the Keplerian, was devised by Johannes Kepler to improve Galileo's telescope. The Galilean telescope had a convex objective but a concave eyepiece. This produced an image the right way up, but limited the magnification and field of view. Though unsuitable for astronomical work, the Galilean system is still used in terrestrial telescopes and opera glasses.

In a refractor the objective lens is achromatic, or colour corrected, by being constructed of a convex lens made of crown glass and a concave lens made of flint glass. The dispersive qualities of the one cancel out those of the other. The eyepiece lens is mounted in a tube which can be moved towards or away from the objective for focusing. The eyepiece itself is made up of two lenses (or lens combinations) – a field lens farthest from the eye and an eye lens. Because the focal length of the objective has to be long, refractors are in general very long for their power.

The Reflector

All reflecting telescopes use a concave mirror to gather starlight, but they differ in the way they manipulate the converging reflected beam. In all except the largest instruments the beam must be reflected and brought to a focus outside the telescope tube, where it can be viewed through an eyepiece.

Far right: A classic amateur refractor with a 10-cm (4-inch) objective. It is equatorially mounted and has a small sighting telescope.

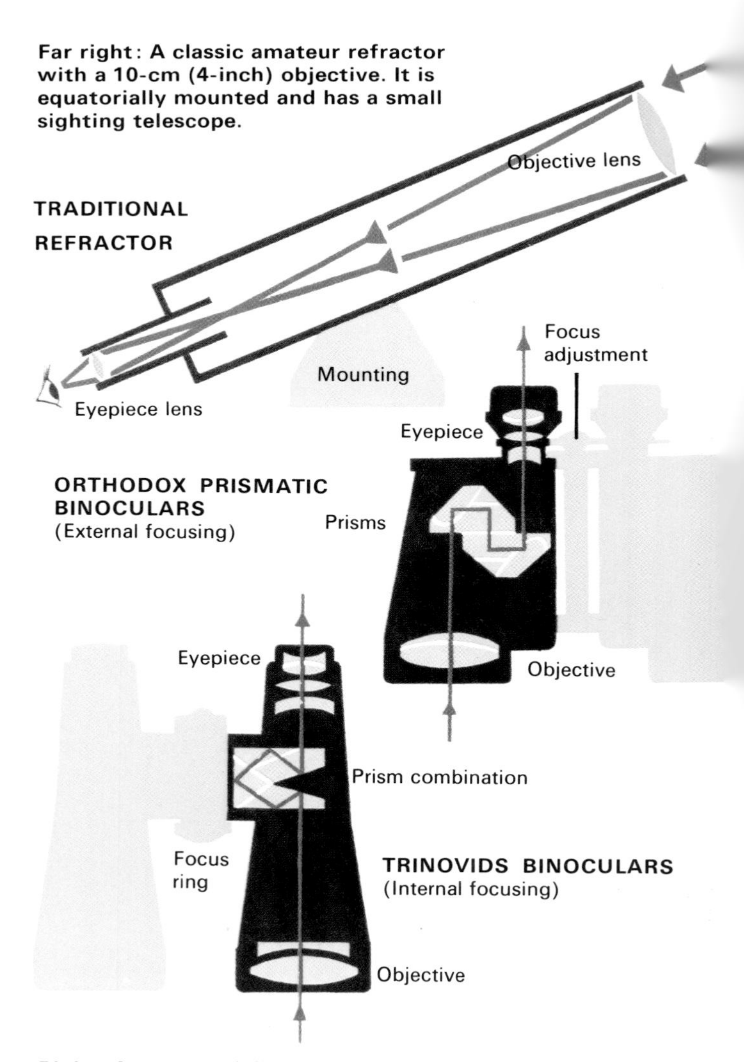

Right: An equatorially mounted 21-cm ($8\frac{1}{2}$-inch) Newtonian reflector, showing how this type provides a comfortable standing-up viewing position. Note the small sighting telescope and the counter-weighted mounting.

USING TWO EYES

Although dwarfed in size and sophistication by their larger relatives – refractors and reflectors – binoculars do have a useful role in astronomical observation, as well as in a more general terrestrial context – sports viewing, bird-watching and so on.

A pair of binoculars is really a special type of twin refractor. Each barrel contains objective and eyepiece lens combinations. It is compact because it contains prisms which 'fold' the light path between objective and eyepiece. The light does a double about-turn as it undergoes total internal reflection in the prisms.

Compared with telescopes, binoculars have very low magnification – but this can be a boon in astronomical work because it results in a wide field of view and a bright image. Astronomers use binoculars particularly to hunt for new comets and novae (exploding stars). Binoculars are rated by their magnifying power and aperture in millimetres. Thus 10 × 50 binoculars have a magnifying power of 10 and an aperture of 50 mm. This is a useful size for general astronomical work.

The problem with any binoculars is holding them steady, which becomes progressively more difficult as the magnification and weight increases. Those with jittery hands will be heartened by the recent development of stabilized binoculars. They contain a gyroscope mechanism coupled to a movable lens. When you 'lock' onto an image, the gyro keeps that image steady no matter how badly you shake.

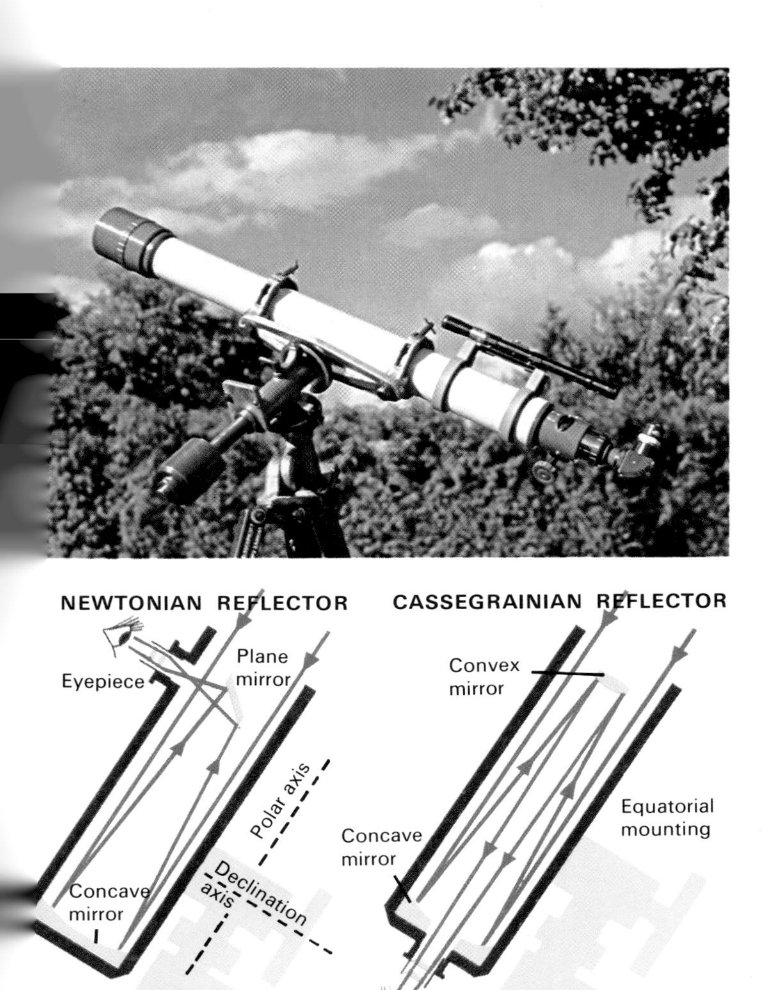

The most common form of reflector is the Newtonian, named after Isaac Newton, who devised it. In the Newtonian light is reflected by the concave mirror (the primary) back up the telescope tube. Near the end of the tube a small plane mirror (the secondary) reflects the converging beam through a right-angle into an eyepiece.

An alternative reflector system is the Cassegrainian. Its secondary mirror is convex and reflects the converging beam back down the telescope tube and through a hole in the primary mirror into an eyepiece. Though a more compact instrument than the Newtonian, the Cassegrainian reflector is more expensive to construct and has a much narrower field of view.

Most of the giant reflectors have both means of manipulating the incident light beam. And in addition they have a Coudé system. This utilizes the convex secondary mirror of the Cassegrainian system and also a third, movable plane mirror, to form an image at the same point, no matter how the telescope moves. Often the Coudé system is permanently linked to a spectrograph for analysing the starlight continuously.

Another important type of reflector used in many observatories is the Schmidt. This uses a spherical rather than a parabolic reflecting mirror and because of this has a much wider field of view. Used alone, the spherical mirror would give rise to distortion in the image, due to spherical aberration. So a thin correcting lens is placed at the mouth of the telescope tube to correct for this defect. A curved photographic plate inside the tube receives the image.

The Schmidt cannot be used for visual observations – it is really a special type of camera. Most large reflectors are in fact used primarily as cameras. Photographic film is more sensitive than the eye and can also 'store' starlight.

Following the Stars

As any casual observer of the heavens knows, the stars wheel overhead during the night. In the northern heavens they appear to arc round Polaris, the pole star. The apparent movement of the stars occurs because the Earth rotates on its axis each 24 hours (actually 23 hours 56 minutes 4 seconds). So to follow the stars for any length of time, an observer must constantly adjust his telescope. He can do this manually or use a motor drive.

But first he must make sure that the telescope is mounted correctly. In the commonest kind of mounting, the equatorial, the telescope is free to rotate about two axes. One axis (the polar) is adjusted to be parallel to the Earth's axis. The other (the declination axis) is at right-angles to the polar. Mounted in this way the telescope simply has to be rotated about the polar axis to follow the stars. By fitting calibrated scales (setting circles) to the axes and zeroing them correctly, the telescope can readily be pointed to any star whose celestial latitude (declination) and longitude (right ascension) are known.

In an alternative and simpler type of mounting, the altazimuth, the telescope is mounted on horizontal and vertical axes. The telescope can then scan simply only in altitude (up and down) and in azimuth – parallel to the horizon. Practically all large observatory telescopes have an equatorial mounting. The notable exception is the Russian 6-m (236-in) diameter reflecting telescope at Zelenchukskaya in the Caucasus, which has an altazimuth mounting to simplify supporting such a huge instrument.

The Giants

The 6-m reflector came into regular use in 1976, displacing the 200-in (508-cm) Hale telescope at Palomar Observatory in California as the world's largest. It is a massive structure, weighing some 840 tonnes. The movable section alone weighs nearly 700 tonnes; yet it can be turned easily because it 'floats' on hydraulic, pressurized-oil bearings. The telescope has such light-gathering power that it can 'see' stars down to the 25th magnitude, or nearly 40 million times fainter than the dimmest we can see with the naked eye. In terrestrial terms this is equivalent to detecting the light from a candle from a distance of 25,000 km (15,000 miles)!

Three other outstanding giant telescopes came into operation in the 1970s: 4-m (157-in) reflectors at Kitt Peak Observatory in Arizona and Cerro Tololo Inter-American Observatory in Chile, and the 3·9-m (154-in) Anglo–Australian Telescope (AAT) at Siding Spring, New South Wales. The establishment of large reflectors in Chile and Australia was particularly notable, since they enabled the far southern heavens to be scanned properly for the first time. This region is a particularly rich part of the sky, containing, for example, the galaxies nearest to us – the Large and Small Magellanic Clouds.

The AAT is more typical of the new breed of giant telescopes than the Russian 6-m reflector, being far lighter and much more sophisticated and flexible in operation. Its total mass is only 326 tonnes.

The 3·9-m diameter mirror is housed near the base of the skeletal steel telescope tube, which is supported in any aptly named horseshoe, which moves on oil-pad bearings. The mirror is a solid disc made from a zero-expansion glass refractory known as Cervit. The upper surface of the disc has a hyperbolic rather than the conventional parabolic curvature and is so accurate that it deviates from an ideal shape by less than one-ten thousandth part of a millimetre. The reflecting mirror surface is provided by about 5 g (0·18 oz) of aluminium vacuum deposited under very careful control.

The mirror is supported from behind by 36 pads, only three of which are fixed. The others are activated by air pistons in which the pressure is varied according to the orientation of the mirror. Side support is provided by 24 counterbalanced mechanical levers, which work either in a push or pull mode again according to orientation. This arrangement virtually eliminates mirror distortion as the telescope changes position.

The AAT, like its other huge cousins, has a variety of focal arrangements. They are achieved by changing the front end of the telescope. In the prime-focus position the observer actually rides in a cage at the top of the telescope tube, some 12·7 m (42 ft) from the main mirror. Strangely enough this does not appreciably affect the quality of the image.

The other three focal arrangements are achieved by bringing into play ancillary mirrors. These mainly downward-facing mirrors are supported from behind by a partial vacuum, which varies with mirror orientation. There are two Cassegrain focal arrangements which reflect the image through a 1-m (3·3-ft) diameter hole in the base of the main mirror. The observer views the image from inside a cage behind the mirror, moving with the telescope.

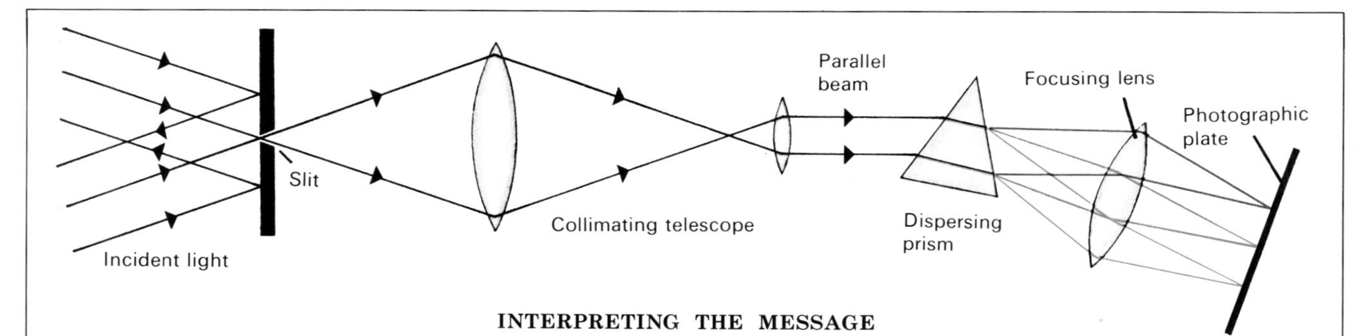

INTERPRETING THE MESSAGE

The only thing that reaches us from a distant star is its light and sometimes other radiation. But from the feeble glimmer of starlight we can often tell the size of the star, how hot it is, what it is made of, how fast it spins and much more besides. A star can be made to yield its secrets by splitting its light into a spectrum. The spectrum of starlight is crisscrossed with a number of dark lines, and it is the position and nature of these lines which give astronomers their information.

The instrument used to make a spectrum, the spectroscope, takes several different forms. Basically they differ in the way the light is dispersed into a spectrum. The traditional spectroscope uses a prism to disperse the light; the more modern way is to use a diffraction grating. A grating consists of closely spaced fine slits (transmission grating) or grooves (reflection grating). Gratings in general gives a much wider spread of colour.

The diagram shows the essential features of a recording spectroscope, or spectrograph. The image from the telescope is allowed to fall on a fine slit set at the focus of a collimating lens. The light emerges as a parallel beam and passes into the dispersing element. The resulting spectrum is then focused onto a recorder, typically photographic film. The visual record is called a spectrogram.

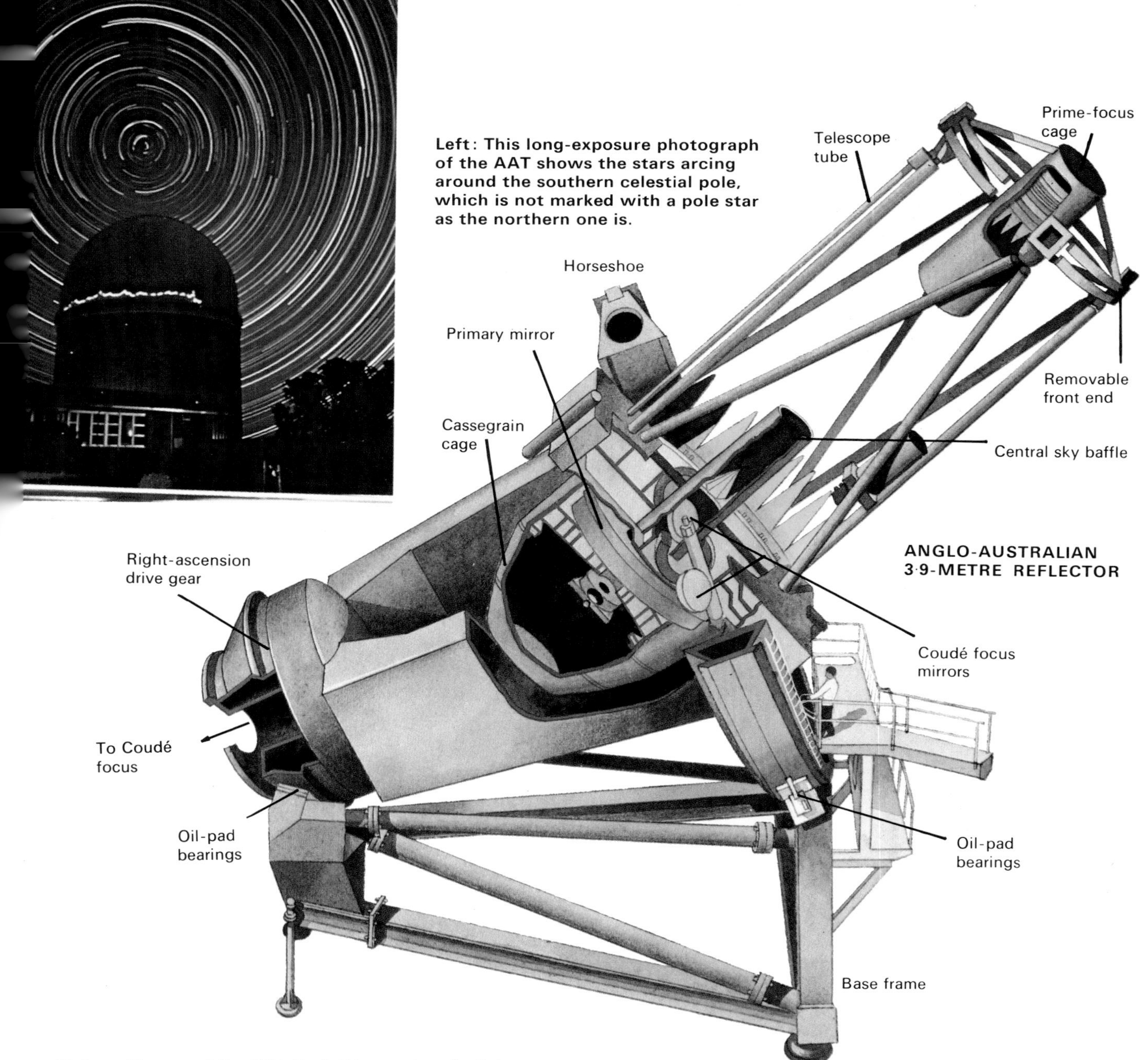

Left: This long-exposure photograph of the AAT shows the stars arcing around the southern celestial pole, which is not marked with a pole star as the northern one is.

Below: Domes of the Kitt Peak Observatory in Arizona. In the foreground is the 4-metre (157-inch) reflector.

Left: A telescopic view of part of the Moon's Mare Tranquillitatis (Sea of Tranquillity) in the region where the Apollo 11 lunar module touched down in 1969. The detail shown is about the best that can be achieved by Earth-based telescopes.

Right: Study of the Sun is carried out with the aid of solar telescopes like this one at Mt Wilson Observatory in California. Mirrors (heliostats) in the dome at the top of the tower reflect sunlight down the central shaft on to an observation table, where a projected image of the Sun's disc can be inspected.

Below: A radio telescope interferometer at the Mullard radio astronomy observatory at Cambridge. The three dish antennae function in concert with others to form a sensitive radio telescope.

Two computers control the operation of the telescope. One controls the telescope's positioning, pointing it to the desired region of the heavens and driving it to follow the stars. The declination and right-ascension motor drives operate through twin pinions and giant spur gears 3·7 m (12 ft) in diameter. (This differs from the worm gearing traditionally used for motor drives.) The position computer also compensates for possible errors due to such things as flexing of the telescope structure and temperature variations. The second computer collects, records and processes data from the instruments coupled to the telescope, such as photoelectric photometers, bolometers, image scanners and spectrographs.

Heavenly Transmitters

In the 15 million degree interior of a star, nuclear fusion reactions provide the energy that the star radiates into space. The star not only pours out this energy as light, it also emits energy at all the other wavelengths of the electromagnetic spectrum, from very short gamma-rays to very long radio waves. The Earth's atmosphere, however, will allow through only a narrow range of wavelengths – those of visible light and radio waves. The rest are absorbed.

It was not appreciated until the 1930s that the heavens could be studied by means of radio waves. But today radio astronomy has achieved almost equal importance with optical astronomy. Our view of the heavens at radio wavelengths reveals a universe much more complex than we once imagined.

Just as ordinary astronomers use optical telescopes to collect and focus light rays, radio astronomers use radio telescopes to collect and focus radio waves. The radio signals from space are so weak that radio telescopes must be very large indeed to detect them. The magnitude of the problem can be put in perspective by the thought that the total energy collected annually by a typical radio telescope is less than the energy of a single snowflake hitting the ground.

Radio Telescope Design

Radio telescopes differ greatly in construction. Most common is the fully steerable dish type. This has a huge parabolic bowl mounted on a supporting tower. The bowl is made of metal mesh or plates and reflects cosmic radio waves to a pick-up unit known as the feed antenna, which is supported at the focus of the parabola. There the weak signals are first strengthened by a preamplifier, then fed by cable to an adjacent building where they are amplified further and detected (sorted out into frequencies). The resulting signals may then be displayed on a moving chart or recorded on digital tapes for subsequent processing by computer.

Largest of the fully steerable dishes is the 100-m (328-ft) instrument at Effelsberg, near Bonn, which is operated by the Max Planck Institute for Radio Astronomy. The dish of this 3000-tonne structure is of homologous design, which means that when shifting orientation it deforms under gravity from one paraboloid to another. This makes it very sensitive and capable of detecting radio wavelengths as short as 1 cm.

Larger than the Effelsberg radio telescope, but with a fixed dish, is the 305-m (1000-ft) radio telescope at Arecibo, in Puerto Rico. The spherical dish is made up of 38,778 perforated aluminium panels supported above a natural depression in the ground. A 600-tonne feed antenna is supported above it by cables from three 76-m (250-ft) high towers. Although the dish is fixed, the radio telescope can be partly steered by moving the overhead feed antenna.

As with many radio telescopes, the Arecibo instrument is used for transmitting as well as receiving. Commonly it is used in a radar mode, sending pulses to and receiving echoes from the Moon and the nearby planets. It has also been used to beam coded messages across the universe, telling any creatures that may intercept it, thousands of years hence, the kind of being who sent the message and where he came from.

Very large dish telescopes are very expensive structures to build. There is, however, a cheaper method of achieving the same kind of sensitivity, or resolution. This is to use a number of smaller dishes acting in concert. The individual antennae are connected to a central receiver and home in on the same cosmic radio source. Depending on their separation apart, their signals will interfere with one another, maybe reinforcing one another or cancelling one another out. Application of this technique (interferometry), which is also done using long, crossed antenna arrays, enables radio sources to be mapped in fine detail.

The most outstanding interferometry set-up is the Very Large Array of the American National Radio Astronomy Observatory located near Socorro in New Mexico. It is made up of 27 separate steerable dish antennae, each 25 m (82 ft) in diameter. They are arranged in a Y-shape with arms measuring 21 km (13 miles) long. Russia also has a very powerful radio telescope of unusual design, the RATAN-600. It consists of 895 movable, tiltable and rotatable

The world's largest fully steerable dish telescope at Effelsberg, near Bonn, the West German capital. Its dish is 100 metres (328 ft) across. It is so sensitive that it can detect water molecules in neighbouring galaxies.

curved reflector plates arranged in a ring 600 m (1970 ft) in diameter. In the centre of the ring is a collector plate comprising 124 flat, tiltable panels. Combinations of plates can be selected and directed by computer to follow a radio source across the sky.

Satellite Science

Viewing the universe at wavelengths other than those of visible light and radio waves was virtually impossible before the advent of the Space Age, though limited success was achieved by launching instruments in high-altitude balloons and sounding rockets. Today instruments can be lofted in satellites above the Earth's atmosphere, where they are able to detect radiation that never reaches sea level.

Orbiting astronomy observatories such as Copernicus and HEAO have made spectacular discoveries in recent years, announcing the presence in our galaxy, for example, of the all-devouring black holes. The launching of the 2·4-m (94-in) Space Telescope from a shuttle in the mid-1980s should result in another gigantic leap forward in our study of the universe. However, satellites have more down-to-Earth uses than this (page 20). They are continuing to revolutionize terrestrial communications, weather forecasting, cartography and many other fields.

But just how can you make an object defy gravity and remain in space, as a satellite? The answer, as Isaac Newton realized, lies in speed. Suppose you are on a high mountain and have a gun that can fire a bullet horizontally at any speed you like. At low speeds the bullet does not travel very far before gravity pulls it back to Earth. As the bullet travels faster and faster, it goes farther and farther before it falls to the ground.

When it travels at a speed of 28,000 km/h (17,500 mph), a strange thing happens. It still falls under the influence of gravity, but the rate at which it falls is the same as the rate at which the Earth curves. In other words the bullet remains the same distance above the ground. It is in orbit as an artificial satellite. The speed of 28,000 km/h (17,500 mph) is called the orbital velocity. If you fire the bullet at a speed of 40,000 km/h (25,000 mph), it can escape from the Earth's gravity completely and vanish into space. This speed is called the escape velocity.

With an ear-splitting roar a Titan-Centaur rocket blasts off from the launch pad at Cape Canaveral, soon to vanish into the heavens. The payload it carries, Voyager 1, is embarking on a decade-long exploration of the outer solar system – first call Jupiter after an 18-month long journey. Standing 49 metres (160 ft) high and weighing 640 tonnes, the rocket is lifted off the launch pad by the twin solid boosters. The first core stage, which burns hydrazine and UDMH, ignites after the boosters are jetisonned.

Rocket Propulsion

Speeds of the order of 28,000 km/h (17,500 mph) are many orders of magnitude greater than those encountered in our ordinary lives. Even supersonic fighters cannot fly much faster than 3000 km/h (2000 mph). Their engines – turbojets – are neither powerful enough to loft things into space, nor are they capable of working in space. They rely on the oxygen in the atmosphere to burn their fuel.

The only engine powerful enough and able to work in space is the rocket. It is completely self-contained, carrying not only fuel but also a substance containing oxygen with which to burn the fuel. Rocket fuel and oxidizer (oxygen-provider) are called propellants. In a rocket the propellants are burned inside a combustion chamber to produce hot gases. The gases expand through a nozzle to form a high-speed jet. Reaction to the backward-facing jet thrusts the rocket forward. This follows from Newton's third law of motion: 'To every action there is an equal and opposite reaction'. The rocket performs better when it is travelling in space, that is, in a vacuum.

Various kinds of propellants are used. Some are solid; others are liquid. They also differ in the energy they give out on combustion. A rocket's power may be expressed either in propulsive thrust (in Newtons or pounds) or specific impulse (thrust per kilo (pound) per second of propellant burned). Specific impulse is increased by increasing the pressure and temperature in the combustion chamber and by lowering the molecular weight of the exhaust gases.

Rocket Propellants

The most powerful kinds of rocket use liquid propellants. Fuel and oxidizer are contained in separate tanks and are pumped at the required flow rate into the combustion chamber where they mix and ignite. The most powerful engines burn liquid hydrogen as fuel and liquid oxygen as oxidizer. These propellants are termed cryogenic because of their low temperature (−253°C and −185°C respectively).

Kerosene is another common fuel, used in combination with liquid oxygen. The five booster engines of the Saturn V Moon rocket used this combination, burning nearly 3 tonnes of propellants per second to achieve a take-off thrust of more than 33 million Newtons (7·5 million pounds). Space launching rockets burn such prodigious amounts of propellants that up to 95% of their take-off weight is accounted for by the propellant.

In liquid-hydrogen/liquid-oxygen and kerosene/liquid-oxygen rocket engines, the vaporized propellants have first to be ignited. In alternative systems two propellants ignite spontaneously when they mix. Such propellants are termed hypergolic. The most widely used hypergolic propellants are chemicals called unsymmetrical dimethyl hydrazine (UDMH) as fuel and nitrogen tetroxide (N_2O_4) as oxidizer. They are used, for example, in the European launching rocket Ariane.

Solid-propellant rockets are much simpler in construction and operation than liquid-propellant rockets. They consist simply of a body shell packed with propellant, which is usually a mixture of a kind of synthetic rubber (such as polybutadiene) and oxidizer (such as ammonium perchlorate). Running lengthwise along the centre of the propellant is a cavity which forms the combustion chamber. An electrical or pyrotechnic igniter starts the propellant burning; burning then proceeds radially.

Solid-propellant rockets are less powerful and less flexible than liquid-propellant rockets and, once set burning, are difficult to control. But solid propellants are more stable and easier to handle than liquid

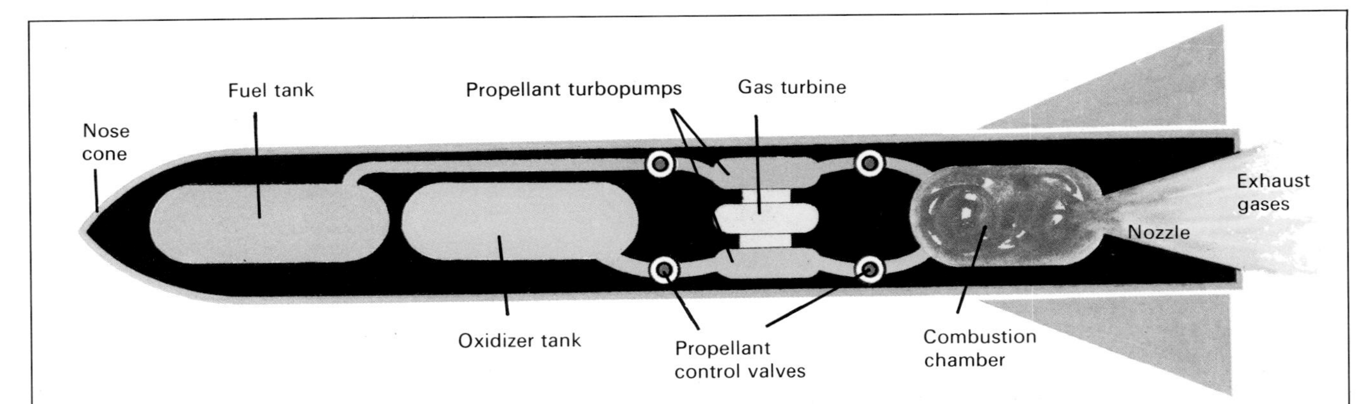

THE LIQUID-PROPELLANT ROCKET

The diagram shows the essential elements of a liquid-propellant rocket whose purpose is to produce a propulsive jet of hot high-speed gases. The fuel and oxidant are pumped from their tanks into the combustion chamber by a pair of turbopumps. These are centrifugal pumps linked to the same shaft and driven by a centrally mounted gas turbine. To prevent cavitation (formation of gas bubbles) in the pumps, the propellant tanks have to be lightly pressurized.

The gas needed to drive the turbine is provided by a generator. This may produce gas by igniting small quantities of the main propellants. Or high-strength hydrogen peroxide may be the source of the gas. It is fed into the generator where it is decomposed by a catalyst into steam and oxygen.

The fuel and oxidant flow via numerous control valves and are sprayed in the required proportions into the combustion chamber through injectors. These atomize the propellants and cause them to mix intimately to form a highly explosive mixture. In hypergolic systems the propellants ignite spontaneously, even at the start. In other systems the atomized propellant mixture must first be fired by an igniter, after which it will carry on burning so long as the propellant flow is maintained. The igniter may bring together a small quantity of hypergolic propellants to provide a starting flame. Or it may work electrically and produce a spark or a red-hot wire.

When the propellants ignite inside the combustion chamber, pressures upwards of 50 atmospheres and temperatures of 3000°C or more may be produced. The high-strength steel alloy normally used to construct combustion chambers cannot withstand such temperatures and has to be cooled.

The most common method is known as regenerature cooling. One of the propellants is pumped, not directly into the combustion chamber but first through the double wall of the chamber, thereby cooling it. Other methods of reducing chamber wall temperature are film cooling and ablative cooling. In film cooling propellant is flowed over the wall surface. In ablative cooling the chamber is lined with material that dissipates heat by boiling away, in the same way that a spacecraft's heat shield does.

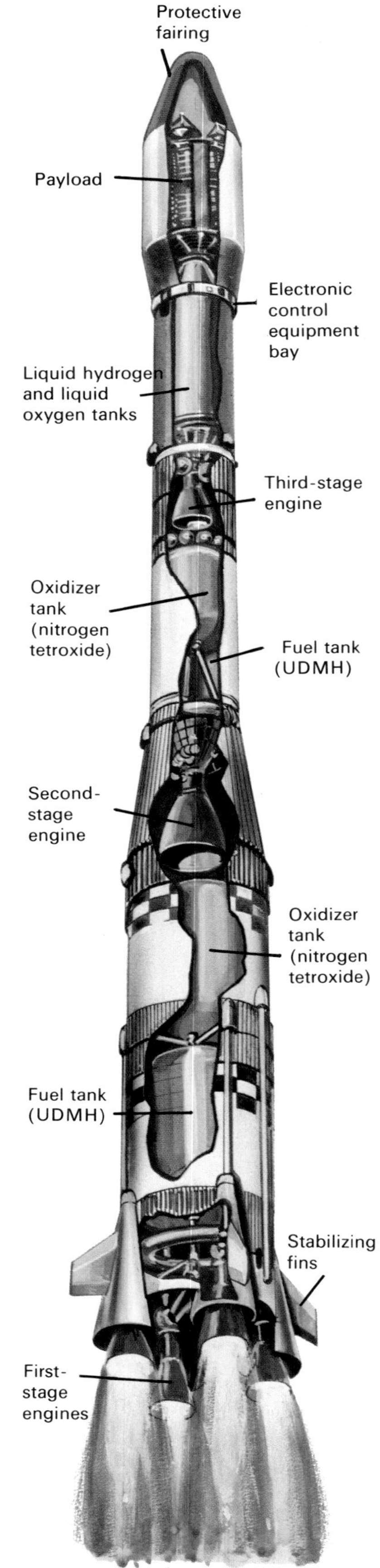

ARIANE, A CLASSIC LAUNCH VEHICLE

The European launcher Ariane is a classic three-stage rocket with a lift-off weight of 207 tonnes and an overall height of 47·4 m (155 ft). Of its weight 90% is accounted for by the propellants, 9% for its structures, and only about 1% for its payload.

The first, or booster stage of the rocket houses four Viking engines, which have a combined lift-off thrust of nearly 2·5 million Newtons (500,000 lb). They burn the hypergolic propellants UDMH and nitrogen tetroxide, which are stored in steel tanks. The combustion chamber of the Viking engine is also made of steel, and its walls utilize film cooling. The exhaust nozzle fitted to the chamber is bell-shaped and lined with refractory graphite.

The main propellants feed the gas generator which drives the propellant turbopumps. The generator also provides the gas to pressurize the propellant tanks and the power for the hydraulic actuators which direct the swivelling of the engines. Swivelling of the engines, which are mounted on gimbals, is necessary to provide directional control of the rocket as it ascends. Aerodynamic stability of the rocket at low altitudes is assisted by the tail fins at the base.

The first stage fires for about 145 seconds, then pyrotechnic cutting cords on the aft skirt of the second stage sever connection with the first stage. Small solid-fuel retro-rockets mounted on the first stage and acceleration rockets on the second stage fire in unison to move the stages apart. Altitude at this point is 52 km (32 miles), speed 6730 km/h (4180 mph).

The single Viking engine of the second stage uses the same propellants as the first stage, stored in aluminium alloy tanks. It is mounted on gimbals and swivelled for directional control over pitch and yaw. Control over roll is effected by means of auxiliary jet thrusters fed by hot gas tapped from the generator.

At an altitude of about 140 km (87 miles) the second stage shuts off and separates in the same manner as the first stage. Just prior to separation the nose fairing is jettisoned, its function of protecting the payload from aerodynamic heating in the dense lower air being over.

As the third stage fires Ariane's velocity is 17,200 km/h (10,700 mph). The third-stage engine is cryogenic, using liquid hydrogen and liquid oxygen as propellants. They are stored in aluminium alloy tanks clad in insulation to prevent them overheating. Gaseous hydrogen is used to pressurize the liquid-hydrogen tank, and helium the liquid-oxygen tank. Gaseous hydrogen is also used for the thrusters which provide roll control for the stage, pitch and yaw being controlled as before by swivelling the engine.

The combustion chamber and the exhaust nozzle are regeneratively cooled by the liquid hydrogen. The cooling channels of the chamber are milled in a copper casting and then covered with an electrolytic deposit of nickel. The same technique has been copied for the space shuttle's main engine. The extended bell-shaped exhaust nozzle is made up of spiral tubes of Inconel, a high-temperature alloy of nickel, chromium and iron. They form the cooling channels through which the liquid fuel flows.

The hardware that controls Ariane from the moment it lifts off the ground is centralized in an equipment bay located at the top of the third stage, beneath the payload. Electronic equipment in the bay controls the sequencing and separation of the stages, guidance, tracking and telemetry (that is, the transmission of digital information between vehicle and ground control).

The inertial guidance and control system is organized around a digital computer and an inertial platform. Accelerometers and gyroscopes sense the velocity changes along the three axes from which the actual speed and direction of the rocket are computed. The computer compares the speed and attitude of the rocket with the instructions contained in its guidance programme. It corrects any deviations as they occur by transmitting appropriate instructions to the hydraulic actuators which orientate the engines.

Ariane's third stage fires for nearly 10 minutes and boosts the vehicle into a 210-km (130-mile) high orbit at a speed of over 35,000 km/h (22,000 mph). About 4000 km (2500 miles) down range the third stage and its payload separate. The former continues in low orbit, while the payload is boosted into an elliptical transfer orbit that will take it to 35,900 km (22,300 miles) altitude. At that altitude the orbit will be circularized, making it geostationary.

The useful payload capacity is large for such a rocket, being some 35 cubic metres (45 cubic yards). This is large enough to house that latest Intelsat communications satellites or two or three smaller satellites, one above the other.

propellants, and can be stored for long periods, unlike liquid propellants. Solid-propellant rockets are nevertheless widely used in space launching vehicles. Large ones are sometimes used as booster rockets to assist take-off, as in the space shuttle. Small ones are used, for example, to separate the stages of a step rocket.

Step by Step

However powerful it might be, a single rocket cannot lift itself and a payload into space. The power-to-weight ratio is never quite high enough. It can be made high enough, however, by constructing a launching rocket from a number of separate rockets joined end to end. This multi-stage or step-rocket arrangement was the brainchild of father of astronautics Tsiolkovsky.

Typically a launching rocket or vehicle has three stages. The first-stage rocket fires until its fuel is expended, and then separates. The second stage rocket then fires and, when its fuel is exhausted, separates in turn. The third stage fires to thrust what remains of the launching vehicle into orbit. By jettisoning unwanted hardware as it goes along, the launching vehicle attains a more than adequate power-to-weight ratio to reach orbit.

Each launching vehicle developed by the space powers differs widely in size, weight and power as well as in the type of engine and the propellant it uses. The Saturn V Moon vehicle impressed by its sheer size and power – 111 m (365 ft) high on the launch pad, with the combined thrust of nearly 50 Concordes. The Titan IIIC planetary-probe launcher is characterized by its large, twin strap-on boosters. The Russian Soyuz launcher is instantly recognized by its four tapered wrap-around boosters.

But whatever their size, shape or rating, step rockets do have much in common in construction and operation. It is worth considering one in detail to illustrate many of the general principles involved (see opposite). The launching vehicle chosen, the Ariane, is a comparative newcomer on the space scene, but it is a classic launch vehicle, based upon proven technology.

Ariane is built by the European Space Agency, a body financed by 10 countries – West Germany, France, Britain, Italy, the Netherlands, Sweden, Spain, Belgium, Switzerland and Denmark – to co-ordinate space activities in Europe. It is designed to have a performance comparable with American launch vehicles and to compete commercially with them, particularly with the space shuttle. Its main purpose is to launch satellites into geostationary orbit, 35,900 km (22,300 miles) above the Earth over the equator. The launching site for Ariane is near Kouron in French Guiana, within a few degrees of the equator, an ideal place for launching geostationary satellites.

The Space Commuter

The United States has now switched its space launching capability from the classic launch vehicle represented by Ariane to the space shuttle. In the classic launch vehicle most if not all of what goes

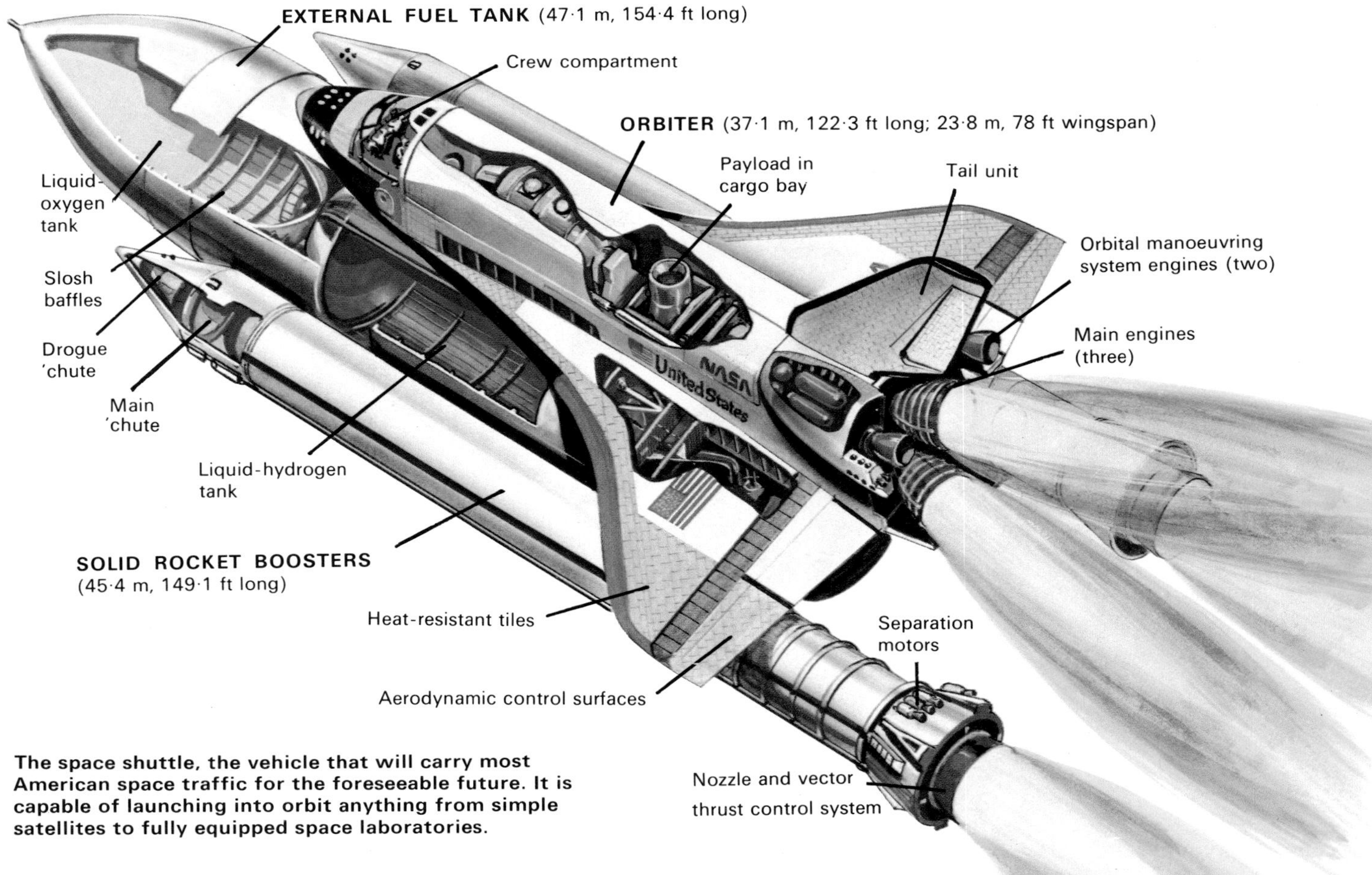

The space shuttle, the vehicle that will carry most American space traffic for the foreseeable future. It is capable of launching into orbit anything from simple satellites to fully equipped space laboratories.

up does not come down, at least in re-usable form. The space shuttle, however, is designed to be re-usable. Much of its hardware can be used time and time again. A shuttle should be able to complete up to 100 missions without a major overhaul. After a mission, it should be ready to return to space in a fortnight.

Another major advantage of the shuttle is that it is manned, thus making it exceedingly flexible. It cannot only launch satellites, but also retrieve those that malfunction. Alternatively it can transport engineers to repair and service satellites *in situ*. And it can carry out all these – and more – activities on the same mission. Last but not least the shuttle can carry a prodigious load – up to 30 tonnes.

The most important part of the shuttle is the orbiter. This houses the crew and payload. It is an interesting vehicle – part rocket, part plane. It has rocket engines and rides into space vertically, but it also has aeroplane wings and tail and lands on a runway. Its length is 37·2 m (122·2 ft) and its wing-span 23·8 m (78·1 ft).

The orbiter starts its journey into space attached to a large tank, which contains the propellants (liquid oxygen and liquid hydrogen) for its three main engines. Strapped to the sides of the tank are twin solid-propellant booster rockets. On take-off all the rockets fire together, with a combined thrust of 30 million Newtons (7 million pounds). When the boosters run out of fuel at a height of about 40 km (25 miles), they separate and parachute back to Earth for recovery and re-use. The orbiter and tank continue on their way, gathering speed all the time. At a height of some 150 km (95 miles), the tank is jettisoned and falls in a ballistic trajectory into the ocean, but is not recovered.

The two engines of the orbital manoeuvring system (OMS) fire next to insert the orbiter into orbit. These engines, which burn hypergolic monomethyl hydrazine and nitrogen tetroxide, are located in pods on either side of the orbiter's tail. The pods also house the thrusters of the reaction control system. These thruster units, in conjunction with one in the nose, provide attitude control (pitch, yaw and roll) for the orbiter in space. They use the same propellants as the OMS engines.

In orbit the orbiter's twin cargo-bay doors are opened. If a satellite is to be launched, this is done by means of a jointed manipulating arm. The arm can also be used to retrieve satellites. Other typical payloads will be instrument platforms that need to be exposed to space, and manned space laboratories, such as Spacelab, which is designed and built by the European Space Agency specifically to fit into the orbiter's cargo bay.

The normal operating crew for a shuttle is three – pilot, commander and mission specialist. For each Spacelab mission the crew is expanded by four other specialists, who are not highly trained astronauts, but scientists and engineers. Whereas the duration of a normal shuttle mission is about a week a Spacelab mission takes up to a month. From start to finish the mission is conducted in a pressurized, 'shirt-sleeve' environment, except when a crew member needs to make a 'space walk', or EVA (extra-vehicular activity). An air lock is provided in the front fuselage for this eventuality.

When the mission is completed, the cargo-bay

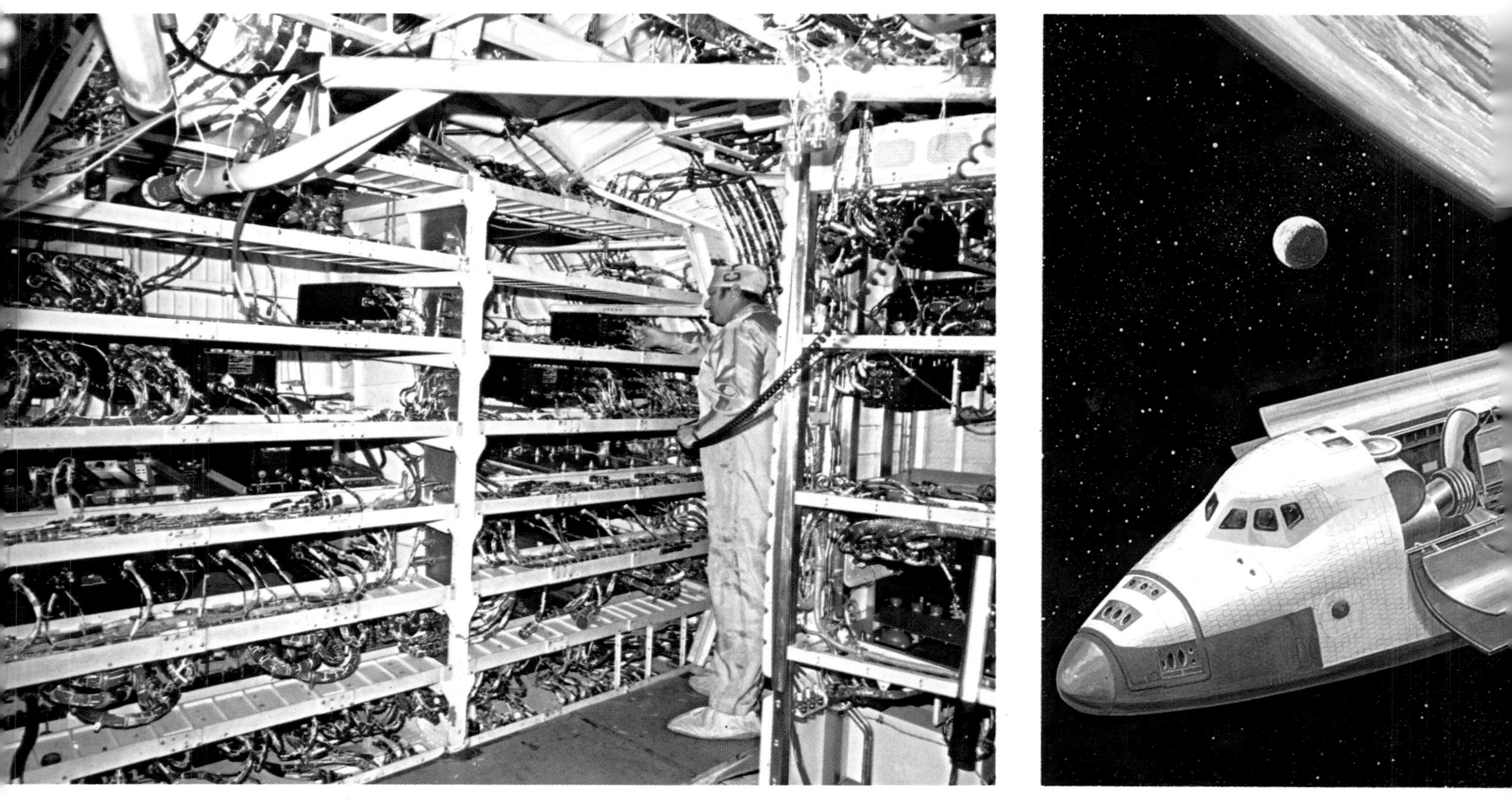

Above: Shuttle Orbiter 101, 'Enterprise' parting company with its carrier aircraft in the final series of approach and landing tests to evaluate the gliding capability of the orbiter at subsonic speeds. The long nose boom fitted on 'Enterprise' for the acquisition of air data will be lacking in operating vehicles. The carrier aircraft is a modified Boeing-747 'jumbo jet'.

Below: In orbit the orbiter opens its cargo-bay doors and exposes the payload to the space environment, in this instance the European space laboratory.

Bottom left: Checking out the seemingly interminable labyrinth of wiring in the avionics bay of Orbiter 102, 'Columbia'.

doors are closed and retrofire by the OMS engines reduces the orbiter's speed to below orbital velocity. The orbiter drops from orbit. As it re-enters the atmosphere at 28,000 km/h (17,500 mph), friction with the atmosphere makes its outer surface glow red hot. This surface is made up of insulating material based on carbon-bonded resins, which protects the aluminium alloy airframe underneath from the high temperatures encountered.

Aerodynamic braking by the atmosphere slows down the orbiter on its 2000-km (1250-mile) glide path from near orbital speed to a landing speed of about 350 km/h (220 mph). The orbiter lands on an orthodox undercarriage on a runway, like a modern airliner. The main shuttle launch and landing site is at the Kennedy Space Centre in Florida, scene of the Apollo launches between 1969 and 1972. The facilities there have been modified and extended for the space-shuttle era. The Vehicle Assembly Building (VAB), one of the biggest buildings in the world – 218 m long, 158 m wide and 160 m high (715 ft by 518 ft by 524 ft) – is back in use. So are the giant eight-track crawler transporters designed to carry the assembled Saturn V rocket on its mobile launcher. The biggest vehicles in the world, measuring 40 m by 35 m (131 ft by 114 ft), they now transport the space shuttle from the VAB to the launch complex.

The Americans are not alone in developing a shuttle. The Russians are also doing so, though few details about it are available. The Soviet space plane, known as Kosmolyot, is, however, believed to resemble the shuttle in size and in mode of operation. ESA might also build a shuttle before 1990.

Many Moons

Every year over a hundred man-made moons, or satellites are launched into orbit, mainly by the United States and Russia. They are used for a variety of purposes – scientific study, military reconnaissance ('spying'), Earth survey, communications, navigation, weather watching. Depending on their purpose, they carry a variety of different instruments and they differ widely in appearance and size. Their shape is purely functional – there is no need for streamlining in the vacuum of space.

To illustrate the variety of satellites, take the American craft Lageos, Landsat and Intelsat IVA. Lageos is a tiny sphere only 60 cm (2 ft) in diameter. Its surface is made of aluminium, but it has a brass core, giving it a weight of 411 kg (906 lb). The surface is covered with 426 reflecting prisms which give it the appearance of a golf ball. It contains no instruments.

Because it is small and dense, Lageos maintains a very stable orbit at 5900 km (3670 miles) altitude. By bouncing laser beams off it and timing the echoes, Earth–satellite distances can be determined to within a few centimetres. By using the stable position of Lageos as a yardstick, geologists can trace, for example, movements of the Earth's crust. Tracing such movements is of critical importance in earthquake research. So stable is the orbit of Lageos that it will remain in orbit for nearly 10 million years.

The Landsat Earth-survey satellite provides a vivid contrast with Lageos. It is about 3 m (10 ft) high and with its 'butterfly' solar paddles extended measures 4 m (13 ft) across. It weighs nearly 1 tonne. Its design is based on that of the successful Nimbus weather satellite. It has two imaging, or picture-taking systems, which scan the Earth's surface at different wavelengths. They record spectral 'signatures' which identify different types of ground features. Landsat also acts as a means of collecting and relaying environmental data from sensing platforms in remote corners of the world.

The Intelsat IVA communications satellite circles in geostationary orbit (that is, it appears fixed in the sky) 35,900 km (22,300 miles) high. It is 7 m (23 ft) tall and is 2·4 m (8 ft) in diameter. From its lofty altitude it can relay telephone, telex and television signals between the continents. It receives signals from one ground station, amplifies them, and retransmits them to another ground station. It can handle 6250 two-way voice circuits and two television channels simultaneously.

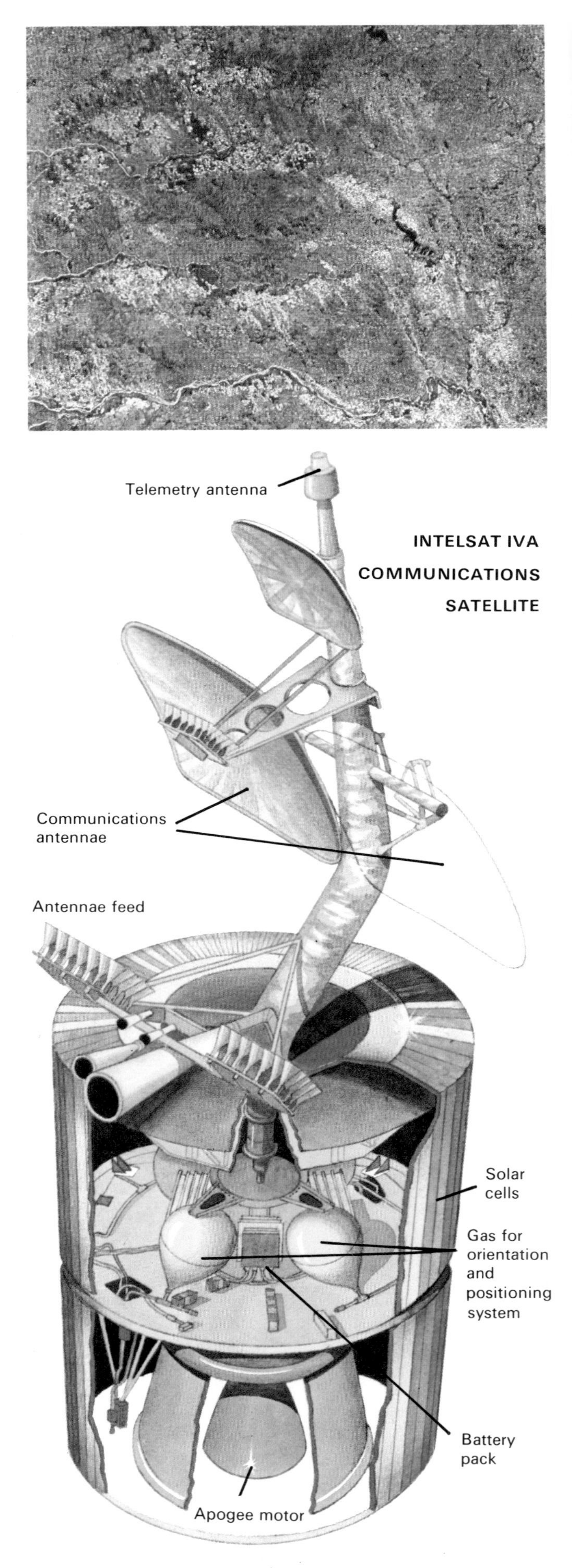

Satellite Systems

Despite their differences most satellites have similar subsystems which enable them to function. These subsystems include instrumentation, communications and telemetry, power and attitude control. They are usually assembled separately as individual modules which are then attached to the basic structure of the satellite. This structure is usually made of aluminium alloy for lightness. It invariably incorporates a means of temperature control to maintain a reasonable operating temperature for the electronics. Temperature control is often effected by means of hinged louvres which open or close as the internal temperature rises or falls.

The instrument module on a satellite may include such things as television cameras, image scanners, radiometers (to measure all kinds of radiation), charged-particle counters, telescopes and magnetometers (to measure magnetism). The communications module is made up of a number of radio receivers and transmitters which receive instructions and transmit data via numerous antennae. Each receiver and transmitter has a different function and operates at a different frequency. Instructions to the spacecraft from ground control come in the form of coded signals. A decoder on board translates them into a form the satellite can use and they are passed, via a distribution unit to the appropriate satellite system. The telemetry ('measuring from afar') system collects data from the satellite's instruments and engineering sensors (thermocouples, pressure gauges, ammeters and so on) and transmits it on instruction back to Earth. The data is first processed by an encoder and then recorded on tape. Only when the satellite comes within range of a ground station does the recorder play back and transmit the data.

Of course, instructions can only be beamed to and received from satellites if their whereabouts in space are accurately known. Such tracking may be done by radar using Earth-based radio pulses. Or it may be done using transponders on the satellite. These are triggered into transmitting signals, and the nature of the triggered signals indicates the range (distance) and range-rate (speed) of the satellite.

Above: Two-views-in-one of a Russian communications satellite of the Molniya 1 series, which form part of the extensive Orbita satellite communications network. The Molniya satellites have a highly eccentric orbit which varies from over 40,000 km (25,000 miles) in the northern hemisphere to only about 600 km (370 miles) in the south.

Top left: A 'false-colour' picture of the American mid-West obtained from a Landsat satellite.

Below: European Space Agency technicians carrying out electronic equipment tests on their weather satellite Meteosat, which was subsequently launched into geostationary orbit.

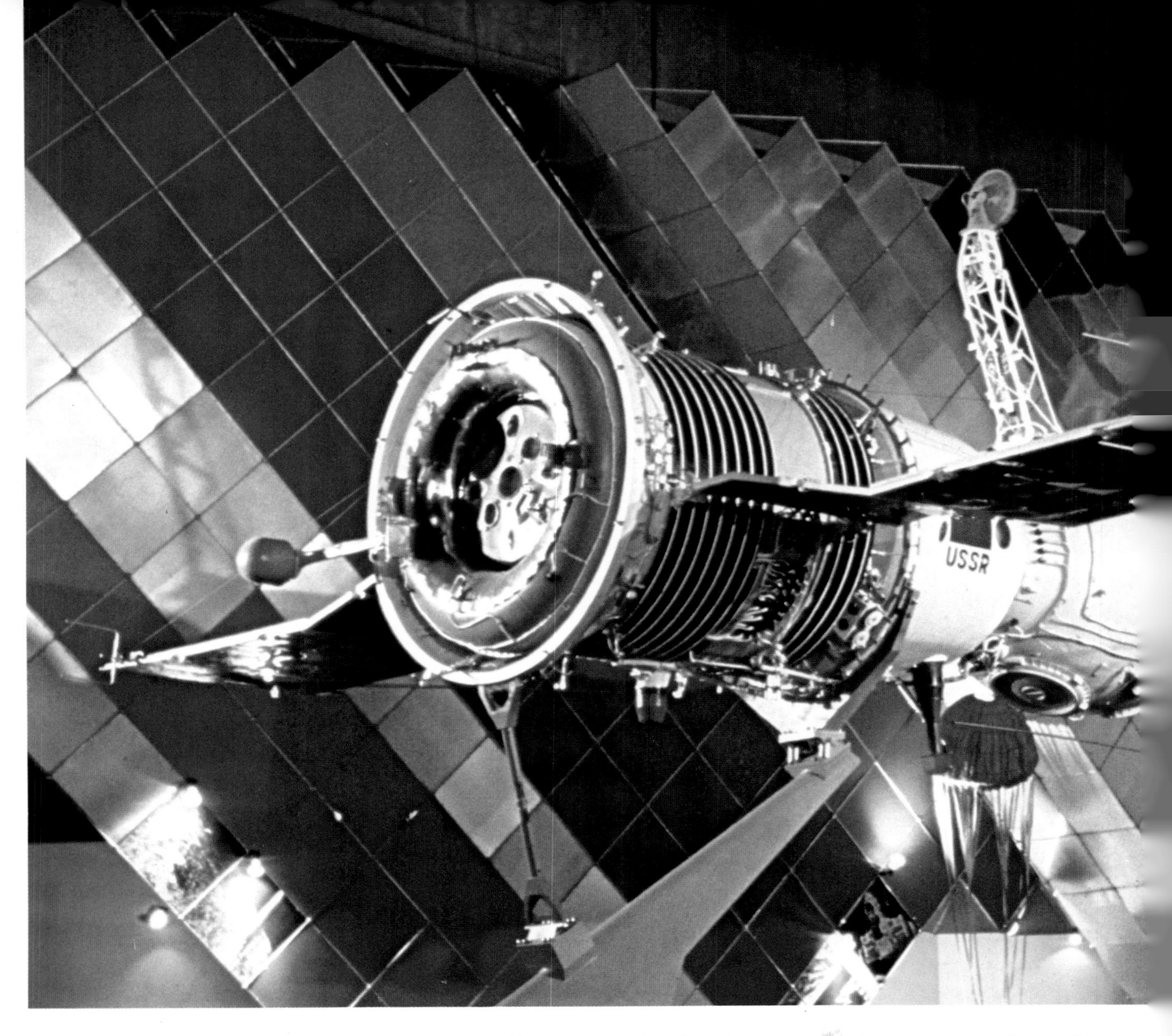

The electricity to power the instruments, radios and other electronic devices is provided by arrays of solar cells. Such cells convert the energy in sunlight directly into electricity. They consist of thin wafers of silicon about 0·15 mm (0·006 in) thick. Tens of thousands of individual cells are required to generate the few hundred watts power a satellite needs. The cells also usually trickle charge a small battery pack, which can provide power when the satellite passes into the Earth's shadow. The solar cells may be mounted on 'paddles' or 'wings' that project from the satellite, or they may cover its body.

In many instances the attitude of the satellite is important. This is particularly true of weather satellites such as Nimbus, Earth-survey satellites such as Landsat, and communications satellites such as Intelsat, which need to point their cameras or directional antennae towards the Earth. To maintain a particular orientation a satellite uses a variety of sensors, thrust units, oscillation dampers and inertia wheels, controlled by computer.

As soon as the satellite has reached orbit, ground control instructs the sensors to 'lock' onto certain of the heavenly bodies – for example, the Sun, the Earth, Polaris or Canopus. Acting on information provided by the sensors, the computer alters the speed of inertia wheels on the three major axes. This causes the satellite to change position and assume the correct orientation. Once this has been done the satellite may be set spinning to help maintain the correct orientation by gyroscopic action. On satellites like Intelsat the directional antennae must be independently mounted to prevent them spinning.

Exploring the Planets

When spacecraft are despatched from the Earth at speeds in excess of 40,000 km/h (25,000 mph) they escape completely from its gravity. Such craft, called probes, are now sent regularly to explore the planets. By the end of 1979 all the planets out to Saturn had been investigated by probes.

Distances in space are so great that probes must travel for months and even years to keep their planetary rendezvous hundreds of millions of kilometres distant. To be assured of success a probe must be boosted away from Earth with incredible precision. If it is a fraction of a degree off course or travelling a few kilometres an hour too slow or too fast, it will miss its target by a wide margin.

To reach an inner planet (that is, Mercury or

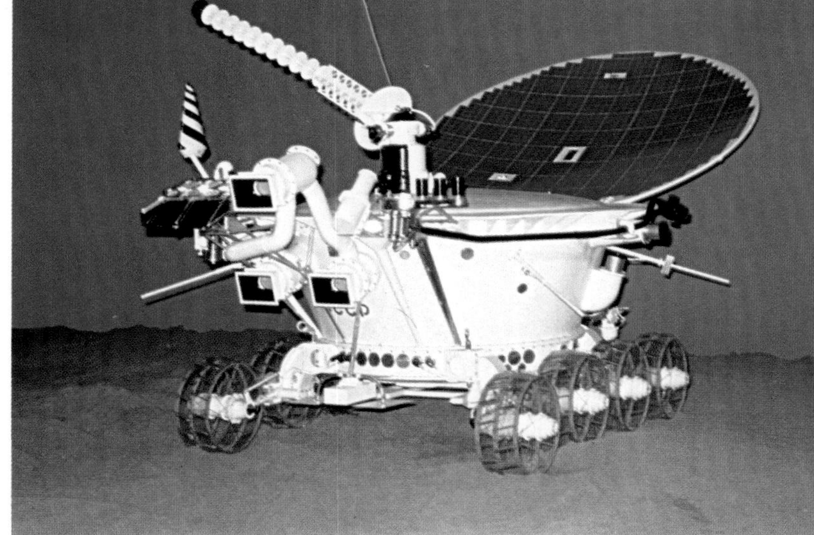

Above: Two Russian Soyuz spacecraft docked together, simulating early flights when Soyuz craft linked up in orbit to make a proto-space station.

Top right: One of the more fascinating pieces of hardware of the Space Age – the 'Moon-stroller' Lunokhod (USSR). The first Lunokhod landed on the Moon late in 1970. It was controlled remotely from Earth.

Venus) a probe is launched in the opposite direction in which the Earth is travelling in its orbit. This ensures that after escape the probe is travelling slower than the Earth. It therefore spirals closer to the Sun, that is, towards the inner planets. Conversely, to reach an outer planet (Mars, Jupiter and so on), a probe must be launched in the direction in which the Earth is travelling. Then it ends up travelling faster and therefore swings away from the Sun towards the outer planets.

In addition to the problem of launching accuracy, there is the problem of communications. Over the vast distances involved in planetary exploration the speed of radio waves (300,000 km/sec, 186,000 miles/sec) cannot be ignored, as it can on Earth. And there can be a considerable time-lag between transmitting inquisitive signals from Earth and receiving telemetric data from the probe. When the Voyager

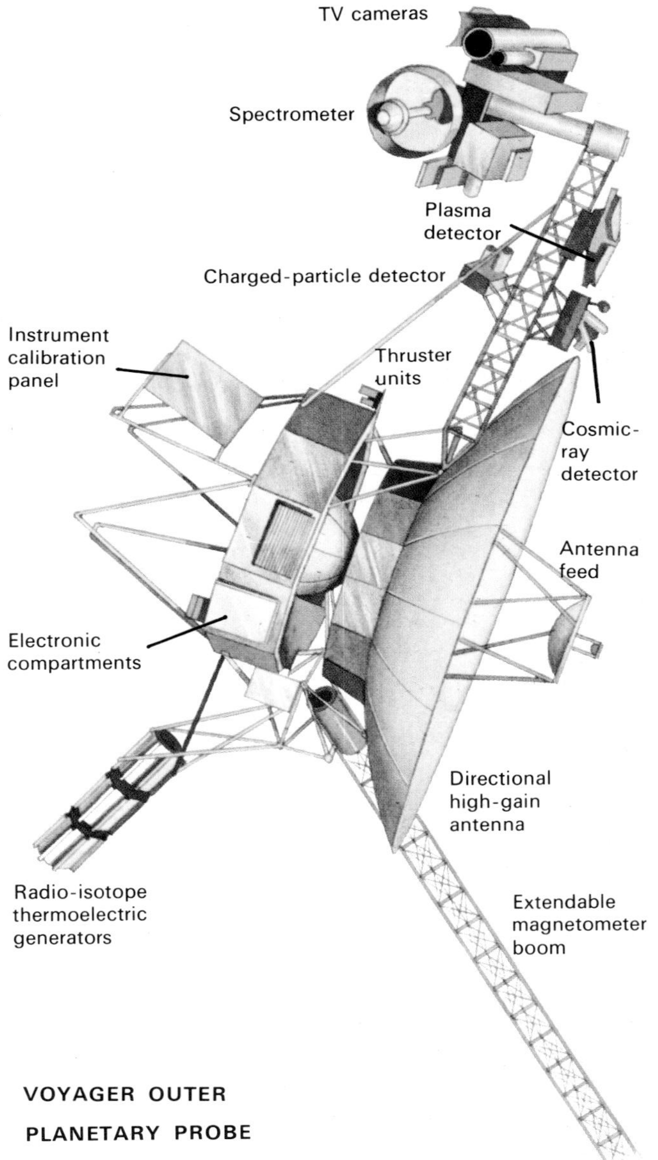

VOYAGER OUTER PLANETARY PROBE

probes encountered Jupiter in 1979, it took over three-quarters of an hour for messages to travel across the 650 million km (400 million miles) of intervening space.

Detecting a signal over such a distance taxes the space scientist's ingenuity to the full. The power of signals from a probe near Jupiter is only something like 10^{-17} of a watt. But with large dish antennae like NASA's 64-m (210-ft) instrument at Goldstone, California, scientists can detect them.

Probes that venture to the fringes of the solar system cannot be powered by solar cells as Earth satellites can, for the Sun's strength at such distances is inadequate. So a nuclear power source is used. It is called a radio-isotope thermoelectric generator (RTG). The Pioneer Jupiter probes, the Viking Mars lander and the Voyager probes to the outer planets all have this kind of power unit.

An RTG unit converts into electricity the heat produced by the decay (breakdown) of a radioactive isotope, usually plutonium-238. The devices it uses to do this work – thermoelectric couples – work on the principle that when the junctions between two dissimilar conductors are maintained at different temperatures, electricity is produced. The RTGs on the Pioneer probes used banks of semiconductor couples composed of lead telluride and a silver, antimony, germanium and tellurium alloy.

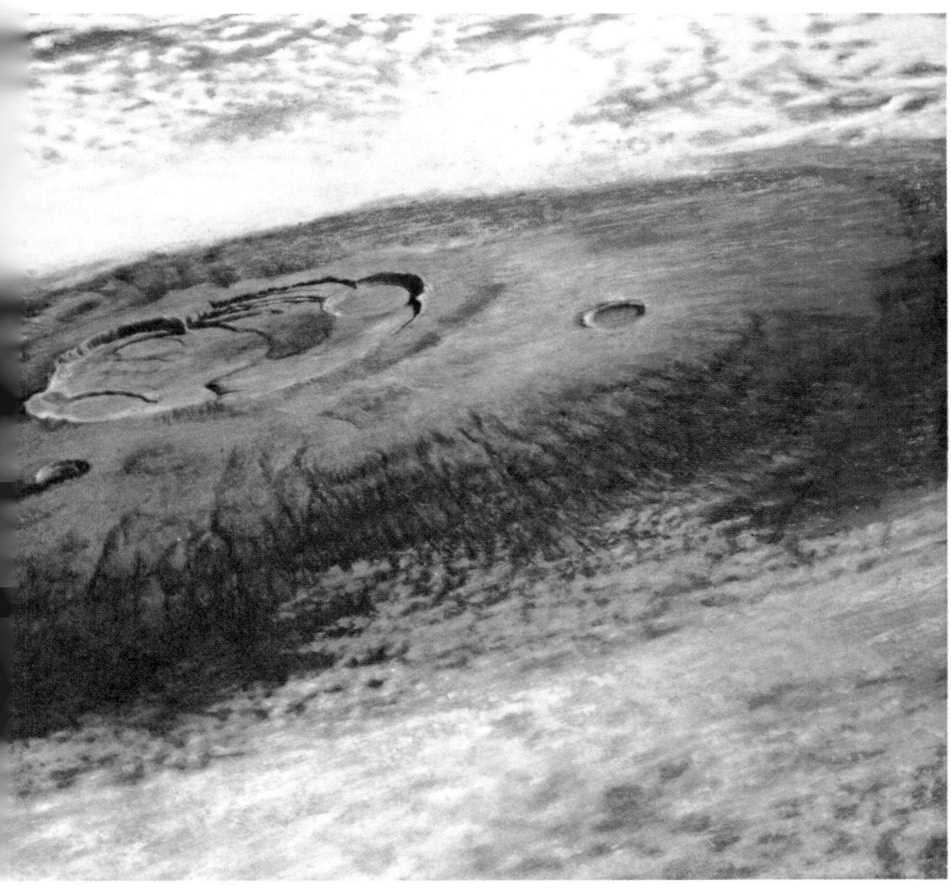

Above: This remarkable picture of Io, one of the 'big-four' Jovian moons, was taken by the space probe Voyager 1 in 1979. This and other pictures revealed the presence on Io, nearest moon to Jupiter, of many active volcanoes.

Left: Any volcanoes there were on Mars have long since become inactive. This picture, taken from Martian orbit, shows the greatest of the dead Martian volcanoes, Olympus Mons (Nix Olympica), wreathed in clouds in mid-morning. The size of Olympus Mons is staggering – 600 km broad at the base, 24 km high, with a summit crater complex 80 km across (370, 15 and 50 miles respectively).

Top left: Close-up of the rust-red surface of the Red Planet Mars, pictured by the Viking 2 probe in 1976. The boulder-strewn, but otherwise featureless landscape stretches to the horizon some 3 km (2 miles) away. Part of the Viking lander is shown, including at the top the high-gain antenna aimed at Earth some 300 million km (200 million miles) away. So great a distance separated the two planets at the time that radio signals took 18 minutes to travel from one to the other.

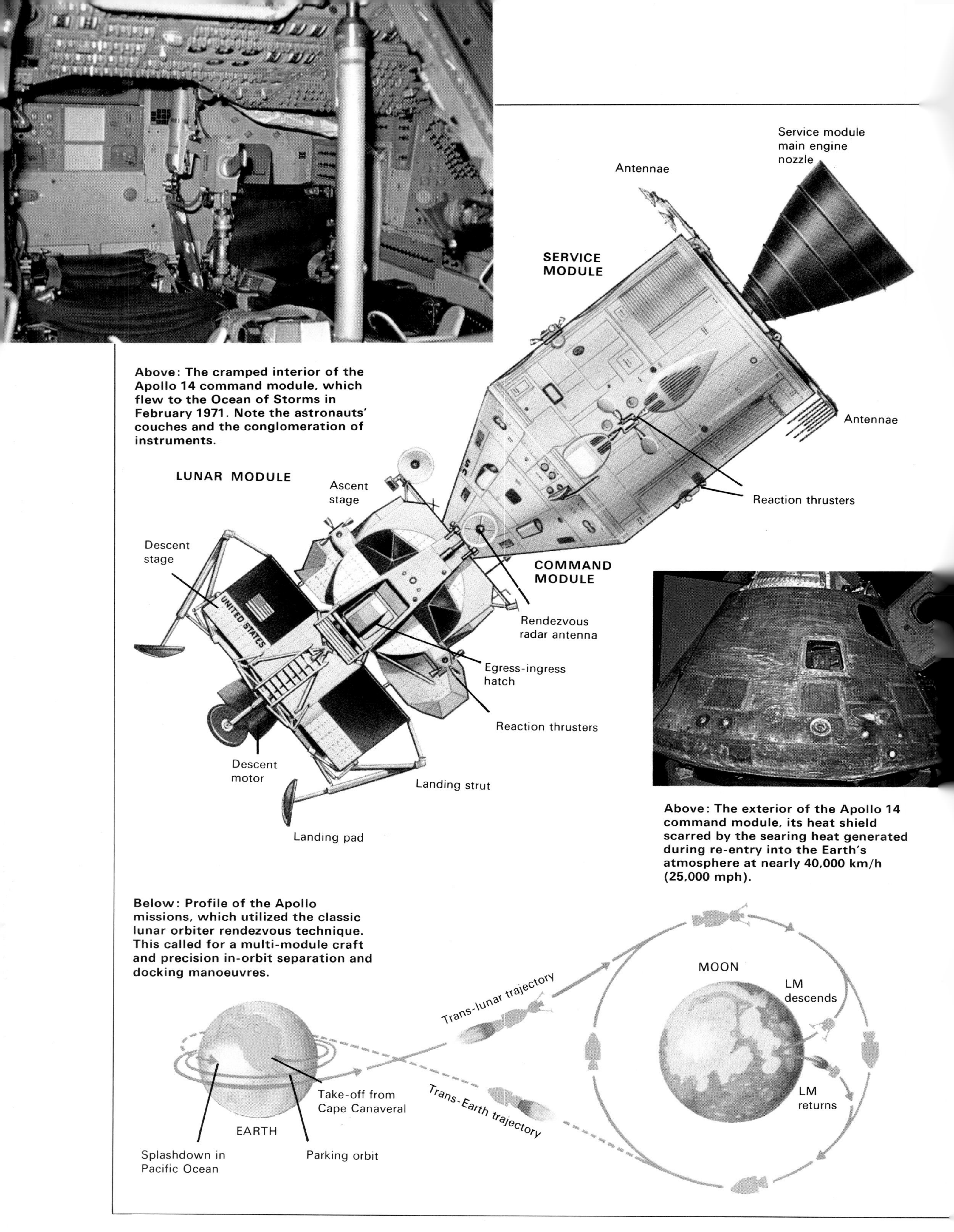

Above: The cramped interior of the Apollo 14 command module, which flew to the Ocean of Storms in February 1971. Note the astronauts' couches and the conglomeration of instruments.

Above: The exterior of the Apollo 14 command module, its heat shield scarred by the searing heat generated during re-entry into the Earth's atmosphere at nearly 40,000 km/h (25,000 mph).

Below: Profile of the Apollo missions, which utilized the classic lunar orbiter rendezvous technique. This called for a multi-module craft and precision in-orbit separation and docking manoeuvres.

MAN'S GREATEST ADVENTURE

In 1961, a few weeks after Yuri Gagarin's epic first space flight, President Kennedy urged that no effort be spared to land an American on the Moon by 1970. At that time the Americans appeared to be lagging behind the Russians in what was coming to be called the space race. By unswerving dedication and the injection of some 25,000 million dollars, American space scientists achieved their goal. They assembled some of the most sophisticated hardware Man has ever created – the Saturn V rocket and the Apollo spacecraft. And they devised most ingenious software (that is, flight programmes) to propel men to the Moon and bring them back safely.

The results obtained from the Moon-landing programme, designated Apollo, exceeded expectations. Six lunar landings took place between 20 July 1969, when Neil Armstrong and Edwin Aldrin in Apollo 11 touched down on the Sea of Tranquillity, and December 1972 (Apollo 17). The Apollo astronauts explored both lowland (mare) and the highland sites, roaming for nearly 100 km (60 miles) and bringing back 385 kg (850 lb) of Moon rock and soil and tens of thousands of stunning photographs and countless reels of magnetic tape carrying all kinds of data about the lunar environment.

The technique chosen to reach the Moon and return, lunar orbiter rendezvous, required the use of a spacecraft made up of several sections, or modules. The complete Apollo spacecraft is shown opposite in its translunar configuration. The three-man crew lived in cramped conditions in the conical command module (CM) which was pressurized. This was linked for most of the time to the cylindrical service module (SM) which contained the bulk of the life-support equipment, rocket motor and propellants. It also contained the fuel cells which provided the spacecraft's power. In these cells hydrogen and oxygen combined to make water, which the crew drank. The command and service module were collectively referred to as the CSM. The lunar module (LM), shown docked with the CSM, was the part of the spacecraft that descended to the Moon's surface. It was made in two parts – an upper and lower – to simplify lunar take-off procedure.

It is worthwhile following the techniques for the Apollo landing because lunar orbiter rendezvous will undoubtedly prove to be one of the great classic space manoeuvres.

Let us turn back the clock and follow an Apollo spacecraft on its journey from Cape Kennedy atop the 111-m (365-ft) tall Saturn V rocket. On the launching pad the CM, surmounted by an escape rocket, is uppermost, with the LM underneath. Saturn V belches flame and boosts Apollo into a parking orbit around the Earth, during which systems are double-checked for the 385,000-km (240,000-mile) trip to the Moon. At this time the 3rd stage rocket is still attached.

The systems check finalized, the 3rd stage re-ignites to boost Apollo into a translunar trajectory at a speed of nearly 40,000 km/h (25,000 mph). Then come a series of manoeuvres to get Apollo into its landing configuration, that is, with the CM mated with the LM. To do this the astronauts detach the CSM from the 3rd stage and do an about-turn by firing the reaction thrusters. They then dock with the LM, still inside the 3rd stage, and pull it clear.

As it pulls away from the Earth, Apollo begins to slow down, since gravity is trying to pull it back. Then, as the Moon looms nearer, lunar gravity begins to accelerate the craft again.

After about 2½ days Apollo is only about 100 km (60 miles) from the Moon and needs to be slowed down before it can go into orbit. The astronauts turn the CSM so that its main engine faces the direction of travel and then fire it. This retro-fire slows down Apollo from some 8000 km/h (5000 mph) to about 5500 km/h (3500 mph), whereupon it goes into lunar orbit.

Preparations now take place for the lunar landing. Two of the astronauts crawl into the LM and separate from the CSM. They manoeuvre into a retro-fire position and fire the LM's engine. The LM drops from orbit and, retro-firing all the time (there is of course no atmosphere to brake them), descends to the surface.

After their stay on the Moon, the astronauts face perhaps the most hazardous part of their dangerous mission – taking-off from the Moon and rendezvousing with their colleague in the CSM circling above them. Any computing or systems error during this phase would spell certain death for them.

They ascend from the Moon in the upper part of the LM, using the lower part as a launch pad. The LM's engine thrusts it into an orbit chosen to intercept that of the CSM. When the two meet, they rendezvous and dock, and the two Moonwalkers rejoin their colleague. The LM is then jettisoned, and at the appropriate point in the orbit, the CSM's engine is fired to boost it beyond lunar escape velocity into a trajectory that will take it to Earth.

As Apollo approaches Earth its speed rises again to nearly 40,000 km/h (25,000 mph) because of the Earth's gravity. Just prior to re-entry into the atmosphere, the SM is jettisoned. With heat shield blazing, the CM plunges through the upper air, slowing all the while. At about 10,000 m (33,000 ft) altitude it is travelling at about the speed of sound. Soon after small drogue parachutes open, followed at 5000 m (16,000 ft) by the three main 'chutes. They lower the CM to a gentle splashdown in the Pacific Ocean. Helicopters are soon on hand to whisk the astronauts to the recovery ship.

Edwin Aldrin, 2nd man on the Moon.

Apollo 16 astronaut and his transport, the remarkable lunar buggy.

Epilogue

A century hence your grandchildren or great-grandchildren could be leading a blissful life in perpetual sunshine in a home among the stars. They would live, with 10,000 others, in a huge habitat, or colony, floating in space some 350,000 km (240,000 miles) from the Earth and at an equal distance from the Moon. The metal and glass used to build the space habitat would have been manufactured on site from ores mined on the Moon. The population would live a self-contained existence, utilizing abundant and inexhaustible solar energy, practising intensive agriculture, and recycling virtually everything so that there would be no waste.

If this seems too hard to swallow, think again. Imagine that you live a century ago, before the invention and application of the petrol engine, the aeroplane, radio and television, and someone had predicted that today there would be machines capable of journeying to the Moon and back. You would have dismissed the predictions as fantastic rubbish of the kind dished out by that French writer Jules Verne, who was currently pioneering a new genre of literature which was to become known as science fiction.

Today's space exploits give credence to the notion that such space habitats can and will be built. An in-depth study of space colonization carried out at Stanford University, California, in 1975 concluded that it offers not only a new challenge but a way out for an expanding world population that is voraciously consuming Earth's limited resources. Man already has most of the technical know-how and resources to colonize space. The major stumbling block is not material, but mental, something that could be termed 'concept shock'. On the other hand the study charged that pursuit of the potential of space should not detract from efforts to conserve terrestrial resources and improve the quality of life on Earth meanwhile.

Boom or Doom?

Until such time as space colonization is feasible, how will mankind fare? People seem to be divided into two camps in their opinions, though of course things in reality are not as cut-and-dried as this. On the one hand are those who believe that the advanced technology that Man has created and will continue to develop will solve each problem as it arises, because it must. They follow the dictum of necessity being the mother of invention. They are prophets of boom – Utopians.

An opposing view is put forward by those who see mankind coming to a sticky end. In technical advancement, they see retrogression. Civilization as we know it today will founder, they say, as Earth's energy and mineral resources fizzle out. If

In the twenty-second century structures like these may be flourishing in space. They are space habitats where tens of thousands of people will live and work. The main habitat is centred on a so-called Bernal sphere, which houses the populace. At either end are ring-shaped agricultural complexes and beyond them industrial modules and panels to radiate away excess heat.

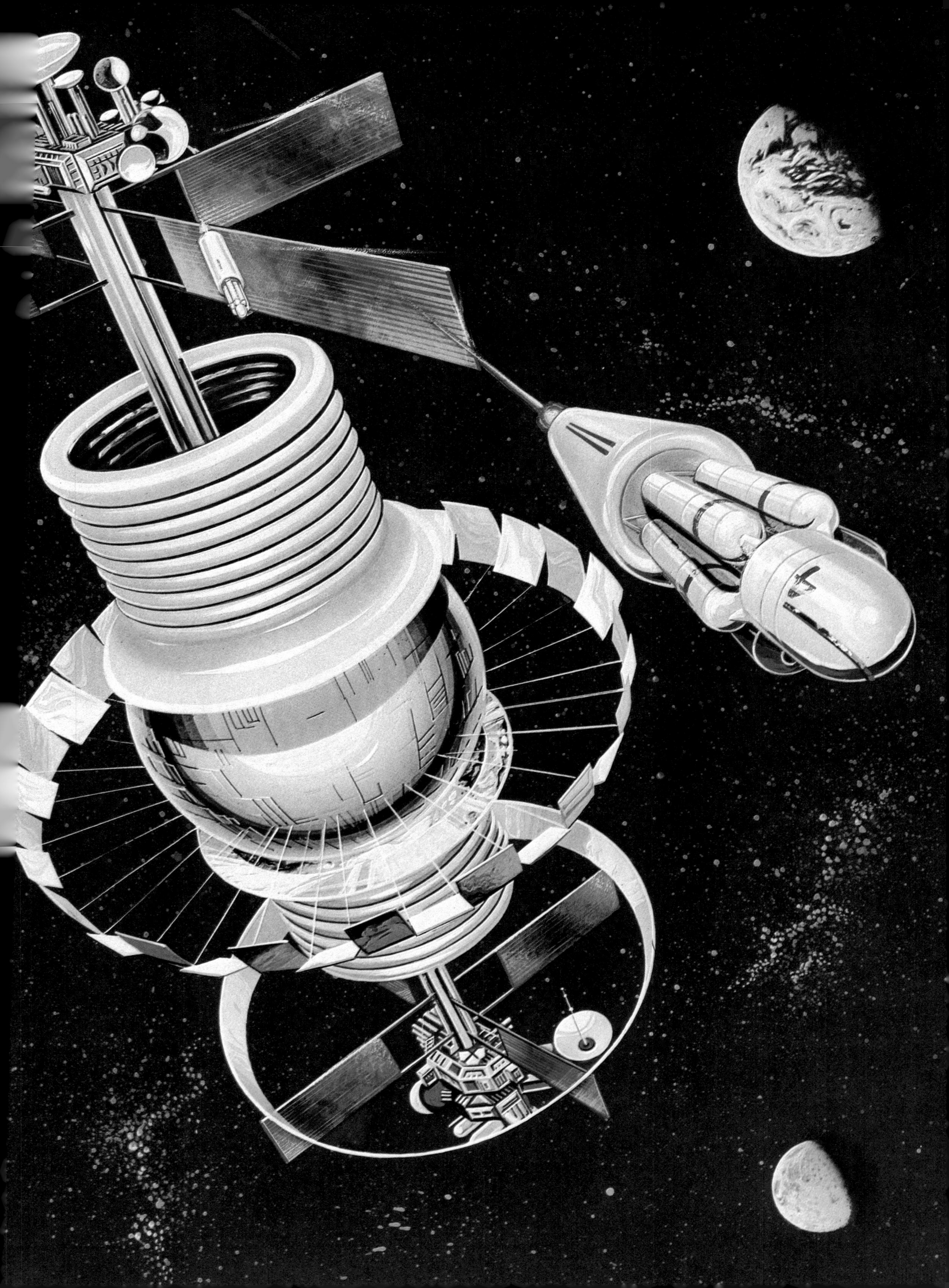

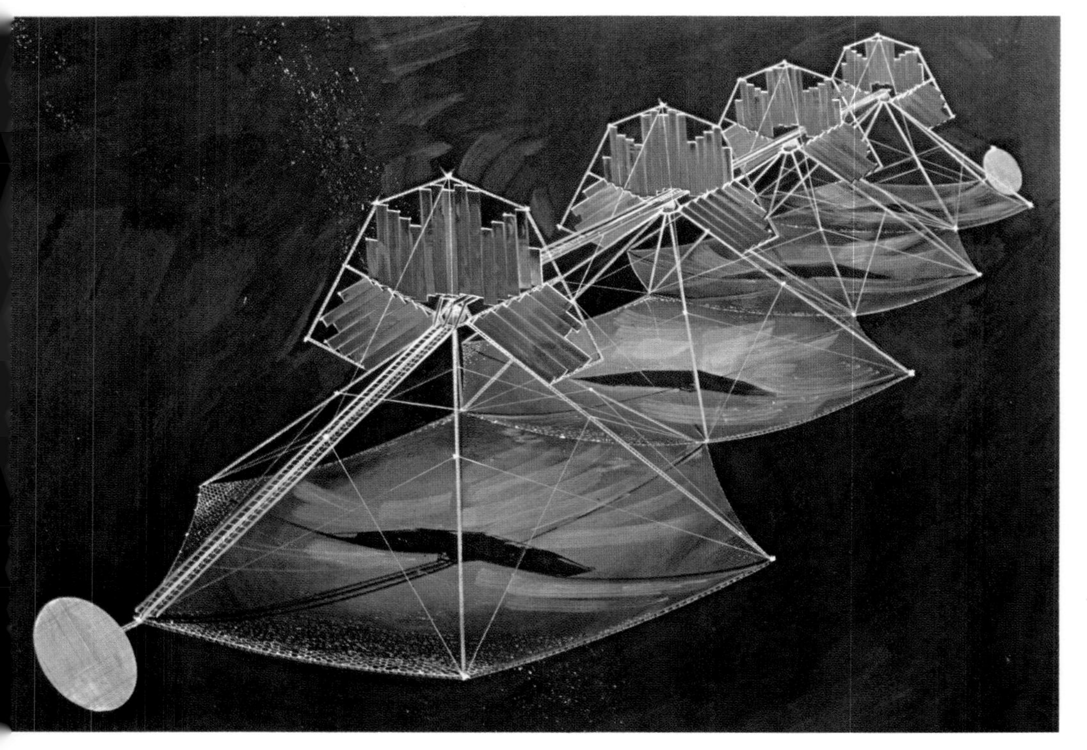

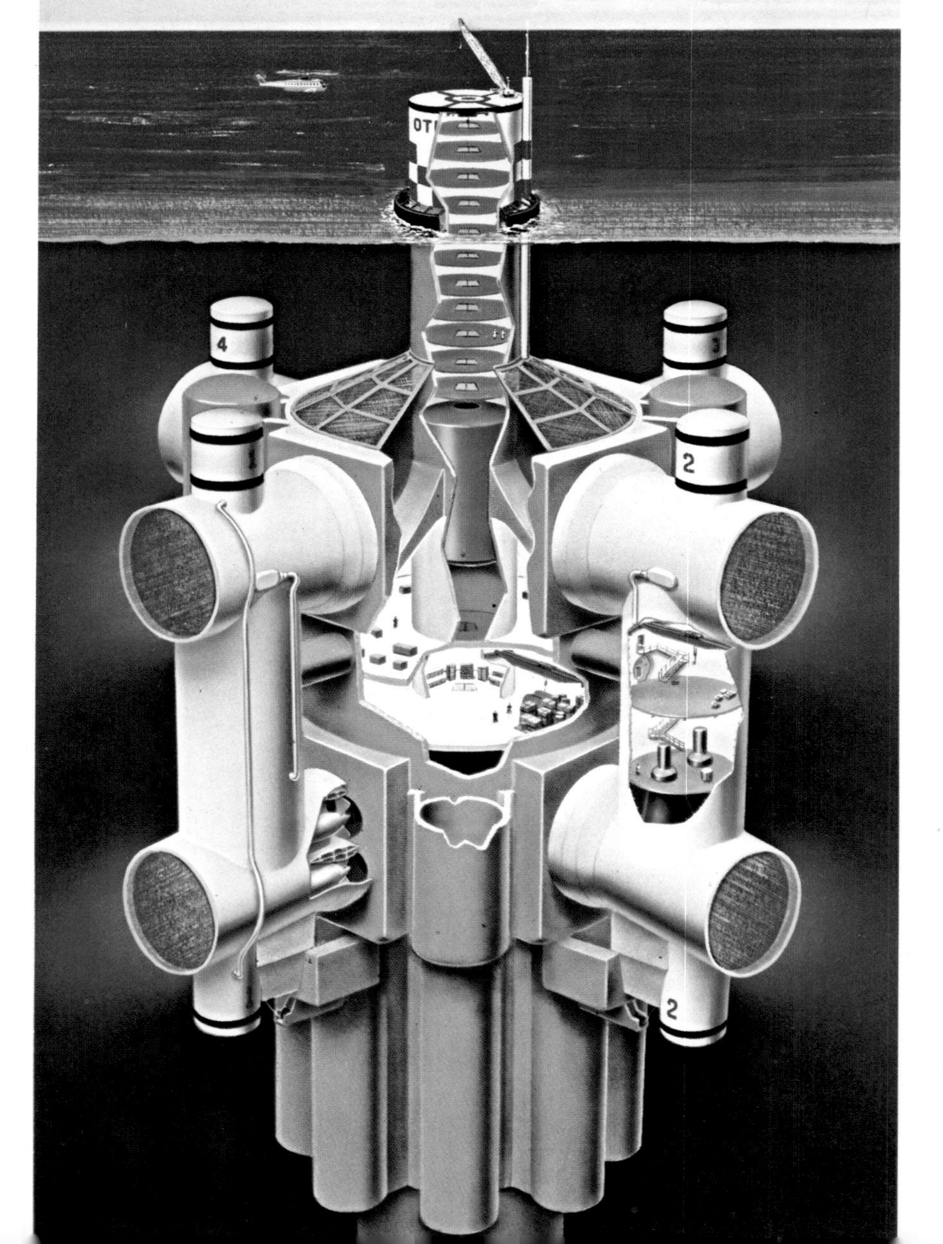

Top left: In the world of tomorrow gigantic satellites will be built. This design is for a four-dish power satellite 25 km (15 miles) long. It will harness solar power to produce electricity and then beam the electricity to Earth via microwaves.

Left: A more down-to-Earth concept for power generation, an OTEC plant floating in tropical waters. It utilizes the temperature difference between hot surface water and the cool water of the deep to drive heat engines.

Above: Glass and concrete will continue to predominate in building construction, though tower blocks will soar to unprecedented heights.

Below: As land regions become increasingly congested, tomorrow's town planners look towards the seas for their salvation. This design for a concrete-and-glass Sea City, put forward by Pilkington Brothers over a decade ago, could provide a blueprint for future offshore development on the continental margins.

mankind does not destroy itself by overpopulation, it will do so in a nuclear holocaust. And there will be no survivors to start a 'brave, new world'. It will be Armageddon. They are the prophets of doom – Luddites.

The probability is that the world will pursue a compromise path, with the excesses of Utopian technocracy being curbed by the prospect of Armageddon. Or perhaps mankind will be saved, at the eleventh hour, by an unpredictable *deus ex machina* (literally 'god from a machine'). Chances are, if Man's past is anything to go by, that this 'god' will *be* a machine, controlled by what is perhaps that most incredible product of Man's inventiveness – the microchip.

The People Explosion

The problems that Man will carry into the twenty-first century stem largely from the rapidly expanding world population. The limited resources our planet has to offer are seemingly inadequate to satisfy the demands of the population at present, let alone in the future. At the beginning of this century, the world population stood at about 1600 million. Currently the population is about 4500 million and will probably double in less than 40 years. If the trend continues unchecked, the population will exceed 10,000 million in the first few decades of the new century.

Ignoring for the moment other problems posed by such a huge population, could the world feed such a multitude? The surprising answer is that, using modern farming methods worldwide, enough food could be grown. And with predictable advances in agricultural technology an even higher population could be sustained.

Where will the population live next century? Far-sighted town planners reckon that society will become almost totally urbanized, living in huge cities interconnected with one another country by country to form what has been called a 'world city'. Development will be most rapid around the coastlines. The isolated small country town and village inland, wasteful of land and energy, will all but disappear. The inland regions will be given over totally to intensive farming or water catchment for the urban populace.

Sky-High Cities

The buildings will not be too dissimilar from those in a modern city of skyscraper tower blocks. And they will be constructed of much the same materials, principally concrete and glass. The main difference will be that the buildings will be very much higher than they are today. Whereas today's tallest structure (the Canadian National Tower in Toronto) is 553 metres (1815 ft) tomorrow's skyscrapers could be 2–3 kilometres ($1\frac{1}{4}$–2 miles) high! According to architects such buildings could be built with existing technology. A 3-km (2-mile) high block towering into the clouds could accommodate a quarter of a million people, who would live, work and play within the confines of its vertiginous walls.

Metals will be used sparingly in the world city, for by the twenty-first century many ore deposits will have been exhausted. Gold, platinum, silver, uranium, lead and copper will have disappeared. Fortunately, aluminium will last for another century and iron for much longer. This is just as well for they are our most essential structural metals, without which we could not build the machinery on which our civilization depends. Recycling metal products will become a vital industry of national and international importance (as perhaps it should today).

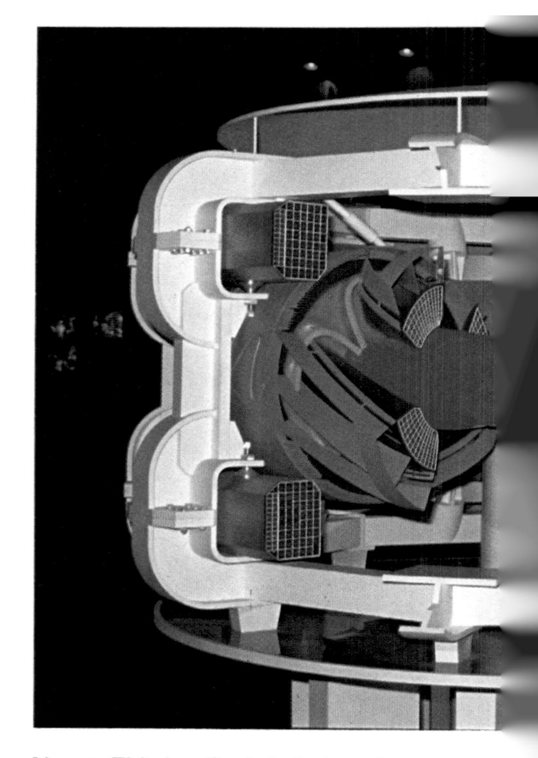

Above: This is a Soviet-designed tokamak, a machine used to investigate ways of achieving controlled nuclear fusion. Nuclear fusion seems to be the only long-term solution to the world's accelerating energy crisis.

Energetic Notions

But we cannot recycle energy. Currently the world as a whole derives the bulk of its energy from four main expendable sources – oil, natural gas, coal and uranium. In less than half a century oil and natural gas will have run out, and supplies of uranium (for nuclear fission) will be scarce. Fortunately coal could last for a further century or more. It is not as convenient a fuel as oil and gas and causes more pollution when it is burned. But it can, at a price, be converted to oil and gas.

A relatively small percentage of the world's energy comes from hydroelectricity. Unlike energy derived from the fossil fuels, hydroelectricity utilizes a renewable resource – flowing water. Hydroelectric schemes using the ebb and flow of the tides are also in operation. But there are relatively few suitable sites remaining for future hydroelectric development. So within a short time span the world has got to come up with some viable alternative to fossil fuels if a catastrophic energy shortfall is to be avoided.

Practically all the alternative-energy schemes under development centre in one way or another on the body that breathes life into our planet: the Sun. The Sun pours into the atmosphere

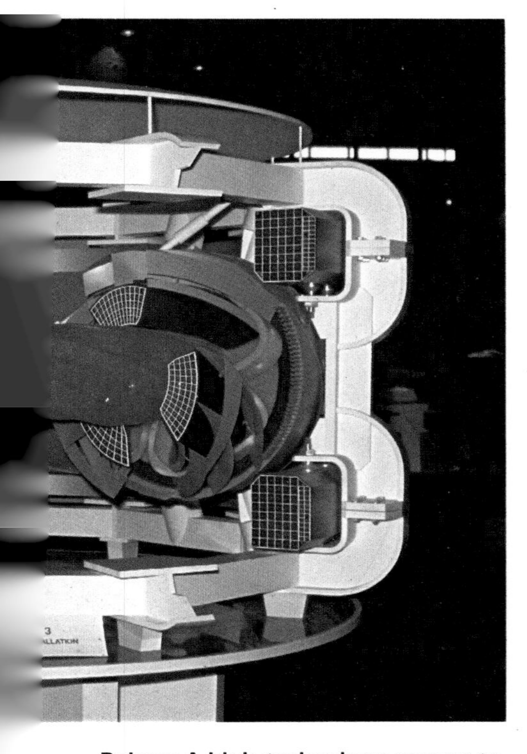

Below: A high-technology answer to tomorrow's transport problems – the high-speed, magnetically levitated monorail. Supported and propelled by magnetic interaction with the track, the maglev vehicle has potential speeds approaching 800 km/h (500 mph). Only in recent years have powerful enough superconducting magnets become available to make magnetic levitation practical. Rivalling the maglev train will be the hover-train, riding on an air-cushion and capable of similar high speeds.

Above: A low-technology answer to tomorrow's transport problems, the wind-powered car. Invented by Michigan aeronautical engineer James Amick, the 'aero-car' can achieve speeds of over 100 km/h (60 mph) in favourable winds.

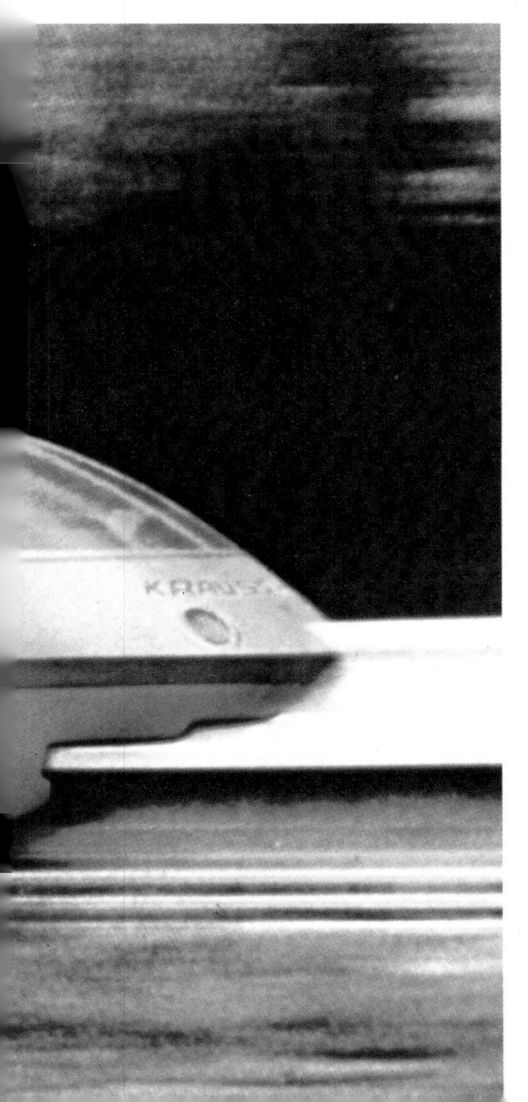

20,000 times more energy than we currently consume. Solar energy not only heats the surface but also powers the weather machine, driving the wind and waves. It is a diffuse source, providing abundant power but over a very large area. How will this be tapped on a large scale? Some people believe that the windmill, in more sophisticated form, could be the answer, while others favour wave-power. One interesting scheme, known as OTEC (Ocean Thermal Energy Conversion), extracts heat from tropical seas. Solar towers are a particularly good bet for sunny climes. They use banks of mirrors to concentrate the heat of sunlight onto a tower-top boiler and use the steam produced to generate electricity. A more grandiose scheme seeks to harness solar energy in space, using a complex of gigantic solar reflectors. The energy produced would be beamed down to Earth by microwave.

With all due respect for their advocates, it seems unlikely that solar energy schemes alone could fill tomorrow's global energy vacuum. Conventional nuclear power – using fission – even if it were developed more extensively, could not do so either. We must learn to tap the very energy source that powers the universe – the fusion of hydrogen into helium. Experiments are already well advanced in nuclear-fusion technology, and fusion reactors could be working before the turn of the century. They could be machines called tokamaks which use ultrapowerful magnets, or machines like Shiva which use laser beams of fantastic power to bring about nuclear fission. If they can be made to work, then Man's problems, energy-wise, will be over, for the hydrogen fuel is readily extracted from the oceans.

Index

(Figures in *italics* refer to illustrations)

Acknowledgements

The author and publishers extend their grateful thanks to the many companies and individuals who have willingly given helpful advice and information and provided illustrative material for inclusion in this book. Particular thanks are due to Anne Lyons and Gael Hayter for picture research. In addition to those provided by the author, the photographs came from the following sources.

(T = Top, B = Bottom, L = Left, R = Right, C = Centre)

Adler 177R; J. Allan Cash 90T, 96B; Almasy 53; Richard Amick 219R; Anglo-Australian Observatory 197T; Balfour-Beatty 58–9T; BBC 137L; Ron Boardman 181B, 189T; Boeing 4–5BL, 130, 131BL, 135, 216T; Boston Museum of Fine Arts 23; Paul Brierley 176B; British Aerospace 101 (inset), 126T & BR; 128C, 129B, 133B, 134B; British Hovercraft Corporation endpapers, 102–3, 118; British Leyland 38, 78–9T, 85R; British Museum 8(inset), 14–15T, 19, 25B, 28T, 31; British Rail 65B, 92, 182T; British Steel 5B, 6–7, 164B, 165L, 167, 174–5; CEGB 50, 63; CERN 180T; Coles Cranes 85B; Coloursport 85; Courtaulds 171R, 172, 173; Gerry Cranham 73B; *Daily Telegraph* 45T, 65T; Raymond Davis 178; EMI 189B; ESA 202, 203B; Fiat 72–3C, 170B; GEC 168; General Electric 181T; General Motors 71BR; Robert Harding Associates 18B; Hirmer-Fotoarchiv 14B; Michael Holford 8–9; IBM 176T; ICI 187BR; JCB 2–3; Jet Ferries 116–7C; JVC 145T, 146T; Krauss-Maffei 218–19B; Lockheed 133T, 216BL; Marconi 188T; Massey-Ferguson 4–5T, 151, 160T; Mercedes-Benz 71BL; John Moss 110, 114B, 174B; NASA 190–1, 198, 200, 204–5, 206T, 210–11, 213; National Coal Board 160BR; *Nuclear Engineering International* 55B, 56–7; P & O 108L; C. A. Parsons 48; Paxman Diesels 33L; Perkins Engines 42T; Pest Control Laboratories 182B; Philips 145BR; Photri 113C; Pilkington Bros 216R; Max Planck Observatory 199; Polaroid 149B; Porsche 71C; Post Office 142TR; RAE Farnborough 126BL, 185; Raleigh 81B; Rank Xerox 137TR; Reid Walker 65L; Rolls-Royce; Ann Ronan 9(inset), 27, 29B; Saab 71T; Scala Museum 26, 30; Seiko 177L; Shell 162TR; Mike St Maur Shiel 109BR; SNCF 43B, 90B, 93, 94, 97; Spalding PR 81T; Spectrum 24–5B, 111; Stanford University 179; Brian Stephenson 96T, 98; Swiss National Tourist Office 51; UKAEA 33BR, 52, 54T, 58B, 150; Union Pacific 88T; John Watney 101TR; White Motors 84B; Young PR 72L; ZEFA 32, 84T, 88B, 92B, 101BR, 106, 108BR, 113T, 170T; Zeiss 186.